Mechanisms
of
Differentiation

Volume II

Modulation of Differentiation
by
Exogenous Agents

Editor

Paul B. Fisher, B.A., M.A., M. Phil., Ph.D.
Director of Neuro-Oncology Research
and Chernow Research Scientist
Columbia University
College of Physicians and Surgeons
New York, New York

CRC Press
Boca Raton   Ann Arbor   Boston

Library of Congress Cataloging-in-Publication Data

Mechanisms of differentiation/editor, Paul B. Fisher.
   p.   cm.
      Includes bibliographical references.
      Contents: v. 1. Model cell culture systems for studying
   differentiation — v. 2. Modulation of differentiation by exogenous
   agents.
      ISBN 0-8493-4947-8 (v. 1). — ISBN 0-8493-4948-6 (v. 2)
      1. Cell differentiation. 2. Cancer cells. 3. Cell
   transformation. I. Fisher, Paul B.
   QH607.M43 1990
   574.87'612 — dc20   90-1547
                                                                          CIP

Direct all inquiries to CRC Press, Inc., 2000 Corporate Blvd., N.W., Boca Raton, Florida, 33431.

© 1990 by CRC Press, Inc.

International Standard Book Number 0-8493-4947-8 (Volume I)
International Standard Book Number 0-8493-4948-6 (Volume II)

Library of Congress Card Number 90-1547
Printed in the United States

These books are dedicated to my wife Marlene, my daughter Danielle, and my son Damien, who have always provided love and support for all of my efforts.

PREFACE

Significant recent advances in cell-culture technology now permit a detailed biochemical and molecular analysis of differentiation in both normal and tumor cells. These studies are important in attempting to understand the complex factors involved in normal growth and development as well as the abnormalities associated with carcinogenesis. Model cell-culture systems indicate that diverse agents, as well as cell shape and the composition of the extracellular matrix, can modulate various programs of cellular differentiation. An induction or inhibition of differentiation, depending on the target cell, has been observed as a consequence of: (a) transformation by viruses or cellular and viral oncogenes; (b) media and cell substrate composition, including the presence or absence of specific hormones, transforming the physiologic growth factors, vitamins, peptides, and ions; (c) exposure to bioresponse modulators, chemotherapeutic agents and cytotoxic drugs; (d) treatment with tumor-promoting agents, such as diterpene phorbol esters and indole alkaloids; (e) exposure to polar solvents, such as dimethyl sulfoxide, hexamethylene bisacetamide and dimethyl formamide; and (f) differentiation-inducing factors produced by cells. Only with an understanding of the role of diverse external and internal stimuli in regulating differentiation and an appreciation of the biochemical and molecular changes associated with differentiation can a clearer insight into important biological phenomena such as repair, regeneration, ageing, and cancer be obtained. In the two volumes of this series on *Mechanisms of Differentiation,* experts in defined areas pertinent to the biochemical and molecular mechanisms of differentiation have contributed review chapters addressing various topics of current interest in this important area of research.

THE EDITOR

Dr. Paul B. Fisher, Ph.D. is Director of Neuro-Oncology Research and a Chernow Research Scientist in the Departments of Neurosurgery, Pathology and Urology, Cancer Center/Institute of Cancer Research, Columbia University, College of Physicians and Surgeons, New York, New York.

In 1974, Dr. Fisher received his Ph.D. from the Waksman Institute of Microbiology of Rutgers University, New Brunswick, New Jersey. From 1968 to 1971, at Herbert H. Lehman College of the City University of New York, he obtained training in genetics in the laboratory of Dr. Joseph W. Rachlin. After obtaining his M.A. in 1971, he entered the Ph.D. program of Rutgers University under the direction of Dr. Vernon Bryson of the Waksman Institute of Microbiology. While at Rutgers University he obtained training in cell biology, virology, and somatic cell genetics. In 1973 he obtained the M.Phil. degree and in 1974, the Ph.D. degree from Rutgers University. His thesis work involved an analysis of the biological properties of methyl ester derivatives of the polyene macrolide antibiotics amphotericin B and nystatin. His thesis research focused on defining the genetic and biochemical basis of polyene antibiotic resistance in mammalian cells. This was a very productive research period and has resulted in 23 publications between 1974 and 1980.

In 1974, Dr. Fisher continued his studies on amphotericin B and nystatin methyl ester as a Busch Postdoctoral Scientist at the Waksman Institute of Microbiology in the laboratory of Dr. Vernon Bryson. From 1975 to 1976, he was a Staff Associate in the Department of Microbiology and Immunology, Albert Einstein College of Medicine, Bronx, New York. During this period he studied the molecular biology of human adenoviruses in the laboratory of Dr. Marshal S. Horwitz. From 1976 to 1980, Dr. Fisher studied the effects of tumor promoters and chemical carcinogens on adenovirus transformation, cellular differentiation and cellular physiology in the laboratory of Dr. I. Bernard Weinstein of the Institute of Cancer Research of Columbia University, College of Physicians and Surgeons. This period was also highly productive and it resulted in 20 papers and 10 reviews between 1978 and 1981. In 1981, Dr. Fisher joined the Department of Microbiology of Columbia University, College of Physicians and Surgeons. During the period of 1981 to 1986, he was involved in independent and collaborative studies on the molecular basis of adenovirus transformation, mechanisms of tumor promoter and carcinogen action, mechanisms of differentiation in normal and tumor cells and enhancement of tumor associated antigen expression by biological response modifiers. In 1987, Dr. Fisher was selected as the Chernow Research Scientist (an endowed research position) in the Department of Urology of Columbia University, College of Physicians and Surgeons and in 1988, he assumed the position of Director of Neuro-Oncology Research in the Department of Neurosurgery of Columbia University, College of Physicians and Surgeons. In 1988, Dr. Fisher also became a member of the Department of Pathology of Columbia University, College of Physicians and Surgeons.

Dr. Fisher has presented numerous papers and chaired sessions at both national and international meetings, as well as presenting lectures at various universities and institutes in both the U.S.A. and in Europe. He has served as a reviewer for the National Cancer Institute, National Science Foundation, Department of Energy, Environmental Protection Agency, Ontario Ministry of Health, and New Jersey State Commission on Cancer Research. He also has served as a reviewer for numerous scientific journals and is presently a member of the editorial boards of *In Vivo* and *Archives for AIDS Research*. Dr. Fisher has published over 150 papers in scientific journals to date. He is currently a member of 18 scientific societies devoted to the fields of cell biology, clinical pharmacology and therapeutics, differentiation, infectious diseases, microbiology, tissue culture, and virology.

His major research interests include: molecular basis of adenovirus transformation; chemical-viral carcinogenesis; mechanisms of interferon action; molecular and biochemical basis of differentiation in normal and tumor cells; mechanisms of tumor progression; genetic determinants of human cancers; and improving and developing new technologies for treating patients with refractile tumors.

VOLUME II CONTRIBUTORS

M. A. Ahmed, Ph.D.
Postdoctoral Research Scientist
Department of Pathology
College of Physicians and Surgeons
Columbia University
New York, New York

Gideon Baum, M.Sc.,
Undergraduate Student
Departments of Molecular Genetics and
 Virology
The Weizmann Institute of Science
Rehovot, Israel

Avri Ben-Ze'ev, Ph.D.
Associate Professor
Department of Genetics
The Weizmann Institute of Sciences
Rehovot, Israel

Ronald Breslow, Ph.D.
S. L. Mitchill Professor
Department of Chemistry
Columbia University
New York, New York

José Luis Rodriguez Fernandez, M.Sc.
Graduate Student
Departments of Molecular Genetics and
 Virology
The Weizmann Institute of Sciences
Rehovot, Israel

Paul B. Fisher, Ph.D.
Director of Neuro-Oncology Research and
 Chernow Research Scientist
Departments of Pathology, Neurosurgery,
 and Urology
College of Physicians and Surgeons
Columbia University
New York, New York

Barbara Gorodecki, M.Sc.
Graduate Student
Departments of Molecular Genetics and
 Virology
The Weizmann Institute of Science
Rehovot, Israel

John W. Greiner, Ph.D.
Staff Scientist
Laboratory of Tumor Immunology and
 Biology
National Cancer Institute
National Institutes of Health
Bethesda, Maryland

Fiorella Guadagni, M.D.
Research Scientist
Laboratory of Tumor Immunology and
 Biology
National Cancer Institute
National Institutes of Health
Bethesda, Maryland

Ludovico Guarini, M.D.
Assistant Professor
Division of Hematology/Oncology
College of Physicians and Surgeons
Columbia University
New York, New York

Duane L. Guernsey, Ph.D.
Associate Professor
Departments of Pathology, Physiology,
 and Biophysics
Dalhousie University Faculty of Medicine
Halifax, Nova Scotia, Canada

Henry Hermo, Jr. D.P.H.
Professor
Department of Biology
Bronx Community College of CUNY
Bronx, New York

Ronald A. Ignotz, Ph.D.
Department of Cell Biology
University of Massachusetts Medical
 Center
Worcester, Massachusetts

Peter A. Jones, Ph.D.
Professor
Department of Biochemistry
University of Southern California
Los Angeles, California

J. A. Kantor, Ph.D.
Cancer Expert
Laboratory of Immunology and Biology
National Cancer Institute
National Institutes of Health
Bethesda, Maryland

H. Phillip Koeffler, M.D.
Professor
Division of Hematology-Oncology
University of California School of
  Medicine
Cedars-Sinai Medical Center
Los Angeles, California

Paul A. Marks, M.D.
President
Memorial Sloan-Kettering Cancer Center
New York, New York

Joan Massagué, Ph.D.
Member and Chairman
Howard Hughes Medical Institute
Memorial Sloan-Kettering Cancer Center
New York, New York

Robert A. Mufson, Ph.D.
Senior Scientist
Department of Cell Biology
Holland Laboratory for Biomedical
  Science
American Red Cross
Rockville, Maryland

Ulrich Nielsch, Ph.D.
Postdoctoral Research Scientist
Department of Urology
College of Physicians and Surgeons
Columbia University
New York, New York

Michio Oishi, Ph.D.
Professor
Institute of Applied Microbiology
University of Tokyo
Tokyo, Japan

Richard A. Rifkind, M.D.
Chairman
Sloan-Kettering Institute
Memorial Sloan-Kettering Cancer Center
New York, New York

Jeffrey Schlom, Ph.D.
Chief
Laboratory of Tumor Immunology and
  Biology
National Cancer Institute
National Institutes of Health
Bethesda, Maryland

Toshio Watanabe, Ph.D.
Institute of Applied Microbiology
University of Tokyo
Tokyo, Japan

Kenji Yamamoto, M.D.
Division of Hematology/Oncology
University of California
Los Angeles, California

VOLUME I TABLE OF CONTENTS

VOLUME II TABLE OF CONTENTS

Chapter 1

MODULATION OF DIFFERENTIATION: A POTENTIAL MECHANISM BY WHICH INTERFERONS INDUCE ANTITUMOR ACTIVITY

Mohammad Almas Ahmed, Ulrich Nielsch, Ludovico Guarini, Henry Hermo, Jr., and Paul B. Fisher

TABLE OF CONTENTS

I. INTRODUCTION

The interferons comprise a family of structurally related proteins which were originally identified as a consequence of their antiviral activity.[1-3] Based on the cell type producing these proteins, they have been referred to as leukocyte (IFN-α), fibroblast (IFN-β), and immune (IFN-γ) interferons, and based on their similar functional and receptor-binding properties, they have been classified as types I (IFN-α/β) and II (IFN-γ) interferons.[4-7] In the case of human leukocyte interferon, multiple species (>16) with at least 70% sequence homology have been identified which have molecular weights of 16 to 27 kDa and pIs ranging from 5.5 to 6.5.[8-12] In contrast, both fibroblast and immune interferon are believed to consist of a single, structurally distinct polypeptide.[12-14] Human fibroblast interferon appears to be a single glycoprotein with an approximate molecular weight of 20 kDa.[13,15] Three forms of human immune interferon have been identified, with apparent molecular weights of 15.5 to 17, 20, and 25 kDa, respectively.[13,15,16] Sequence analysis of these three proteins indicates differences in the extent of glycosylation of a single polypeptide species.[17] Using recombinant DNA technologies, many of the genes coding for several human IFN-α species, IFN-β, and IFN-γ have been molecularly cloned and expressed in bacteria and yeast, resulting in large quantities of homogenous preparations of recombinant interferons for structural and biochemical studies.[11-16]

Studies employing both natural and recombinant interferons indicate that these proteins can induce a wide spectrum of changes, in addition to antiviral activity, in appropriate target cells (Tables 1 and 2).[18-21] A property of both natural and recombinant DNA-derived human interferons of potential clinical importance is the ability of these proteins to inhibit the growth of tumor cells, both *in vitro* and *in vivo*.[15,20] When applied to the clinical setting, certain interferon species have displayed some inhibitory activity toward specific human tumors.[22-25] However, in the majority of situations, interferons have resulted in only a minimal or negligible activity in inhibiting the progression of the malignant state in humans.[22-25]

Although the mechanism(s) by which interferons induce their antitumor activity is not presently known, possible insights into this process have come from investigations indicating that these agents are potent modulators of cellular differentiation (Table 3)[18,26-31] and immunological functions.[19,21,30,32] In addition, recent studies indicate that the antitumor activity of various interferons can be enhanced by combining these agents with other compounds which can modulate tumor cell growth (Table 4). An important group of compounds which represent potentially useful tools in suppressing tumor growth are those agents which can induce terminal differentiation in tumor cells and/or alter the antigenic profile of tumors, resulting in an increased surface expression of tumor-associated antigens.[18,19,21,30] In the present review, we will address the potential relationship between the ability of interferons to modulate cellular gene expression and their effects on cell growth and differentiation. Studies from our laboratory which have been designed to exploit the ability of interferon, alone and in combination with other agents, to alter tumor cell growth and induce cellular differentiation will also be discussed.

II. EFFECTS OF INTERFERON ON DIVERSE PROGRAMS OF DIFFERENTIATION

A. MYOGENESIS

Techniques are currently available which permit the *in vitro* growth of myogenic muscle satellite cells from both normal and diseased human skeletal muscle.[33-37] When grown under the appropriate conditions, muscle satellite cells can be induced to undergo both morphologic changes, formation of multinucleated myotubes, and biochemical changes — creatine kinase (CK) isoenzyme transition from CK-BB (found in unfused myoblasts) to CK-MM (which

TABLE 1
Inhibitory Effects of Interferons on Cellular Activities

1. Virus growth in cells
2. Growth of intracellular parasites (*Toxoplasma* and *Plasmodium*)
3. Growth of mammalian cells in monolayer, liquid, and agar suspension (normal, tumor, and embryonic cells)
4. Tumorigenicity
5. Differentiation (adipogenesis, erythrogenesis, melanogenesis, and myogenesis)
6. Myocardial beat
7. Enzyme synthesis and secretion (glucocorticoid-inducible enzymes [glutamine synthetase, glycerol-3-phosphate dehydrogenase, and tyrosine aminotransferase], hepatic cytochrome P-450-linked monooxygenases, ornithine decarboxylase, and plasminogen activator)
8. DNA synthesis (both basal and growth factor-induced)
9. Lipid-associated changes (acetate to lipid, high-density lipoproteins, high-density lipoprotein cholesterol, leucine to fatty acids, and total cholesterol)
10. Toxin binding (cholera and diptheria toxins)
11. Hormone and growth-factor binding (thyrotropin and epidermal growth factor)
12. Cell locomotion (membrane ruffling and saltatory movements of intracellular granules)
13. Cell fusion (membrane lipid fluidity induced by Sendai virus)
14. Syncytium formation induced by Moloney murine sarcoma virus
15. Murine leukemia virus release
16. Bovine papillomavirus transformation of mouse cells
17. Redistribution of cell-surface components (receptor mobility)
18. Membrane transport (including thymidine uptake and uridine release)
19. Unsaturated fatty acid content
20. Synthesis of pp60^{src} and expression of the transformation-related phenotype
21. Synthesis and expression of the p21 c-Ha-*ras* gene product
22. Expression of the transformed phenotype in Ha-*ras*-transformed cells
23. Expression of growth-related genes (including *myc* and *fos*)
24. Salmonella/Shigella invasiveness
25. T-lymphocyte proliferation and production of leukocyte migration inhibitory factor
26. B-lymphocyte proliferation and IgG; IgM and IgE antibody production
27. Delayed hypersensitivity responses (including afferent pathway, efferent pathway, and graft-vs.-host reaction)
28. Maturation of human monocytes into macrophages
29. Hematopoietic cell (bone marrow) proliferation
30. Natural killer (NK) cell target cell sensitivity
31. Antigen expression
32. Granulocyte migration

develop at the time of myoblast fusion), which normally occur during myogenesis *in vivo*.[33] The addition of either crude natural or recombinant human interferon (IFN-α) (100 to 5000 units per ml) to human muscle satellite cells *in vitro* results in an acceleration in myotube formation and CK isoenzyme transition, without a change in the proliferative capacity of myoblasts.[38] The specificity of the IFN-α effect was demonstrated by the inability of heat-inactivated IFN-α, trypsin-treated IFN-α, or mouse L-cell interferon to induce changes in differentiation in human muscle cell cultures. In contrast to the effect of interferon, tumor promoters such as *12-0-tetradecanoyl-phorbol-13-acetate* (TPA) and teleocidin inhibit differentiation in human muscle cells and they antagonize the stimulatory effect of interferon.[36,38] The specificity of the TPA effect is suggested by the inability of structural analogs of this diterpene phorbol ester which are devoid of *in vivo* tumor-promoting activity to alter the pattern of differentiation in human muscle cell cultures.

A different response to interferons is observed in chicken myoblasts and the murine myoblast cell line, MM14DZ, which are inhibited in their program of differentiation by interferon.[39-41] Studies on the effect of interferon on chicken muscle cell differentiation employed impure preparations of crude interferons, which did not permit a distinction

TABLE 2
Stimulatory Effects of Interferons on Cellular Activities

1. Anchorage-independent growth in methylcellulose and agar cultures
2. Experimental metastasis and tumor formation
3. Enzymatic changes (induction of aryl hydrocarbon hydroxylase, conversion of cyclooxygenase to prostaglandins [E2a and F2a], induction of protein phosphokinase, induction of 2-5A synthetase, synthesis of tRNA methylase, and tryptophan degradation)
4. Nucleotide and protein changes ($2'$-$5'$-A_n synthesis, cAMP, cGMP, phosphorylated proteins — including P1 [67K], eIF-2, ODC, and P2 [37K] (a subunit peptide chain initiation factor [eIF2]) — serum amyloid A
5. Induction of new mRNA species and unique proteins (many with undefined functions)
6. Differentiation in human skeletal muscle, melanoma, B-lymphoid, epidermoid carcinoma, keratinocyte, promyelocytic leukemic and macrophage cells, as well as murine erythroleukemia, murine macrophage, murine myeloid leukemia, and rat pheochromocytoma cells
7. Concanavalin A binding to cells
8. Beat frequency of myocardial cells
9. Rigidity of the plasma membrane bilayer
10. Estrogen receptors
11. Abundance of submembraneous microfilaments (bundles of actin filaments and long filaments of fibronectin)
12. Excitability of cultured neurons
13. Early increase in membrane lipid fluidity
14. Proportion of saturated acyl side chains in cellular phospholipids
15. Gamma-radiation sensitivity
16. Actinomycin D sensitivity
17. Tumor necrosis factor growth inhibition and cytotoxicity
18. Tumor cytotoxicity when combined with chemotherapeutic or other biological response-modulating agents
19. Morphologic reversion of bovine papillomavirus transformed mouse cells
20. Morphologic reversion of Ha-*ras* transformed rodent cells
21. Antiviral activity
22. Production of interferon (priming)
23. Macrophage functions (phagocytosis, spreading, cytotoxicity against tumor cells, cytotoxicity toward virus-infected cells, and production of macrophage-activating factor)
24. Expression and shedding of histocompatibility antigens
25. Expression and shedding of tumor-associated antigens detected by monoclonal antibodies
26. Induction and increased expression of intercellular adhesion molecule 1 (ICAM-1)
27. T-lymphocyte (cytotoxicity and suppression)
28. B-lymphocytes (antibody production)
29. Natural killer cell cytotoxicity (antibody dependent)
30. Killer cell cytotoxicity (antibody dependent)
31. IgE-mediated histamine release by basophils

between a direct effect of interferon or a contaminant-inducing inhibition of differentiation.[39,40] In a more recent study by Multhauf and Lough,[41] purified murine IFN-β was employed and cultures treated with 20 to 2000 units per ml of IFN-β for 5 d exhibited a dose-dependent inhibition in differentiation, as indicated by a reduction in myotube formation and CK activity. Studies employing human, chicken, and murine myoblast cultures indicate an important temporal kinetics for interferon to exert its effect on myogenesis. When cultures have initiated fusion, interferon has no effect (human[38] and chicken[39]) or stimulates myogenic traits when applied after myotube formation (chicken[40]). Although further studies are required, these observations suggest the intriguing possibility that the effect of interferons on myogenesis will be dictated by the status of differentiation of the target myogenic cell population. In this context, various interferons may function as normal regulators of the process of myogenesis during muscle cell development *in vivo*.

The mechanism by which interferon alters differentiation in myogenic cells and the reason for the differential response between human versus chicken and murine myoblasts is

TABLE 3
Modulation of Cell Differentiation by Interferon

Cell culture system	Program of differentiation	Inducer
Induction of Differentiation		
Human B-lymphoid (Daudi Burkitt lymphoma-derived cell line)	Plasmacytoid	Spontaneous
Human epidermoid carcinoma (A431)	Epithelial	Spontaneous
Human histiocytic lymphoma	Macrophage and granulocyte	Spontaneous
Human keratinocytes	Epithelial	Spontaneous
Human macrophage	Macrophage	Spontaneous and colony-stimulating factor-1
Human melanoma	Melanogenesis	Spontaneous
Human promyelocytic leukemia (HL-60)	Macrophage and granulocyte	*12-0-tetradecanoyl-phorbol-13-acetate* (TPA), 12,13-*trans* retinoic acid (RA), and dimethyl sulfoxide (DMSO)
Human skeletal muscle	Myogenesis	Spontaneous and Ca^{2+}-induced
Murine erythroleukemia	Erythroid	DMSO
Murine macrophage	Macrophage	Spontaneous
Murine myeloid leukemia	Macrophage and granulocyte	D-factor, lipopolysaccharide or polyinosinic acid
Rat pheochromocytoma (PC12)	Neuronal	Nerve growth factor
Inhibition of Differentiation		
Chicken embryo muscle	Myogenesis	Spontaneous
Human myeloid cells	Granulopoiesis	Spontaneous
Murine B-16 melanoma	Melanogenesis	Spontaneous or α-Melanocyte-stimulating hormone (MSH)
Murine erythroleukemia	Erythroid	DMSO
Murine myoblast	Myogenesis	Spontaneous
Murine 3T3 cells	Adipogenesis	Spontaneous or insulin

Modified from Fisher, P. B., Hermo, H., Jr., Pestka, S., and Weinstein, I. B., *Pigment Cell 1985: Biological, Molecular, and Clinical Aspects of Pigmentation,* University of Tokyo Press, 1985, 325.

not known. Similarly, the mechanism underlying the ability of interferon to induce differentiation in human muscle cells, whereas TPA and teleocidin inhibit differentiation in human muscle cells as well as many other cell types, remains to be elucidated. A possible mechanism underlying the differential response of human cells to interferon and TPA may involve the induction — following the binding of these ligands to their cell-surface receptor, type I (IFN-α/β) receptor for leukocyte and fibroblast interferons[3-5,15] and protein kinase C[43-45] for TPA — of different transmembrane signals, which results in either the induction or abrogation of the appropriate gene transcriptional and expression changes required for differentiation.[46] Recent studies have resulted in the identification and molecular cloning of a series of genes, including MyoD,[47,48] myogenein,[49,50] *myd*[51] (not yet cloned), and Myf-5[52] (which is different from both MyoD and myogenein, but its relationship to *myd* remains to be determined), which can induce a myogenic phenotype when transfected and expressed in mammalian cells. Studies determining the effect of interferon on the expression of these muscle determinant genes should prove especially rewarding in defining the biochemical and molecular basis of interferon and TPA modulation of muscle cell differentiation in different myoblast cultures.

TABLE 4
Strategies for Enhancing the Clinical Utility of Interferon

Construction of molecular hybrids formed between different cloned interferon genes which produce novel hybrid
interferon proteins. These new molecules may exhibit enhanced antitumor activity and/or unique immunomodulating activities.
Direct modification of the structure of cloned interferon genes by site-specific mutagenesis and genetic recombination
techniques, thereby creating interferon molecules which display enhanced stability, antitumor activity, and/or
immunomodulating activities.
Combining specific unmodified recombinant interferons (or hybrid interferon proteins) and interferon plus other
cytokines, resulting in enhanced antitumor activity and/or immunomodulating activities.
Combining interferon with differentiation and/or immunomodulating compounds, resulting in a synergistic suppression of tumor cell growth and/or an increased expression of tumor-associated antigens.
Combining interferon with cytostatic and/or cytotoxic chemotherapeutic agents, thereby enhancing the therapeutic
index of these compounds in specific tumors.
Administering interferon prior to injecting monoclonal antibodies to upregulate the surface expression of tumor-
associated antigens. This approach can be utilized to increase the utility of monoclonal antibodies for the detection
and also the therapy of specific human cancers.

B. MELANOGENESIS

Melanogenesis involves the conversion of tyrosine into melanin and is regulated by a
single enzyme, tyrosinase.[53,54] In our initial studies on the effect of interferon on melanogenesis, we employed the C3 clone of mouse B-16 melanoma cells, which produces low
levels of melanin during active proliferation, but when cultures reach confluency and stop
proliferating, they begin to produce large amounts of melanin.[53,55] By growing B-16 cells
in the presence of α melanocyte-stimulating hormone (α-MSH), $5 \times 10^{-7} M$, the onset of
melanogenesis and the level of melanin production is increased.[36,55,56] Incubation of B-16
cells with 0.03 to 30 units per ml of crude natural mouse L-cell interferon results in a dose-
dependent inhibition in both spontaneous and α-MSH-induced differentiation.[56] When mouse
L-cell interferon is combined with TPA, which inhibits differentiation in B-16 cells when
applied alone,[55] a synergistic inhibition in both spontaneous and α-MSH-induced melanogenesis in B-16 cells results.[56] The inhibitory effect of both agents occurs in the absence of
growth suppression and is apparently mediated by a delay in the peaking of tyrosinase in
treated cultures. Mouse L-cell interferon and TPA, alone or in combination, can only
transiently block B-16 differentiation, as indicated by the eventual terminal differentiation
of cultures, even in the continuous presence of these agents. The mechanism by which
interferon and TPA inhibit differentiation in B-16 cells is apparently different, based on
three types of evidence: (1) Interferon is able to inhibit B-16 differentiation when added to
B-16 cultures at a later time point than TPA; (2) The inhibitory effect of interferon on α-
MSH-induced differentiation of B-16 cells is greater than that of TPA; (3) The combination
of interferon plus TPA is more effective in inhibiting differentiation than either agent alone.
The molecular cloning of the mouse and human tyrosinase genes[57,58] and the development
of mouse and human tyrosinase monoclonal antibodies[59] will now permit studies to be
conducted to determine if interferon and TPA inhibit melanogenesis in B-16 mouse melanoma
cells by altering mRNA transcription, steady-state mRNA, processing, location, and/or the
enzymatic activity of tyrosinase.

The human leukocyte interferons comprise a multigene family consisting of >13 non-
allelic human α genes, many of which have been cloned, sequenced, and expressed in
Escherichia coli.[12,14,15] The ability to analyze the specific role of defined regions within the
human leukocyte proteins in mediating diverse biological effects of these molecules on target
cells has been aided by the isolation of multiple cDNA and genomic recombinants of these
genes.[12,14,15] By enzymatically cleaving the different leukocyte cDNA or genomic clones
containing common restriction endonuclease sites with the appropriate restriction enzymes
and religating complementary DNA segments, a series of novel hybrid leukocyte interferon

DNA constructs and proteins have been synthesized and hybrid proteins have been purified to homogeneity.[12,14,15,60-65] Hybrid leukocyte interferons formed between IFN-αA and IFN-αD have proven especially informative in studying the structure/function relationship of human leukocyte interferon in regulating differentiation in the heterologous murine B-16 cell culture system.[62,66] In contrast to IFN-αA or IFN-αD, which do not display significant antiviral activity on mouse cells, two hybrid constructs, IFN-αA/D (Bgl) and IFN-αA/D (Pvu), generate recombinant proteins which display antiviral activity when applied to mouse cells.[62] IFN-αA/D (Bgl) consists of amino acids 1 to 62 from IFN-αA and amino acids 64 to 166 from IFN-αD, while IFN-αA/D (Pvu) consists of amino acids 1 to 91 from IFN-αA and amino acids 93 to 166 from IFN-αD. Other hybrid constructs, including IFN-αD/A (Bgl), IFN-αD/A (Pvu), and IFN-αA/D/A (Bgl-Pvu), do not induce antiviral activity in mouse L-cells. In addition to inducing antiviral activity in mouse L-cells, IFN-αA/D (Bgl) also: (1) inhibits both spontaneous and α-MSH-induced differentiation in mouse B-16 melanoma cells, in a manner similar to that of mouse L-cell interferon;[56,66] (2) acts synergistically with TPA in inhibiting B-16 differentiation in a manner analogous to that of mouse L-cell interferon;[56,66] (3) induces antitumor activity in nude mice carrying human tumor xenografts;[67] and (4) enhances tumor-associated antigen expression in MCF-7 human breast carcinoma and Colo 38 human melanoma cells with a potency similar to that of IFN-αA, even though IFN-αD is essentially devoid of this property in MCF-7 cells.[19,68,69] In contrast, IFN-αA/D (Pvu) does not inhibit differentiation in B-16 cells, suggesting that amino acids 64 to 91 from IFN-αD represent an important domain in this hybrid protein responsible for the differentiation-modulatory activity of this hybrid protein construct in B-16 mouse melanoma cells.[66]

The observations described above indicate that by genetically engineering the structure of the leukocyte interferon genes, novel proteins can be produced which differ in their properties when compared with their unmodified parental molecules. These unique proteins can be employed to directly test the effect of human interferons on the growth of human tumor cells in nude mice.[67,70,71] In addition, since certain hybrid constructs retain antiviral activity but lose differentiation repression activity in heterologous mouse cells, these proteins should prove useful in defining the functional domains of the interferon protein which induce specific receptor-mediated responses in appropriate target cells. Since new interferon molecules can be generated by constructing molecular hybrids, not only between different leukocyte interferon species, but also between different classes of interferon (e.g., IFN-α/IFN-β, IFN-α/IFN-γ, and IFN-β/IFN-γ) and between interferons derived from different animal species (e.g., human/mouse, human/rat, etc.), it may be possible by employing this approach to identify interferons with unique properties; for example, (1) molecules with high antitumor activity but devoid of negative adverse biological effects; (2) molecules with an increased capacity to induce terminal differentiation in tumor cells; (3) molecules with potent antiviral activity but without negative side effects; and (4) molecules with enhanced immunomodulatory activity (including an increased ability to stimulate NK cell activity and/or augment tumor-associated antigen expression on tumor cells). These novel interferon proteins could prove to be important reagents in both the diagnosis and treatment of cancer in humans as well as valuable molecules to analyze specific biological properties in target cells.

The effect of natural murine IFN-α and IFN-β and recombinant IFN-γ, alone and in combination with α-MSH, on the differentiation of a recently derived 7,12-dimethylbenzanthracene-induced JB/MS melanoma has been evaluated by Kameyama et al.[59] As was observed with B-16 melanoma cells,[55,56] treatment with α-MSH increased the rate of melanin synthesis 25-fold and tyrosinase activity 6-fold after a 4-d treatment. Although the three types of interferons did not directly alter melanogenesis in JB/MS melanoma cells, the interferons synergistically increased melanin production when used in conjunction with α-MSH. By fluorescence-activated cell-sorter analysis in conjunction with tyrosinase-specific antibodies, evidence was presented indicating that the stimulation of melanogenic activity by α-MSH and interferon plus α-MSH may have occurred by the activation of preexisting enzyme. A potential mechanism by which the combination of interferon plus α-MSH increases melanogenesis in JB/MS melanoma cells was suggested by Scatchard analyses of α-MSH receptors on interferon-treated cells. All three interferon preparations increased (~2.5-fold) the number of α-MSH receptors on cells relative to untreated controls. Further studies are required, however, to determine if the ability of interferon to stimulate α-MSH-induced differentiation in JB/MS melanoma cells is unique to this cell type or if it represents a general phenomenon in early-passage murine melanoma cells and if other murine melanoma cell cultures which display a similar response to interferons plus α-MSH also express increased numbers of α-MSH receptors.

In contrast to murine B-16 melanoma cells, and the JB/MS melanoma cell line exposed only to interferon, treatment of specific human melanoma cultures with interferon or TPA results in the induction of differentiation, as demonstrated by an increase in melanin synthesis.[31,72-77] The most effective interferon preparation in inducing growth suppression as well as inducing melanogenesis was IFN-β.[73] When IFN-β was combined with mezerein, a compound which shares a number of *in vitro* properties with the diterpene phorbol ester tumor promoters, including its potent cell differentiation-modulating activity,[78-82] a potentiation of both growth suppression and melanogenesis was observed. This effect was apparent even in melanoma cells displaying an innate relative resistance to either preparation when employed alone.[73,75] As will be discussed later in this review, this system is proving valuable for defining the specific antigenic and gene expression changes which occur concomitant with differentiation and growth inhibition in human melanoma cells and should permit the identification and molecular cloning of gene(s) which may control the melanogenesis phenotype.

C. ADIPOGENESIS

When maintained at confluency for 1 to 3 weeks, specific clones of both Swiss 3T3 (3T3-L1) and Balb/C 3T3 (clone A31T) cells undergo adipogenic conversion which is characterized by the cytoplasmic accumulation of triglycerides.[83-87] This *in vitro* adipogenic conversion is similar in a number of ways to the process of adipogenesis which occurs *in vivo,* including cessation of cell growth, distinctive morphological changes, induction of new enzyme synthesis, acquisition of increased insulin membrane receptors, and altered metabolism which can be further modified by hormones.[86] Keay and Grossberg[88] demonstrated that crude and partially purified mouse interferon could inhibit the number of Swiss mouse 3T3-L1 cells which accumulated lipids in response to treatment with insulin. In this model cell-culture system, interferon also reduced the levels of triglycerides, cholesterol, and cholesterol esters produced in response to insulin treatment. The specificity of the mouse interferon effect on adipogenesis in 3T3 L-1 cells was suggested by the inability of inactivated (trypsin-treated or antibody-inactivated) mouse interferon or heterologous interferons (human leukocyte, human fibroblast, or chicken interferons) to inhibit differentiation. Less than one unit of crude mouse fibroblast interferon inhibited the incorporation of $[1\text{-}^{14}C]$ acetate into lipids, whereas the incorporation of labeled leucine into trichloroacetic acid-precipitable protein was unaffected. Similarly, interferon did not alter DNA content in insulin-treated 3T3 L-1 cells. These results indicate that interferon can specifically inhibit insulin-induced adipogenesis in Swiss mouse 3T3-L1 cells without inducing a generalized suppression in the metabolic activity of treated cells.

Cioe et al.[89] analyzed the effect of mouse interferon and TPA, administered alone or in combination, on adipogenesis in BALB/C 3T3 (clone A31T) mouse cells. A minimum dose of 25 units per ml of crude mouse interferon reduced the number of lipid-positive cells developing both spontaneously and as a result of treatment with insulin. In order to induce complete suppression of adipogenesis in BALB/C 3T3 (clone A31T) cells, a concentration of 200 units per ml of crude interferon was required. These results are at variance with those of Keay and Grossberg,[88] who demonstrated that less than one unit per ml of crude mouse interferon could inhibit insulin-induced adipogenesis in Swiss 3T3 (3T3-L1) cells. Several explanations are possible for the different dose response to interferon reported in the studies of Keay and Grossberg[88] vs. Cioe et al.,[89] including differences in the cell types used and/or the types of interferon preparations employed. Cioe et al.[89] also demonstrated that: (1) interferon was able to inhibit adipogenesis if added at day 0 or 8 d after confluency when cells had already begun to convert to adipocytes; (2) the inhibitory effect of interferon on adipogenesis was fully reversible, even after 14 d of exposure to interferon; (3) the interferon effect on differentiation was specific, since no antidifferentiation effect was observed when heterologous human interferon, heat-inactivated mouse interferon, or trypsin-treated mouse interferon was used in place of untreated mouse interferon; (4) the simultaneous treatment of BALB/C 3T3 (clone A31T) cells with TPA and interferon resulted in a greater inhibition of adipogenesis than when either agent was used alone, e.g., the combination of an ineffective dose of TPA (1.6×10^{-9} M) and a weakly effective dose of interferon (25 units per ml) resulted in a level of inhibition of adipose conversion similar to that observed with 100 units per ml of interferon; and (5) the mechanisms by which TPA and interferon inhibit adipogenesis are different, e.g., cultures treated with 10^{-7} or 10^{-8} M TPA often escape from inhibition and differentiate even in the presence of fresh TPA, whereas 200 units/ml of interferon or TPA (10^{-9} M) plus interferon (25 to 100 units/ml) results in a complete suppression of adipose conversion.

In more recent studies, Taylor et al.[90] have determined the effect of mouse fibroblast interferon and alpha-fluoromethyl ornithine (DFMO), the inhibitor of ornithine decarboxylase (ODC), added alone or in combination, on differentiation in Swiss 3T3 (3T3-L1) cells. Both agents inhibited adipogenesis in a dose-dependent manner and the combination of mouse fibroblast interferon and DFMO resulted in an additive antidifferentiation effect. The mechanism by which these agents inhibited differentiation and affected cellular processes was different, as indicated by: (1) the ability of putrescine, the product of ornithine decarboxylation, to reverse the inhibitory effect of DFMO on differentiation, whereas this agent had little effect on the antidifferentiation activity of fibroblast interferon; and (2) the ability of DFMO to reduce the antiviral activity of fibroblast interferon in both undifferentiated and differentiated 3T3-L1 cells, whereas DFMO by itself had no detectable effect on the replication of encephalomyocarditis virus in 3T3-L1 cells. These studies also suggest that the antiviral action of interferon does not involve the regulation of polyamine metabolism by ornithine decarboxylase.

Significant recent advances have resulted in a clearer understanding of the molecular basis of adipogenic conversion of fibroblast cells.[91] By employing newly isolated molecular probes for identifying specific changes in gene expression which occur concomitant with adipogenesis, future studies will permit a more detailed molecular analysis of the mechanism by which interferon alters both spontaneous and insulin-induced adipogenic conversion of cultured fibroblast cells.

D. HEMATOPOIESIS

A large body of literature has accumulated describing the effects of interferon on hematopoiesis in both normal and tumor cells.[15,18,20,26,28-31] In this part of our review, we will discuss those studies investigating the effect of interferon, used alone and in combination with differentiation-modulating agents, on hematopoiesis in leukemic cells. We emphasize these systems because they not only provide information pertinent to the mechanism of action of interferon, but also suggest that interferon in combination with other agents may prove useful clinically in the treatment of hematopoietic malignancies which may be refractile to either agent employed alone.

When grown in a medium containing 1.5% dimethyl sulfoxide (DMSO), or other chemical inducers of differentiation such as hexamethylene *bis*-acetamide (HMBA) or hemin, Friend erythroleukemia (FEL) cells undergo erythroid differentiation (from the proerythroblast to the normoblast stage of erythropoiesis), as indicated by an increased incorporation of iron into heme, increased synthesis of hemoglobin, and an increased production of globin mRNA.[92-97] FEL cells are immortal erythroid precursor cells, blocked in their differentiation at the proerythroblast (late CFU-E) stage, which are chronically infected with Friend murine leukemia virus (F-MuLV) and spontaneously release viral particles.[92] An increase in the intracellular accumulation of viral antigens, with a blockage of virus shedding, is observed in FEL cells treated with interferon.[98-101] When FEL cells are treated with interferon, they exhibit a dose-dependent biphasic response, i.e., at interferon doses between 200 and 500 units/ml, hemoglobin and globin mRNA synthesis are enhanced,[99,101-103] whereas at higher doses of interferon ($\geq$1000 units/ml), erythroid differentiation is inhibited.[103-106] Treatment of FEL cells with higher concentrations of interferon is also associated with a prolongation of the G_1 and G_2 phases of the cell cycle and growth suppression.[107] The biphasic effect of interferon on FEL differentiation is not a consequence of impurities in the interferon preparations, since similar effects on antiviral activity and differentiation are apparent when using purified preparations of interferon.[106]

The role of interferon in modulating erythroid differentiation has been studied extensively by Rossi and collaborators[101,103-106,108-113] using FEL cells sensitive (S) and resistant (R) to various interferon-induced biochemical changes. Earlier studies employing this system indicated that the ability of high doses of interferon (1000 units/ml) to inhibit differentiation of FEL-S cells was (1) temporally dependent, i.e., addition of interferon 24 h following DMSO exposure resulted in optimal inhibition, whereas addition of interferon 2 or 3 d following DMSO exposure resulted in an attenuated response, and (2) related to an inhibition in the transcription and translation of globin mRNA, as opposed to an effect on the assembly of globin into hemoglobin or heme production.[104-106] In the case of FEL-R cells, selected by growth in interferon-containing medium, these cells were found to be resistant to the induction of the antiviral state against lytic viruses (mengo and vesicular stomatitis virus) as well as a retrovirus (F-MuLV), even when exposed to levels of interferon which were 1000-fold higher than those eliciting a protective effect in FEL-S cells.[108] When exposed to low doses of interferon (15 to 480 units per ml), FEL-R cells were resistant to the differentiation-enhancing effect of interferon observed in FEL-S cells, whereas high concentrations of interferon (5000 to 20,000 units per ml) did not alter vesicular stomatitis virus yield or F-MuLV production, but they did inhibit DMSO-induced differentiation in FEL-R cells to an extent similar to that observed in FEL-S cells.[108] These findings suggest that the mechanism by which interferon induces an antiviral state is different from that involved in inhibiting differentiation in FEL cells. These observations are also consistent with the large body of evidence indicating that the mechanisms which underly the antiviral, antiproliferative, and immunomodulatory actions of interferon may be different.[15,18-20]

More recent studies employing FEL-R cells which are resistant to growth inhibition by α/β-interferon[108,110,112] indicate that when these cells are exposed to α/β-interferon they do not develop an antiviral state, show enhanced HLA Class I (H-2) antigen expression,[114] or display an induction of 2-5A synthetase.[109,111,112] However, these α/β-interferon FEL-R cells are sensitive to IFN-γ[111] and they bind IFN-β.[108] In contrast to their resistance to α/β-interferon, upon exposure to DMSO, all variant clones differentiate in a manner similar to that of FEL-S cells. Affabris et al.[112] have determined the effect of recombinant IFN-α, IFN-β, and IFN-γ on the growth and DMSO-induced differentiation in FEL cells. Treatment with murine recombinant IFN-β enhanced DMSO-induced FEL differentiation, whereas IFNα (recombinant) and IFN-γ (both recombinant and natural) inhibited DMSO-induced FEL differentiation. By employing FEL-S and FEL-R cells, evidence was presented indicating that the mechanisms underlying the stimulatory effect of IFN-β and the inhibitory effect of IFN-α on DMSO-induced FEL differentiation occur via different mechanisms. Although the response of FEL cells to the different interferons is obviously complex, this system represents an interesting model for defining the mechanism by which interferons induce growth suppression and the relationship between this response and modulation of differentiation. In addition, these studies emphasize the diversity of cellular responses which can be induced when IFN-β or IFN-α interact with the same cell-surface receptor molecule.

Analyses of terminal differentiation of human myelomonocytic cells have most frequently employed the promyelocytic HL-60,[115] the promyelocytic ML3,[116] and the histiocytic U937[117] leukemia-derived human cell lines. Growth of myelomonocytic cells in TPA,[118,119] 1,25-dihydroxyvitamin D_3,[120] or leukocyte products present in conditioned medium from phytohemmaglutinin-stimulated leukocytes[121] induces them to differentiate into cells with traits of monocytes/macrophages. In contrast, when HL-60 cells are grown in medium containing DMSO[122] or retinoic acid (RA),[123] they differentiate along the myeloid lineage. A number of studies have indicated that IFN-α and IFN-β can inhibit the growth of HL-60 cells without inducing these cells to differentiate.[124-129] In contrast, IFN-γ has been shown to induce monocytoid differentiation in HL-60 cells, as indicated by the expression of monocyte-specific surface antigens, nonspecific esterase, receptors for the Fc fragment of IgG, and the ability to mediate antibody-dependent cell-mediated cytotoxicity.[126,127,130-132] Similarly, IFN-γ induces U937[126,132-136] and ML3[126,137] cells to differentiate into monocyte-like cells. Of potential clinical importance has been the observation that the combination of specific interferons with differentiation-modulating agents, including retinoic acid,[125,127-129,134] DMSO,[138] TPA,[125,138] tumor necrosis factor,[135,137] lymphotoxin,[137] actinomycin D,[128] and 1,25-dihydroxyvitamin D_3,[134,135,139-142] results in an additive or synergistic induction of differentiation in human myelomonocytic leukemia cells. In a number of cases, the differentiation-enhancing effect of interferon was not only apparent when using IFN-γ, which induces differentiation when used alone, but was also evident when using IFN-α or IFN-β, which do not modulate differentiation independently.[125,127-129,138,142] The potential increased effectiveness of the combination of interferon and retinoic acid in suppressing growth and inducing differentiation in fresh leukemic blast cells from patients with acute myelogenous leukemia (AML) has been demonstrated by Gallagher et al.[142] The combination of IFN-αA and retinoic acid was more effective than either agent employed alone in inhibiting the growth of a majority of AML patient-derived cells, as determined using a myeloid leukemic blast cell clonogenic assay. The combination of agents was also more effective in inducing terminal differentiation of AML patient-derived cells. In addition, an increased inhibition of clonal growth and/or differentiation by retinoic acid alone was observed in two of five cases following *in vivo* treatment with IFN-αA. These observations suggest that the combination of interferon with modifiers of differentiation may prove valuable in enhancing the effectiveness of therapy in specific patients with myelogenous leukemia.

A major problem in cancer chemotherapy is the development of tumor cell populations which are not only resistant to the originally employed chemotherapeutic agent, but which also display a cross-resistance to other structurally diverse chemotherapeutic agents.[143-147] This phenomenon, commonly referred to as multidrug resistance (MDR), is one mechanism by which tumor cells appear to evade destruction by chemotherapy and may be a major factor in the ineffective treatment of patients with recurrent tumors following a second course of chemotherapy. In **typical MDR,** overexpression of a gene which encodes a 170 kDa mol wt glycoprotein, called the P-glycoprotein, has been found in the membrane of multidrug-resistant cells.[143-147] The P-glycoprotein has been characterized biochemically as an integral plasma membrane glycoprotein which spans the entire lipid bilayer and appears to function as an efflux transport pump similar to that found in bacterial membranes.[143-147] In this context, interferons may prove useful in the treatment of specific tumors which display this phenotype.

We have recently inserted a cloned human MDR gene, mdr1, into a human glioblastoma multiforme cell line G18 and selected cells for resistance to colchicine, referred to as G18[MDR].[148] Stably transfected G18[MDR] cell clones have been obtained which overexpress the mdr1 gene and which are resistant to the toxic effects of colchicine. By continued growth in gradually increasing concentrations of colchicine, we have also obtained subclones of G18[MDR] which have increased mdr1 mRNA expression, referred to as G18[MDR-AMP], and which are resistant to >30-fold higher levels of colchicine than G18[MDR] and >100-fold higher levels of colchicine than parental G18 cells. When grown in the presence of IFN-β or IFN-γ, growth suppression of G18, G18[MDR], and G18[MDR-AMP] cells was similar, and the combination of IFN-β plus IFN-γ was more effective in suppressing growth, as was originally observed in other glioblastoma multiforme cell lines,[149,150] than either preparation employed alone. In recent studies employing the G-18 series, we have additionally found that the ability of interferons, other immunomodulators such as tumor necrosis factor and various interleukins, and chemotherapeutic agents to enhance and/or induce the expression of specific cell-surface molecules identified using monoclonal antibodies and fluorescence-activated cell-sorter analysis is different.[151] Further studies are required to ascertain the relevance of these findings with respect to the use of specific bioresponse modulators and immunomodulators, alone or in combination with each other and with classical chemotherapeutic agents, in the management of tumors exhibiting an MDR-phenotype.

The studies described above were discussed in order to introduce the concept of differentiation-resistant HL-60 promyelocytic leukemia cells and the use of interferons to overcome the differentiation-resistance phenotype. By growing HL-60 cells in gradually increasing concentrations of TPA, cis or trans retinoic acid (RA), or DMSO, a series of resistant variants were obtained which were no longer responsive to either the growth-suppressive or differentiation-inducing effects of the respective selection compound when receiving doses of agents which induced these effects in unselected parental HL-60 cells.[138] Although these HL-60 variants were not tested for the MDR-phenotype, they were useful in determining if interferon could alter their resistance phenotype. When grown in the presence of IFN-αA or IFN-β, a dose-dependent inhibition of growth was observed without a concomitant alteration in the status of differentiation of HL-60 cells.[125,138] Similarly, IFN-αA or IFN-β were only growth suppressive and not differentiation inducing in the resistant variants, HL-60[TPA], HL-60[RA], or HL-60[DMSO].[138] When treated with appropriate doses of inducers of growth suppression and terminal differentiation in parental HL-60 cells, HL-60[TPA], HL-60[RA], or HL-60[DMSO] were not altered in these parameters. In contrast, growth in the presence of IFN-αA or IFN-β (1000 units/ml) in conjunction with each inducer in both the parental and the appropriate variant cell line resulted in a synergistic antiproliferative effect, whether measured by suspension culture growth or clonigenicity in soft agar.[125,138] In addition, administration of interferon together with one of these agents reduced the concentration of TPA, RA, or DMSO necessary to induce terminal differentiation in resistant cells, as indicated by their ability to reduce NBT dye.[138] Even though the combination of interferon and the appropriate inducer was able to induce terminal differentiation in a proportion of the variant cell population, full restoration of responsiveness to TPA, RA, or DMSO was not achieved in the variants, since considerably higher concentrations of the appropriate inducer were necessary to produce cellular differentiation than for the parent cells.[138] Since TPA, RA, and DMSO presumably display different mechanisms of action, it is possible that the antiproliferative effect of interferon facilitates in a general manner the ability of resistant leukemic cells to embark upon a differentiation program. The ability of interferon to increase the sensitivity of both sensitive and resistant HL-60 cells to inducers of terminal differentiation suggests that a combination of agents, i.e., interferon plus a differentiation inducer, may be more beneficial than either agent alone in the treatment of both drug-sensitive and drug-resistant leukemias.

E. OTHER MODEL DIFFERENTIATION SYSTEMS

Although not studied as extensively as the systems described above, interferons have also been found to modify additional programs of differentiation. In this section of our review, we will discuss some of the modifications induced by interferon, alone and in combination with other agents, in the differentiation of human hairy cell leukemia,[152] rat pheochromocytoma (PC12),[153] human teratocarcinoma (embryonal carcinoma),[154] human keratinocytes,[155,156] and human epidermoid carcinoma (A431)[157] cells.

Of the human malignancies studied to date, hairy cell leukemia, characterized by pancytopenia and an accumulation in bone marrow, spleen, and often peripheral blood of incompletely differentiated preplasmatic B lymphocytes with low proliferation, have proven responsive to long-term administration of IFN-α. Treatment of hairy cell leukemia patients with this interferon results not only in a decrease in the hairy cell mass, but also restores granulocytes, macrophages, platelets, and erythrocytes to normal levels. Recent *in vitro* studies by Michalevicz and Revel[152] indicate that IFN-α or IFN-β can alter the differentiation pattern of hematopoietic colonies derived from the peripheral blood of hairy cell leukemia patients, resulting in the enhanced formation of myeloid and monocytic elements, as well as the increased formation of normoblasts and megakaryoblasts, whereas the number of lymphoid cells decreased. In contrast, IFN-γ did not induce such a response in *in vitro* hairy cell leukemia hematopoietic colony cultures. These observations may have relevance to the mechanism by which specific interferons induce remission and relieve pancytopenia in hairy cell leukemia patients. In addition, this system may prove valuable in understanding the mechanism by which interferons modulate hematopoietic differentiation.

The effect of both natural and recombinant IFN-γ on the growth and differentiation of the rat pheochromocytoma cell line PC12 in the presence or absence of nerve growth factor (NGF) has recently been evaluated.[153] IFN-γ induced a reversible inhibition of proliferation of PC12 cells, whereas antimitotic agents, such as cytosine-arabinofuranoside (AraC), colchicine, mitomycin, and hydroxyurea, or removal of serum resulted in a mitotic arrest which was followed by cell death. An interesting observation was that NGF alone did not alter the level of $2',5'$-A synthetase, whereas IFN-γ induced an average 4.4-fold enhancement in $2',5'$-A synthetase activity and the combination of IFN-γ and NGF resulted in a marked (5- to 18-fold) increase in $2',5'$-A synthetase activity in PC12 cells. In addition, IFN-γ-treated PC12 cells responded more rapidly to NGF with respect to the development of neuronal outgrowth. Although further studies are required, the present investigation suggests that the combination of NGF and IFN-γ can enhance neuronal differentiation, and this approach could theoretically prove useful in inducing neuronal differentiation in tumors of neuroectodermal origin.

In the case of teratocarcinoma cells, interferons were only able to induce a partial response, and evidence has been presented that the extent of the response to these agents is developmentally regulated.[154] Human IFN-γ was able to induce Class I (HLA-A, B, and C) antigens and β_2-microglobulin in two human embryonal carcinoma cell lines and a yolk sac carcinoma cell line, whereas IFN-α and IFN-β were less effective inducers of these surface molecules. In contrast, no antigenic induction was observed with a gestational choriocarcinoma cell line. In addition, none of the interferon preparations induced growth inhibition, expression of Class II (HLA-DR) antigens, resistance to vesicular stomatitis infection, or expression of $2',5'$-A synthetase in any of these cell cultures. IFN-γ was also ineffective in inducing differentiation or potentiating the induction of differentiation when combined with RA in a pluripotent embryonal carcinoma cell line. The effect of the status of differentiation on the responsiveness of embryonal carcinoma cells to interferon was suggested by the observation that (1) IFN-γ induced a greater increase in the surface expression of Class I (HLA-A, B, and C) antigens and β_2-microglobulin in RA-induced differentiated than in undifferentiated cells, (2) both IFN-α and IFN-β induced $2',5'$-A synthetase only in differentiated cells, and (3) IFN-α induced a small resistance to vesicular stomatitis infection in aged cultures of differentiated cells. It is not known whether the differential response of human teratocarcinoma cells observed in this study reflect a potential role for interferon in the normal physiology of human development or whether these changes are reflective of pathological states of abnormal development.

The effect of recombinant IFN-α[155,156] and IFN-β[155] on the growth and differentiation of human keratinocyte cultures have also been determined. When cultured in serum-free hormone-supplemented medium, IFN-α or IFN-β inhibited the growth of keratinocytes, as indicated by cell counts, total protein, and the appearance of stained colonies.[155] Growth inhibition of keratinocytes by interferon was reversible following removal of interferon and was substantially blocked by the simultaneous addition of cholera toxin and interferon to the medium. In contrast to the effect of IFN-α and IFN-β on normal human keratinocytes, the growth of a human epidermal carcinoma cell line was less affected by interferon. IFN-α and IFN-β also reversibly promoted terminal differentiation of human keratinocytes, as assessed by the desquamation rate of cells from the colony surface and by the proportion of total cells having cornified envelopes. Similarly, studies by Stadler et al.[156] indicate that IFN-αA inhibits DNA synthesis, as indicated by ^{3}H-thymidine incorporation, and promotes terminal differentiation, as indicated by a decrease in the amount and synthesis of noncovalently bound keratins and an increase in the disulphide cross-linked keratins, in human keratinocyte cultures. Both of these studies support the hypothesis that interferon may function as a physiologic regulator of epidermal growth *in vivo* with properties of a negative growth factor or chalone.

Treatment of the human epidermoid carcinoma cell line, A431, with IFN-γ, but not with IFN-α or IFN-β, results in a rapid induction of morphological alterations and cell death.[156] This effect was associated with an increased expression, both RNA and protein, of c-Ha-*ras* and the epidermal growth factor receptor. By employing a three-dimensional *in vitro* tumoroid cell culture system, it was further demonstrated that keratin expression was increased in A431 cells by IFN-γ. These results suggest that IFN-γ can specifically kill A431 cells, and this effect is associated with an induction of terminal differentiation. Further studies are required to determine how general this effect of IFN-γ is on other epidermoid carcinoma cell cultures.

III. MODULATION OF GENE EXPRESSION BY INTERFERONS

A. DIFFERENTIAL INDUCTION OF NEW PROTEINS BY INTERFERONS

A wide assortment of proteins, with molecular weights ranging between 15 and 105 kDa, can be induced by exposing cells to interferons.[158-164] At the present time, the function of the majority of these proteins remains unknown, since in many cases they are detected only by gel electrophoresis. However, the identity and putative functions of several interferon-inducible genes is known, including the MX protein in mice which is associated with resistance to influenza viruses,[165] oligo(A) synthetase which induces the synthesis of oligonucleotides of the general structure pppA(2'-5A)n,[166,167] class I and II histocompatibility antigens and β2-microglobulin,[168-170] a GTP-binding protein,[171] and metallothionein.[172] It appears that some of the interferon-inducible proteins can also be induced by other mechanisms. For example, Bersini et al.[173] demonstrated that some of the interferon-inducible proteins can also be induced by interleukin 1a, interleukin 1b, and tumor necrosis factor. Similarly, metallothionein gene expression is not only regulated by interferons, but also by heavy metals,[174] glucocorticoid hormones,[174-177] interleukin,[178] and by activation of protein kinase C.[179]

The pattern of protein induction in a particular cell type in response to exposure to interferon depends on the type of interferon. Interferon-α and β (type I interferons) cause the induction of a common set of proteins, while interferon-γ (type II interferon) induces a set of proteins which partially overlap with those induced by interferons-α and -β.[159] For example, the 6-16 gene[180,181] is specifically induced by type I interferons, while HLA-DR[182] and β2-microglobulin[183] are preferentially induced by interferon-γ. Furthermore, Weil et al.[159,160] detected 12 proteins by 2D-gel electrophoresis that were induced by interferon-γ, but not by interferon-α or -β. The differences in protein induction between interferon-α and -β vs. interferon-γ have been explained by the observation that interferon-α and -β interact with a common membrane-localized receptor, while interferon-γ recognizes a distinct receptor.[4] An example of the functional importance of having two receptor systems for regulating gene expression is suggested by the observations of Sen and Rubin.[184] They demonstrated that both interferon-α and interferon-γ can induce the synthesis of a 56- and 67-kDa protein; however, the induction of these proteins is transient in response to continuous application of interferon-α, while the response is maintained in the presence of interferon-γ. Furthermore, by measuring the mRNA levels for these proteins, it was shown that after 24 h at a time point when the synthesis of proteins occurred only in response to interferon-γ, *in vitro* translatable mRNAs were present after both interferon-α and interferon-γ treatments. This demonstrated that the synthesis of interferon-inducible proteins is regulated both transcriptionally and translationally by interferons and that the nature of such regulation may depend on the receptor with which the interferon interacts.

Although the mechanism(s) by which interferons induce their differential induction of specific cellular proteins is not presently known, several studies employing protein synthesis inhibitors indicate that the regulation of these processes may be different for type I vs. type II interferon.[185-188] Protein synthesis is required for IFN-α and IFN-β, but not IFN-γ, to mediate induction of tumoricidal activity in the murine macrophage.[185] In contrast, only IFN-γ appears to utilize polypeptide mediators in establishing the antiviral state[186] and inducing (2'-5')oligoadenylate synthetase enzymatic activity[188] in human cells. Similarly, IFN-α and IFN-β increase the expression of histocompatibility gene products and a melanoma-associated antigen located in the cytoplasm and the plasma membrane of human melanoma cells by a process not requiring protein synthesis, whereas IFN-γ requires *de novo* protein synthesis for its antigen-enhancing activity.[188] These observations suggest that specific responses, such as induction of enhanced antigen expression, induction of (2'-5')oligoadenylate synthetase, and development of an antiviral state, which are induced by type I and type II interferon in human cells, are regulated by different mechanisms which are dependent on preexisting cellular factors (type I interferons) or require the synthesis of new cellular proteins (type II interferon).

B. INDUCTION AND CLONING OF CELLULAR GENES INDUCED BY INTERFERONS

Using recombinant DNA technology, it has been possible to study interferon-inducible genes in more detail than by separating proteins by gel electrophoresis. Using cDNA cloning techniques, a considerable number of interferon-inducible genes have been cloned, some of which are listed in Table 5. The kinetics for the induction of interferon-inducible genes is highly complex and depends both on the gene under investigation and on the cell line employed. For example, ISG-54 and ISG-56 are maximally transcribed within 2 h of continous interferon treatment in human fibroblasts, and this is followed by a decline in transcription.[193] The increased transcription is reflected in the accumulation of the respective mRNAs within 1 h, and peak mRNA levels are reached by 6 h. This increase and peaking in mRNA levels is followed by a decline in mRNA levels to less than 10% of peak levels within the subsequent 24 h.[190] It appears that the decline in the transcription of ISG-54 and ISG-56 is due to repression of gene expression, since it can be delayed or prevented by blocking protein synthesis with cycloheximide. Other examples of genes which are transiently induced by interferon include gene 11-25, metallothionein II, and 2'-5'oligo(A) synthetase.[180] In contrast to the above genes, a number of interferon-inducible genes are induced for up to 40 h. These include mRNAs encoded by genes 1-8, 9-27, 6-16, 8-27, and HLA class

TABLE 5
Interferon-Inducible Genes

Gene		Ref.
ISG-15		189
ISG-54	(IFN-IND 1)	190
		191
ISG-56	(IFN-IND 2; mRNA 561)	192
		193
		194
1-8		195
6-16		195
6-26		195
8-27		195
9-27		195
11-25		195
Metallothionein-II		195
2A		195
10Q		195
HLA-DRα		196
Mx		197
2'-5'oligo(A) synthetase		198
P56		184
P67		184
C202		199
C203		200

I,[180,195] and genes 2A and 10Q.[195] The induction of these genes is a primary response, since inhibition of protein synthesis does not prevent the induction of the mRNAs.[195] These results suggest that the induction of the genes mentioned above does not rely on the synthesis of new transcription factors, but, rather, is dependent on the modification of preformed transcription factors. However, the precise mechanism by which the binding of interferons to their receptors results in the induction of specific transcriptional changes in cells remains to be determined.

C. EFFECT OF INTERFERONS ON GROWTH REGULATORY GENE AND ONCOGENE EXPRESSION DURING DIFFERENTIATION

Oncogenes and their cellular homologues, proto-oncogenes, encode proteins which profoundly affect a large number of cellular processes, including the regulation of normal cellular proliferation, differentiation, and development. In addition, specific oncogenes have been directly implicated in the induction of cellular alterations associated with the transformed, tumorigenic, and metastatic phenotypes. In situations in which interferons induce a suppression of the growth of transformed cells and a loss of oncogenic potential in tumor cells, possible cellular targets for their action may be activated oncogenes and/or proto-oncogenes. With an increasing awareness of the function(s) normally controlled by proto-oncogenes and the cellular alterations which occur following their activation into oncogenes, a clearer picture of normal as well as abnormal cellular physiology is emerging. Oncogene proteins display a significant structural and functional homology with cellular growth-control elements; for example, (1) the *ras* family of oncogene products is similar to the GTPase/ GTP binding proteins that are involved in signal transduction from cell-surface receptors to adenyl cyclase;[201] (2) the *erb*-B oncogene is a truncated form of the epidermal growth-factor receptor (EGFR);[202] and (3) the *sis* oncogene product is similar to the B-chain of platelet-derived growth factor (PDGF).[203]

Oncogenes can be broadly classified into three categories: (1) those oncogenes whose products act as ligands for growth factor binding, e.g., sis oncogene; (2) those oncogenes

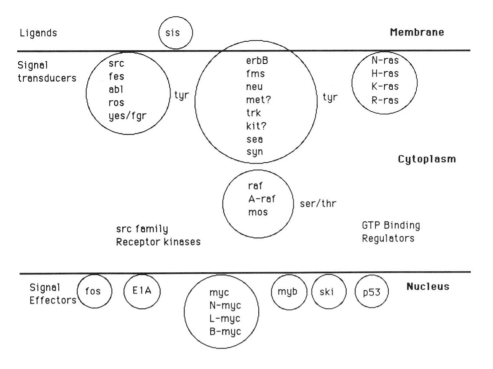

FIGURE 1.   Oncogenes grouped according to their location, amino acid sequence homology, and position in the signal transduction pathway of growth factors. (Modified from Rapp, U. R., Storm, S. M., and Cleveland, J. L., *Cancer Rev.,* 9,34, 1987.)

whose products act as transmembrane receptors and signal transducers, including members of the *src* gene family (*erb*-B, *fms, neu, met, trk,* c-*fgr,* c-*yes,* and c-*syn*) and the *ras* gene family (N-*ras,* H-*ras,* and K-*ras*); and (3) those oncogenes whose products are nuclear-acting proteins, e.g., *myc, fos, p53, myb,* and adenovirus *E1A.* Many of the nuclear on-cogenes display distinct patterns of regulation when exposed to specific growth factors. Although the list of oncogenes and their functions continue to grow, Figure 1 presents a simplified view of some of the different classes of oncogenes and their location in the cell and putative functions.[204]

The mechanism by which proto-oncogenes exert their influence on normal cellular differentiation and the manner in which these genes are regulated during spontaneous and induced differentiation is not well defined. Similarly, the role of interferon in modifying the expression of specific proto-oncogenes and inducible cellular genes during modulation of differentiation is just beginning to be analyzed on a molecular level. As indicated in Tables 1 and 2, interferons induce a large number of changes in target cells, including both inhibitory and stimulatory effects. It is important, therefore, to bear in mind that although interferon may alter the expression of a specific oncogene, this does not permit one to easily distinguish between a direct effect or a secondary effect of interferon (as a consequence of alterations in cellular physiology) on differentiation in the target cell. However, even with this reservation, several experimental models have been developed to study the effect of interferons on differentiation, and they are beginning to provide important insights into this process (Table 3).[18-20,26-31,97,205-207] In some systems, interferons have the capacity to inhibit differentiation, whereas in other systems they exert the opposite effect. The reason for these apparently conflicting experimental results is not known, but it may simply reflect differences in the transmembrane signals generated following binding of appropriate differentiation-modulating compounds to receptors in the cell membrane.[46] In one case, the binding of a specific molecule, such as interferon, to cell-surface receptors may function as a direct inducer of differentiation, resulting in initiation of the cascade of gene changes involved in differentiation, whereas in other situations the binding of interferon may actually inhibit the ability of a normal inducer of differentiation to exert its effect.[25,39,56,73] In this section of our review, we discuss the effects of interferons on the expression of some of the key proto-oncogenes and growth regulatory genes which may contribute to the differentiation process.

1. Autocrine Functions of Interferons

In order to understand the role of interferons in modulating the function of oncogenes and growth regulatory genes during differentiation, the growth regulatory function(s) of the interferons have to be defined within the context of normal physiologic conditions. Initially, interferons were regarded as proteins whose production was induced by external stimuli such as viral infection. However, it is now well accepted that interferons are produced under both normal and pathological conditions, and they are involved in cell proliferation, differentiation, and maintenance of the organism. It is also now apparent that some species of leukocyte and fibroblast interferons are produced constitutively by cells. Therefore, it is important to identify the endogenous inducers of interferons which operate *in vivo* and the role of interferon in normal cell growth, differentiation, and development of the organism.

Because of the variety of effects exerted by interferons on cell growth and differentiation, it is now realized that interferons may be producing their effects by interacting with other cytokines, growth factors, and regulatory gene products. An interaction between interferon and growth factors would be anticipated, since these molecules regulate cellular proliferation. However, only recently has it become apparent that growth factors enhance the expression of IFN-β_2, which exerts a negative feedback effect on the mitogenic response produced by growth factors. IFN-β_2 is identical to human fibroblast B-cell differentiation factor (BSF-2).[208] Two forms of IFN-β_2 have been recognized, and these have been designated IFN-β_{2a}[209-211] and IFN-β_{2b}.[212,213] For the present review, we will employ the general term IFN-β_2. An IFN-β_1 has also been identified which is structurally distinct from IFN-β_2.[214,215] IFN-β_1 is produced from a 0.9-kb mRNA transcript, whereas IFN-β_2 is synthesized from a 1.3-kb mRNA transcript, and the translational products of these two genes display only some amino-acid sequence homology.[214,215] Furthermore, only IFN-β_2 has been shown to have a negative feedback effect on the mitogenic response induced by growth factors. Tumor necrosis factor (TNF), which has mitogenic activity in human fibroblasts,[216,217] specifically induces the expression of the human IFN-β_2 gene and not the IFN-β_1 gene.[207] A negative feedback effect of IFN-β_2 on cell growth is suggested by the fact that antibodies against IFN-β_2 enhance the mitogenic effect of TNF. Similarly, recent studies by Chen et al.[218] indicate that IFN-β_2 selectively suppresses the growth of human breast carcinoma and leukemia/lymphoma cell lines at doses which stimulate DNA synthesis and growth in B-cell hybridomas and plasmacytomas. PDGF and calf serum also induce IFN-β_2 and not IFN-β_1 expression in quiescent human fibroblasts.[219] In addition, IL-1 has been found to induce IFN-β_2 in human fibroblast cells,[220] and IL-1 itself is induced by TNF.[221] Although EGF stimulation does not appear to induce production of IFN-β_2 in quiescent human fibroblasts, treatment of these cells with IFN-β_2 antibodies prolongs the EGF stimulation of DNA synthesis in human and Syrian hamster fibroblasts.[222] Furthermore, the effect of IFN-β antibodies could be reversed by the simultaneous addition of IFN-β_2.[222] In addition to growth factors, synthetic diacylglycerols, such as 1,2 dioctanoylglycerol (DiC8) and 1-oleolyl-2-acetylglycerol (OAG), strongly enhance IFN-β_2 mRNA expression in human fibroblasts.[223] The enhanced expression of IFN-β_2 by synthetic diacylglycerols can be blocked by a protein kinase C- and cyclic nucleotide-dependent protein kinase inhibitor, 1-5-isoquinolinesul-phonyl-2-methylpiperazine dihydrochloride (H7). Furthermore, the antiviral activity displayed by DiC8 can be blocked by anti-IFN-β antibodies.[223] These findings indicate a relationship between PKC activation and IFN-β_2 expression, and they also may explain why IFN-β_2 expression can be induced by PDGF, IL-1, and TNF. In summary, these results support the existence of a complex autocrine feedback loop in cells involving IFN-β_2 which functions as a negative feedback regulator of the mitogenic response normally induced by growth factors. At the present time, however, this autocrine mechanism of growth regulation has only been associated with IFN-β_2.

Because IFN-β_2 can be induced by a number of cytokines in a variety of cells, it is likely that IFN-β_2 may exert different functions, depending on the cell type and the conditions under which it is produced. Constitutive production of IFN-β_2 has been detected in a number of different cells, including human fibroblasts,[224] T lymphocyte cell lines,[225] and monocytes/macrophages.[226] It is likely that constitutive production of IFN-β_2 in normal cells is probably involved in the regulation of cell proliferation. The constitutive production of other interferons has also has been found, but no autocrine mechanism has been detected. These observations support the hypothesis that specific interferons are a component of the complex cytokine network operating in the organism, and they may play an integral role in the regulation of normal cellular differentiation and development.

2. c-*myc* Gene

The *myc* family of oncogenes has been extensively studied since v-*myc* was first identified in the avian myelocytomatosis virus (MC 29).[227] The initial identification of the v-*myc* oncogene was followed by the discovery of its cellular homologues, including c-*myc*, n-*myc*, and L-*myc*.[228] The two open reading frames of the c-*myc* gene predict that it will encode a protein of 49 kDa in size. However, a protein with an apparent molecular weight of 65 kDa has been indicated by electrophoresis.[229] This suggests that there is extensive posttranslational processing of the c-*myc* protein, which may occur in a similar manner in different organisms.

The levels of c-*myc* protein are relatively constant during the various stages of the cell cycle in proliferating cells,[230] while the levels in nonproliferating cells are very low or below detectable levels.[231] Following stimulation with cell type specific mitogens or growth promoting agents, a several fold induction of c-*myc* mRNA is seen in nonproliferating cells.[232] In addition, it has been demonstrated that proliferating cells, when induced to differentiate, display a rapid inhibition of c-*myc* expression.[233,234] An interesting observation was that IFN-α- or -β-induced inhibition of c-*myc* expression stimulated by PDGF was diminished in Balb/c 3T3 cells after treatment with the protein synthesis inhibitor cycloheximide.[235] This finding suggests that interferon-induced inhibition of c-*myc* expression is dependent on new or continued protein synthesis.[235] In another study, it was shown that HL-60 cells induced to differentiate, either with IFN-γ or 1,25 dihydroxy vitamin D3, displayed a parallel decrease in c-*myc* mRNA expression.[236] However, these findings do not unequivocally demonstrate that the reduction in c-*myc* expression is responsible for differentiation, since the two phenomena may be separable and c-*myc* expression may be a consequence rather than a cause of differentiation in HL-60 cells.

In IFN-sensitive Daudi cells, human IFN-α markedly reduced the level of c-*myc* mRNA, whereas in interferon-resistant Daudi cells the levels of c-*myc* mRNA did not significantly change following IFN-α treatment.[237] The downregulation of c-*myc* in the interferon-sensitive Daudi cells is an early event in these cells and precedes inhibition of growth by several hours, suggesting that this change is not a consequence of IFN-α-induced cessation of Daudi cell growth. It has been suggested, based on studies of the transcription rate and half-life of c-*myc* mRNA in interferon-treated cells, that decreased levels of c-*myc* may not be a prerequisite for the inhibition of cell proliferation observed in Daudi cells following IFN-α treatment.[29] In several additional studies, both IFN-α and IFN-β have been found to reduce c-*myc* mRNA in interferon-sensitive Daudi cells by as much as 75%.[234,238,239] This reduction in c-*myc* mRNA in interferon-treated Daudi cells did not depend on new protein synthesis, since it occurred in the presence of the protein synthesis inhibitor cycloheximide.[240] The reduction observed in c-*myc* mRNA levels in interferon-sensitive Daudi cells may be the result of a reduction in the half-life of the mRNA rather than a reduction in the rate of c-*myc* transcription.[239] This conclusion is supported by the observation that treatment of Daudi cells with either IFN-α or IFN-β does not alter the transcriptional activity of c-*myc*, as monitored by nuclear "run-on" experiments.[241] However, these studies do not exlude the possibility that posttranscriptional modification may contribute to the reduction in c-*myc* mRNA levels in interferon-treated Daudi cells.

In SH-SY5Y human neuroblastoma cells induced to differentiate with TPA, a downregulation in c-*myc* expression was accompanied by a rapid and transient increase in the expression of c-*fos*.[242] The upregulation of c-*fos* expression may not be important for differentiation, since the addition of dioctoglycerol (DiC8) induced a rapid increase in c-*fos* expression which was not accompanied by downregulation of c-*myc* or the induction of differentiation. The authors speculated that since both TPA and DiC8 activated protein kinase C (PK-C), activation of PK-C is not sufficient to cause differentiation in SH-SY5 cells. They also suggested that it was the downregulation of c-*myc* expression and not the induction of c-*fos* which is associated with differentiation of SH-SY5Y cells. Phorbol esters activate PK-C, but the role of PK-C in the downregulation of c-*myc* observed in SH-SY5Y cells is not clear from this study, since DiC8 which had activated PK-C to the same extent at 30 min as TPA had no effect on c-*myc* RNA expression.[242] DiC8 also failed to induce differentiation in HL-60 promyelocytic leukemic cells.[243] Similarly, bryostatin, an effective activator of PKC, inhibited TPA-induced differentiation in HL-60 cells.[244] In a separate study, similar findings were reported for human leukemia cells induced to differentiate to macrophages by phorbol esters.[245] However, in quiescent fibroblasts[246] and lymphocytes,[247] TPA has been found to induce both c-*myc* and c-*fos* mRNA. These results suggest that in different systems TPA can have either an inductive or suppressive effect on c-*myc* and c-*fos* expression.

A large number of studies have addressed the effect of interferons on c-*myc* expression during differentiation, and for the sake of brevity, only a small number of studies will be discussed. In several cell culture systems, a repression of c-*myc* expression has been observed following terminal differentiation.[248-253] These results suggest that irreversible suppression of c-*myc* expression is necessary for the withdrawal of cells from the cell cycle and terminal differentiation. However, in other systems, this correlation between c-*myc* expression and terminal differentiation has not been observed.[249] For example, in rat skeletal myoblasts, c-*myc* gene expression is inducible in biochemically and terminally differentiated myotubes, and the transient expression of c-*myc* does not suppress the differentiated phenotype.[249] It is possible, however, that suppression of c-*myc* expression at a specific point in the differentiation program of a cell may be a necessary event for terminal differentiation. In a recent study, a plasmid carrying an antisense human myc DNA sequence and the gene encoding *E. coli* xanthine/guanine phosphoribosyl transferase (Ecogpt) was introduced into the HL-60 cell line.[254] In these transfected HL-60 cells, high expression of antisense *myc* RNA reduced *myc* expression at both the transcriptional and translational level.[254] The suppression of endogenous *myc* in HL-60 cells was accompanied by a decrease in cell proliferation and monocytic differentiation. From this study, a role for c-*myc* in HL-60 cell proliferation and an involvement in differentiation is suggested. However, in most situations where a direct correlation has been found between inhibition of c-*myc* expression and the induction of differentiation, it is often difficult to ascertain if a change in c-*myc* expression is the primary mediator of differentiation. For example, in a recent study, it has been proposed that c-*myc* acts as a molecular switch which either directs cells toward a pathway culminating in terminal differentiation or in continued proliferation.[255] In this study involving forced expression of c-*myc* in 3T3-L1 preadipocytes, it was found that cells overexpressing c-*myc*, in contrast to unmodified 3T3-L1 preadipocytes, reentered the cell cycle from the G_0/G_1 stage when challenged with 30% serum and also failed to differentiate. The differentiation block could be reversed by high expression of an antisense *myc* RNA, indicating that the differentiation block was due to the increased constitutive expression of c-*myc*. These results suggest that 3T3-L1 preadipocytes enter the G_0/G_1 stage after treatment with differentiation inducers, whereas cells expressing high levels of c-*myc* are blocked from entering this stage of the cell cycle. Therefore, c-*myc* is acting as a molecular switch which either directs a cell to continued proliferation or to terminal differentiation. In several cell systems, it has been found that repression of c-*myc* expression occurs concomitantly with differentiation.[248-252] If c-*myc* can act as a molecular switch for all cell systems, then interferon-induced suppression of c-*myc* at a specific stage of the cell cycle could be used to direct cells to differentiate.

In other systems, where IFN-α inhibits cell proliferation, a direct association between changes in oncogene expression and induction of differentiation has not been definitively demonstrated. For example, there have been several reports indicating the IFN-α inhibits proliferation in HL-60, U937, and Friend erythroleukemia cells without inducing changes in c-*myc* expression.[234] These observations suggest that either c-*myc* may not play a key role in differentiation or that there may be separate pathways leading to differentiation and only one or a few are active in any specific cell type. This may explain the discrepancies observed in demonstrating a direct correlation between changes in c-*myc* expression and induction of differentiation by diverse agents in different target cells.

3. c-*fos* Gene

The cellular homologue of v-*fos*, c-*fos*, has been detected in the genome of various vertebrates, including human and mouse.[256] The product of the c-*fos* oncogene is a 55-kDa nuclear protein that is complexed to a cellular protein, referred to as p39,[257] and possibly DNA. It is now known that several different stimuli cause a transient increase in the expression of c-*fos*, e.g., EGF, PDGF, c-AMP, TPA, Vit D3, and the *ras* p21 protein, in various cell types. In addition, c-*fos* can be activated by the addition of a transcriptional enhancer or the disruption of the 3′ noncoding region;[258] deletion of the 3′ noncoding region is believed to stabilize the mRNA. The transient upregulation of c-*fos* expression is associated with signal transduction occurring as a consequence of binding of interferon to specific receptors. Induction of c-*fos* has been reported to be associated with differentiation,[246,250,257,259,260] while (as discussed briefly above) in other studies no such correlation has been found.[245]

Einat et al.[235] determined the effect of IFN-β and IFN-γ on c-*fos* gene expression and cell growth of quiescent Balb/3T3 cells and observed that the addition of interferon could reduce by 50 to 80% the increase in c-*fos* normally observed after the addition of either PDGF or serum. In addition, interferon reduced the mitogenic response of Balb/3T3 cells following treatment with PDGF. This was not due to interferon interfering with PDGF binding to receptors, since the deoxyglucose uptake associated with PDGF binding was not changed. Furthermore, the mRNA levels for ornithine decarboxylase, β-actin, and c-*myc* were also reduced by interferon, and these genes are thought to be part of the mechanism that makes cells progress from G_0 to G_1. Thus, interferon, by repressing the expression of these and other competence genes and thereby counteracting the effect of PDGF, determines the proliferative status of the cell.

Rapid induction within 15 min and peaking at 30 min of c-*fos* mRNA occurs in F9 embryonal carcinoma and NIH 3T3 cells following treatment with either IFN-α/β or IFN-γ.[261] This change in c-*fos* mRNA is transient, since message levels decline rapidly and return to baseline levels after 2 h.[261] Both types I and II interferon induced c-*fos* expression with similar kinetics in NIH 3T3 cells. In contrast, the interferons did not effect the expression of c-*myc*, although the expression of c-*fos* and c-*myc* are often coupled in this cell line. The induction of c-*fos* by the interferons exhibited kinetics similar to those obtained with a variety of receptor-mediated stimuli, such as PDGF, CSF-1, and EGF.[262,263] It has also been shown that compounds which increase c-AMP levels stimulate c-*fos* expression. Based on these and other observations, it has been suggested that c-*fos* expression induced by the interferons is associated with signal transduction elicited by binding of the interferons to their receptors.[20] A combination of the protein synthesis inhibitor cycloheximide with IFN-γ resulted in an amplified induction of c-*fos*.[261] The kinetics of cycloheximide versus IFN-γ induction of c-*fos* are different, suggesting that the mechanisms of induction by these two agents are also different. In addition, induction of c-*fos* mRNA has been reported in other cells, including T-cell hybridomas, fibroblasts, and epithelial cells derived from murine embryos.[261] The failure of F9 cells to differentiate after interferon-induced c-*fos* expression is in agreement with a previous study on F9 cells[264] and also with a study which indicated no correlation between c-*fos* expression and monocytic differentiation.[265]

A direct role of c-*fos* in mediating differentiation has not been easy to demonstrate, and it has been difficult to directly distinguish between a cause and effect relationship. For instance, in monocyte differentiation, it has been found that c-*fos* expression can be dissociated from differentiation.[265,266] In PC12 neuroblastoma cells induced to differentiate by nerve growth factor, an increase in c-*fos* expression has been observed, but again there is no evidence proving that this change in c-*fos* expression is essential for the induction of differentiation.[267] However, in F9 teratocarcinoma cells containing either a normal c-*fos* gene or c-*fos* constructs under the transcriptional control of a metallothionein promoter, colonies of differentiated cells were obtained even in the presence of increased c-*fos* expression.[268] Although these F9 teratocarcinoma cells displayed some of the markers characteristic of endodermal differentiation, their defined embryological status remained obscure. The differentiation potential of c-*fos* cannot be assessed from this study, since differentiation of F9 cells can also be achieved with other oncogenes, including the ras oncogene from EJ carcinoma cells.[268] Since differentiation in F9 cells can also be induced by genes other than c-*fos*, it is unlikely that c-*fos* is performing a unique function required for differentiation. It has been suggested that this phenomenon may be a property of the F9 carcinoma cells and not a general property of all differentiating cells., i.e., an aberrant type of differentiation may be occurring in this cell line.

Currently available scientific evidence does not indicate a definitive role of c-*fos* in the control of cellular differentiation, although correlations with cell proliferation or differentiation have been observed. The majority of experimental data suggest that c-*fos* is not a primary gene involved in differentiation, but rather, is most likely involved in the regulation of cell proliferation, which is often directly related to the process of differentiation. The variety of stimuli which induce c-*fos* expression indicates that it is involved in a number of different growth regulatory pathways, some of which may be lineage specific or stage specific. Based on studies employing two-dimensional gel electrophoresis, a role for the product of the *fos* gene in the differentiation of preadipocytes to adipocytes was suggested.[269] In the maturation from preadipocyte to adipocyte, major changes in at least 100 proteins were observed, of which 40% represented newly synthesized polypeptides.[270] From this complex array of differentiation-associated gene changes, several cDNAs encoding differentiation-dependent proteins were identified, molecularly cloned, and sequenced.[271,272] One of these proteins was a homologue of myelin P2 protein, which was designated adipocyte P2 or aP2.[272] In the 5′ flanking region of aP2, two fat-specific elements (FSEs) were identified, FSE1 and FSE2, which were believed to be important for aP2 regulation.[273,274] Recent studies indicate that the *fos* protein is present in protein complexes that bind to the FSE2 sequence and closely related sequences.[275] *Fos* protein binds to the sequence TGACTCA,[275,276] which has been identified as the consensus for binding of the transcription factor AP-1[277] and the yeast transcription factor GCN4.[278] AP-1 can produce a positive or negative effect on specific transcription enhancers and it is thought to play a role in enhancer elements stimulated by TPA. Recently, the transcription factor AP-1 has been identified as the product of the proto-oncogene jun.[279,280] It is now believed that AP-1 and the *fos* protein form a complex which can bind to DNA in a sequence-specific manner.[281] Furthermore, it has been reported that *fos* protein directly modulates the jun function by means of a heterodimer of *fos* and *jun* proteins,[282] and the "leucine zipper" domain of *fos* is necessary for the binding of this heterodimer to DNA.[282] The complex binds to the DNA through the AP-1 component, while the *fos* component, by its close proximity, influences the DNA binding capacity of the protein complex.[283] In this manner, the *fos* protein regulates the expression of the aP2 and possibly other genes. These results emphasize the complexity of gene expression changes which may be mediated by the interaction of *fos* with additional specific DNA binding proteins. The role of the *fos* protein in mediating differentiation will have to inevitably account for these complex interactions and will await a further understanding of the mechanism by which the *fos* protein (in conjunction with additional proteins) mediates their transcriptional changes in defined genes in target cells.

4. *ras* Gene Family

Studies in the last decade or so have demonstrated that two rat sarcoma-inducing genes originally identified as the transforming principle of the Harvey (referred to as H-*ras*-1) and Kirsten (referred to as K-*ras*-2) strains of rat sarcoma virus are often associated with tumor formation in mammalian species, including approximately 10 to 15% of human tumors (reviewed in Reference 284). In the case of human tumors, ras oncogenes have been identified in carcinomas of the bladder, breast, colon, kidney, liver, lung, ovary, pancreas, and stomach; in hematopoietic tumors of lymphoid (acute lymphocytic leukemia, B-cell lymphoma, Brukitt's lymphoma) and myeloid (acute and chronic myelogenous leukemias, promyelocytic leukemia) lineage; and in tumors of mesenchymal origin such as fibrosarcomas and rhabdomyosarcomas (reviewed in Reference 284). *ras* oncogenes have also been variably found in other human tumors, including gliomas, keratoacanthomas, melanomas, neuroblastomas, and teratocarcinomas. The members of the *ras* gene family include H-*ras*, K-*ras*, and N-*ras*,[284,285] each of which encodes a protein of 21 kDa[286-289] which is localized to the inner side of the plasma membrane and anchored to the lipid bilayer by a covalently attached palmitic acid at the carboxy terminus.[290] A series of *ras*-related genes that share 35 to 55% sequence homology with the *ras* gene family have been identified, including R-*ras* (found in human and rodents), *Dras3* (found in fruit flies), *ral* (found in primates), and *YP2* (found in yeast) (reviewed in Reference 284). It remains to be determined, however, if any or all of these ras-related genes are members of conserved gene families. Unless specifically stated, the remaining discussion of *ras* oncogenes refers to the direct members of the *ras* gene family (H-*ras*, K-*ras*, and N-*ras*) and not the *ras*-related genes. The *ras* proteins bind guanine nucleotides (GTP and GDP),[291,292] have a GTPase activity,[293,295] and display a significant sequence homology with G proteins.[284,296] These observations and their membrane localization suggest that the *ras* proteins may contribute to the process of transmembrane signaling. The G proteins play a critical role in the signal transduction pathways from membrane-bound receptors to adenylate cyclase.[296] The exact nature of the involvement of *ras* proteins with signal transduction from receptors is not presently known, and the cellular target(s) for these proteins also remain to be elucidated. Recently, it has been suggested that *ras* proteins may form a link between growth factor receptors and phospholipase C,[297,298] a key enzyme in the phosphoinositol breakdown pathway. With further research, a clearer picture of the role of *ras* proteins in signal transduction should become evident. Abnormal forms of p21 have reduced or no GTPase activity and as a consequence display a constitutive activation of the *ras* protein.[299] Constitutive activation of p21 leads to the loss of normal regulation of cell proliferation.[284] Also, the microinjection of either the human or mouse p21 oncogene protein in human or 3T3 cells causes a loss of normal regulation of growth control and the appearance of a transformed morphology.[300,301]

It is apparent that the ras oncogene can profoundly modify cellular phenotypes, including morphology, biochemical processes, proliferative capacity, and differentiation (reviewed in

Reference 284). For example, in PC12 phaeochromocytoma cells, infection with *K-MSV* (or K-v-*ras*) or *H-MSV* (or H-v-*ras*) induces differentiation by mimicking the effect of nerve growth factor (NGF).[302] The v-*src* and v-*mos* oncogenes also mimic the effect of NGF, but not as markedly as v-*ras*.[303] In contrast, normal *ras* p21 protein do not induce the differentiation of PC12 cells into neuron-like cells. Differentiation could be induced in PC12 cells by injection of oncogenic *ras* proteins[304] and could be blocked by injection of anti-*ras* antibodies.[305] The temporal kinetics and requirement for RNA and protein synthesis to induce neuronal differentiation in PC12 cells is similar for the oncogenic *ras* proteins and NGF.[306] These findings suggest that both the oncogenic *ras* proteins and NGF may employ the same transmembrane signaling mechanism in modulating PC12 differentiation. The mechanism by which oncogenic ras proteins induce differentiation in PC12 cells appears to be different than the mechanism by which c-AMP induces PC12 differentiation. This hypothesis is supported by the observation that microinjection of PC12 cells with antibodies against p21 *ras* proteins suppresses neurite formation induced by NGF but not by cAMP.[305]

The effect of interferon on c-*ras* expression has been studied in a number of different experimental systems. In human bladder carcinoma cells (RT4), IFN-β results in a reduced expression of the c-H-*ras* gene, but not from the c-K-*ras* gene.[307] The effects on gene expression occurred before inhibition of tumor growth was observed, indicating that the reduced gene expression was not due to cessation of growth. In this study, it was reported that IFN-β treatment also altered the phosphorylation of an approximately 21 kDa protein, but they presented no evidence which indicated that it was the product of the *ras* gene. In a recent study, Giacomini et al.[308] determined the effect of recombinant IFN-α, IFN-β, and IFN-γ on the level of Ha-*ras*-1 mRNA and protein in the human melanoma cell line Colo 38. In this model system, only IFN-γ induced a dose (20 to 200 units per ml)- and time (48 to 96 h)-dependent reduction in the accumulation of Ha-*ras*-1 mRNA and in the synthesis of the Ha-*ras*-1-encoded p21 protein. As was observed in RT4 cells, downregulation of this proto-oncogene was observed prior to the antiproliferative effect of interferon. However, in contrast to RT4 cells, IFN-β, even at concentrations as high as 1000 units per ml, did not affect the total amounts of Ha-*ras*-1 products in Colo 38 cells. These studies indicate that different interferons will induce different effects on cellular proto-oncogenes, depending on the target cell, and that downregulation of proto-oncogene expression precedes interferon-mediated suppression of cell growth.

In a study employing NIH 3T3 cells transformed with a cloned human c-H-*ras* gene, IFN-β treatment resulted in a reduced level of c-H-*ras* mRNA,[309] and with continued IFN-β treatment, transformed cells acquired a revertant or nontransformed phenotype. When the application of IFN-β was discontinued, the levels of c-H-*ras* increased slowly and only a small proportion of the revertant cells reacquired a transformed phenotype. With time, the vast majority of cells not continuously exposed to IFN-β did not regain the transformed phenotype, even though their levels of c-H-*ras* were elevated.[310] These persistent revertant cells displayed relatively high levels of p21 protein, compared to the original revertants and similar to the transformed cells, but they retained a normal cellular phenotype.[310] Based on these observations, Samid et al.[310] concluded that overproduction of p21 in the persistent revertants was insufficient to induce these cells to become retransformed.

The N-*ras* gene has been found in a number of human tumors, including glioblastomas and neuroblastomas, and has been identified as the transforming gene.[311,312] A substitution in the p21 protein has been shown to be responsible for the activation of the transforming function.[312] In human Daudi cells, N-*ras* expression has been shown to be regulated both qualitatively and quantitatively by IFN-β.[313] Furthermore, the reduction in N-*ras* mRNA levels occurred prior to cessation of growth, indicating, in a manner analogous to interferon-induced reductions in c-Ha-*ras* expression in RT4 and Colo38 cells,[307,308] that the reduction is not due simply to cessation of growth. It was also demonstrated that the reduction in N-

ras expression in Daudi cells was dependent on the synthesis of new protein, since cycloheximide abrogated the regulation of N-ras by IFN-β.

Further studies are required to determine if a direct association exists between the induction of differentiation in specific target cells and an altered expression of endogenous ras genes. Even if a correlation is observed, however, it may be difficult to directly determine if this change in Ha-*ras* proto-oncogene expression is a cause or a consequence of the differentiation process.

5. Other Oncogenes

A potential association between the altered expression of other oncogenes (in addition to c-*myc*, c-*fos*, and *ras*) and differentiation has been demonstrated.[114-131] For example, the c-*src* oncogene appears to contribute to neuronal differentiation.[314] Immunocytological studies indicate that, in neural retina and cerebellum, pp[60]-c-*src* is detectable only at the onset of differentiation in neurons and at a time when neuron proliferation stops.[315,316] Furthermore, the expression of c-*src* persists in terminally differentiated neurons for extended periods. Similar results have been obtained in a study analyzing c-*src* expression in primary cultures of neurons or astrocytes, where a 15- to 20-fold increase in *src* protein was found in comparison with fibroblasts.[317] These studies suggest that c-*src* may play a role in neuron differentiation, but they do not indicate whether c-*src* is involved directly in the induction of differentiation or if it provides a function required by differentiating cells. However, the persistence of pp[60]-c-*src* in differentiated neurons supports its latter role in neuronal differentiation. In *Drosophila*, a gene similar to v-*src* has been isolated and, although initially expressed in all embryonic tissues, after 8 h of development it is confined to the neural tissues and smooth muscle cells of the gut.[318] The expression of the c-*src* gene in *Drosophila* corresponds to the pattern of expression of c-*src* in vertebrates and correlates with cellular differentiation rather than proliferation, suggesting that its function has been evolutionarily conserved. IFN-β has been shown to induce a reduction in cellular mRNA hybridizable to v-*src* in RT4 carcinoma cells.[307] A decrease in the level of a protein similar in size to pp[60] v-*src* was also observed, but there was no evidence presented as to whether this protein was antigenically related to the viral protein.[307] An earlier study also showed that interferon treatment of Rous sarcoma virus-transformed cells resulted in both reduced synthesis of pp[60]-src the *src* gene product, and decreased expression of the transformed phenotype.[319]

Another oncogene with a putative role in differentiation is c-*fms*,[320] which is believed to encode the receptor for the growth factor, colony-stimulating factor (csf-1). The c-*fms* oncogene is a member of the *src* gene family of related oncogenes and encodes a tyrosine kinase.[321,322] In undifferentiated HL60 cells, the expression of c-*fms* is low, but when these cells are induced to differentiate by treatment with TPA or Vit D3, c-*fms* expression is increased.[323] Furthermore, the *in vivo* level of c-*fms* expression in mature macrophages is high, as is the number of receptors, compared with myeloid precursor cells, supporting its potential role in differentiation.[324] It is likely, therefore, that c-*fms* expression during macrophage differentiation may be a consequence of differentiation rather than a cause of it. In addition, a role for the regulation of c-*fos* by c-*fms* has also been proposed.[324] These observations suggest that c-*fms* expression may be an important component of monocyte/macrophage differentiation.

A reduction in the level of mRNA complementary to the v-*mos* oncogene has been observed in Daudi cells following treatment with an antiproliferative dose of interferon, but not with a dose of interferon inducing only antiviral activity.[325] These observations suggest that different levels of interferon may be required to modulate the expression of defined sets of genes which regulate the diverse effects of interferon. Further studies are required, however, to ascertain whether an altered expression of v-*mos* is a general phenomenon associated with interferon treatment and growth suppression.

A nuclear-acting oncogene with a putative role in differentiation is c-*myb*. Friend erythroleukemia cells (FELC) can be induced to differentiate with HMBA, and the induction can be blocked with TPA.[78,94,326] A rapid reduction in c-*myb* levels occurs 1 h after HMBA treatment, and c-*myb* expression remains depressed even after 48 h.[327] In contrast, under the same conditions, c-*myc* mRNA levels were dramatically reduced within 1 h, but they returned to control levels within 24 h of treatment. Furthermore, in cells permanently commited to differentiate but as yet undifferentiated (HTLC2), the levels of c-*myb* expression were low in comparison to parent cells, suggesting that the decrease in c-*myb* expression is one feature of FELC commitment to differentiation. Further evidence for this conclusion comes from a study in which proto-oncogene expression in normal, preleukemic, and leukemic murine erythroid cells was examined.[328] High levels of c-*myb* were observed in both stages of erythroleukemia, but not in normal erythroid cells. Induction of terminal differentiation by DMSO in the erythroleukemic cell line, 745A, resulted in an absence of c-*myb* transcripts by 5 d post-treatment. The loss of c-*myb* transcripts could not be attributed to exposure to DMSO, since c-*myb* transcripts were found at high levels in TFP10 erythroleukemic cells not terminally differentiated by DMSO after 5 d of treatment.[328] These results suggest that c-*myb* levels observed in erythroleukemic cells may be related to their undifferentiated state and not simply to changes in cell proliferation. Furthermore, other studies using erythroleukemic cells treated with HMBA have indicated that persistent suppression of c-*myb* gene expression is critical for permitting these cells to proceed to commitment to terminally differentiate.[329] A contributing role of c-*myb* in erythroleukemic cell differentiation is evident from these studies, and it has been proposed that c-*myb* is involved in the control of cell proliferation at a specific stage of hematopoietic maturation.[330,331] Further studies are required to determine if the alterations observed in c-*myb* expression as a consequence of induction of differentiation in FELC and F745A cells represent a general phenomenon or whether this effect is cell-lineage specific or even specific for differentiation only in erythroleukemia cells.

D. MECHANISM(S) OF GENE REGULATION BY INTERFERONS

After binding to their specific receptors, interferons induce a rapid transcriptional activation of a number of genes which may mediate the variety of effects elicited by these molecules. Based on a preponderance of experimental evidence, it is now widely accepted that there are two types of interferon receptors, each specific for the two major classes of interferons:[332] type I (IFN-α and IFN-β) and type II (IFN-γ).[4,333-335] A third type of IFN receptor has been postulated, based on the evidence that IFN-β can be displaced by IFN-γ.[336] The number of interferon receptors can vary from a few hundred to more than 50,000 on some cell types, although the majority of cell types have 1000 to 2000 receptors. There is now evidence which indicates that there may be subtypes of type I and type II interferon receptors which are functional in different cells.[337] For instance, there is a difference in the molecular weight of type II receptors from monocytes and HeLa cells, and the HeLa cell receptors do not display downregulation following exposure to type II interferon.[338] It has also been reported that murine type I interferon receptors respond differently to alpha and beta interferon.[339] Based on the complexity of the alpha interferon gene family, it may eventually be demonstrated that there is a diversified family of interferon receptors in target cells which are involved in specific interferon-induced changes in cells.[340]

As briefly discussed above, the earliest event in interferon action appears to be the interaction of these molecules with high-affinity cell-surface receptors specific for types I or type II interferon. Although not well defined, secondary events in interferon action most likely involve a transmembrane signaling process which results, very rapidly, in induction of the transcription of specific cellular genes. Interferon responsiveness of a gene is conferred by an interferon-responsive sequence (IRS), and a consensus for this sequence has been identified on many interferon-inducible genes (Table 6). The IRS for type I interferons is recognized by three nuclear factors: B1, B2, and B3.[191] DNA binding protein B1 is also

TABLE 6
Sequence Homology between Interferon Response Elements

Gene	Sequence	Ref.
Human 2'-5'oligo(A) synthetase	TGAGGAAACGAAACC	341
Human ISG-15	AGGGAAACCGAAACT	189
Human ISG-54	GGGAAAGTGAAACT	189
Human 6-16	GGGAAAATGAAACT	342
Mouse C202	GGGAAATTGAAAGC	199

present in uninduced cells, while B2 and B3 are present only in interferon-treated cells. Factor B3 is induced rapidly and then declines, while factor B2 is induced after a lag period of 1.5 h, and this factor reaches maximal levels after 2 to 4 h. Furthermore, the induction of factor B3 does not require *de novo* protein synthesis, while the induction of factor B2 can be blocked by cycloheximide.[191] This two-stage binding of nuclear factors correlates with the two-phase regulation in the transcriptional cycle of ISG-54 and ISG-56,[190,193] which involves transcriptional activation of ISG-54 and ISG-56, and this phase does not require *de novo* protein synthesis. Transcriptional activation of these genes is followed by repression of transcription, and this phase depends on ongoing protein synthesis. Thus, Levy et al.[191] speculated that B3 may act as an inducer while B2 may function as a repressor of transcription of interferon-stimulated genes. However, this interpretation may be simplistic, since some genes are induced in response to interferons and, thus, B2 may also act as an inducer of some genes or, alternatively, additional control mechanisms must exist. The complexity of the mechanisms by which interferons regulate gene expression is emphasized by the observation of McMahon et al.[163] that some genes (e.g., 6-26, 2A, MT11, 2'-5'A) are equally induced by IFN-α in wild-type and in interferon-resistant Daudi cells, while other genes (e.g., 1-8, 9-27, 6-16) are induced only in wild-type Daudi cells. This observation suggests that there are at least two transduction pathways for IFN-α. Similarly, Kusari et al.[343] demonstrated that mRNAs ISG-56, 6-16, 1-8, 2A, and 6-26 are induced in both fully interferon-responsive Hela cells and in partially responsive RD-114 cells. In contrast, 2'-5'oligo(A) synthetase is induced much more effectively by IFN-α in HeLa cells than in RD-114 cells. Other evidence for multiple transduction pathways for the interferon response is based on the observation that the induction of 2'-5'oligo(A) synthetase by IFN-α in HeLa cells is not blocked by cycloheximide,[344] while the induction of ISG-56 can be blocked with cycloheximide.[345] The transduction mechanism(s) for IFN-γ appears to be at least partially separate, since the induction of 2'-5'oligo(A) synthetase in HeLa cells by IFN-γ can be blocked by cycloheximide, while the response to IFN-α is unaffected. Furthermore, ISG-56 mRNA in HeLa cells is induced by IFN-α, but not by IFN-γ.[345] The signaling pathways do at least in part overlap, since the cycloheximide blockage of the induction of ISG-54 mRNA by IFN-α can be prevented with IFN-γ, which by itself cannot induce ISG-56 mRNA.

The precise mechanism of communication between interferon receptors and the nucleus remains unclear, but it seems likely that a second messenger should be involved, because internalization of interferon[346] or proteolytic fragments of interferons do not appear to act as the signal for the interferon response.[346] The search for the second messenger has not lead to any conclusive evidence. It has been shown that interferons may raise cAMP[347,348] and GTP levels[349] in some cells, however, no conclusive link between changes in second messenger and changes in gene expression have been demonstrated. Further studies are required to define on a biochemical and molecular level the mechanism by which interferon-induced transmembrane signaling processes induce transcriptional changes in specific target cells, resulting in the diversity of interferon-induced changes in cells (see Tables 1 and 2).

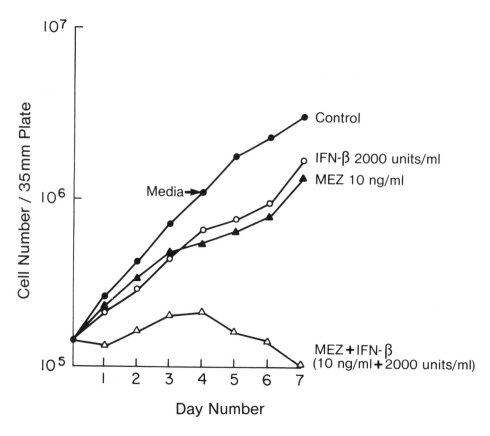

FIGURE 2.   Growth of HO-1 cells in the presence of IFN-β (2000 units per ml), MEZ (10 ng/ml), or IFN-β (2000 units/ml) + MEZ (10 ng/ml). Cells were seeded at 1.5 × 10⁵/35 mm plate, and cell numbers were determined in triplicate by Coulter counter every 24 h. Fresh media containing the indicated compounds were added 4 d postplating. Replicate values varied by <10%. (●) control; (○) IFN-β (2000 units/ml); (▲) MEZ (10 ng/ml); (△) IFN-β (2000 units/ml) + MEZ (10 ng/ml). (From Fisher, P. B., Prignoli, D. R., Hermo, H., Jr., Weinstein, I. B., and Pestka, S., *J. Interferon Res.*, 5, 11, 1985. With permission.)

IV. MODULATION OF GENE EXPRESSION AND THE ANTIGENIC PHENOTYPE OF HUMAN MELANOMA CELLS AS A FUNCTION OF CHEMICALLY INDUCED DIFFERENTIATION

In previous studies, we have demonstrated that the combination of the antileukemic compound mezerein (MEZ) and human interferon results in enhanced growth suppression and an induction of terminal differentiation in human melanoma cells.[31,73,75,76] Of the human interferons tested, recombinant IFN-β was the most effective interferon preparation, alone or in combination with MEZ, in inhibiting the growth and inducing differentiation of human melanoma cells.[73] The effect of the combination of IFN-β and MEZ on the growth of the human melanoma cell line HO-1 is shown in Figure 2. As can be seen, when treated with the combination of IFN-β and MEZ, a complete suppression in growth is observed as early as 24 h, whereas either IFN-β or MEZ, alone, induce only a partial suppression in growth. When treated with MEZ, or MEZ plus IFN-β, a distinct morphological change, characterized by the appearance of dendritic-like projections, is induced in human melanoma cells (Figures 3 and 4). Morphological changes in HO-1, as well as other human melanoma cells, exposed to MEZ or IFN-β plus MEZ were apparent as early as 8 h post-treatment. Employing the

HO-1 human melanoma cell culture model system, we have identified several discrete phases in the induction of differentiation by IFN-β plus MEZ in human melanoma cells.[76] Treatment of human melanoma cells with specific concentrations of IFN-β plus MEZ for 72 or 96 h results in a **reversible commitment to differentiate,** i.e., removal of these agents results in a gradual resumption of proliferation and a return to the undifferentiated phenotype of the original untreated tumor cell. In contrast, exposure to higher doses of IFN-β plus MEZ or continued exposure for 7 d to doses of these agents, which result in a reversible commitment to differentiate after 72 or 96 h treatment, results in an **irreversible induction of terminal differentiation,** i.e., cells remain viable but they lose their proliferative capacity. Employing the HO-1 cell line and other human melanoma cell lines, we have begun to analyze both the gene expression[76] and antigenic changes[77] associated with the induction, reversible commitment, and terminal stages of human melanoma cell differentiation.

A. EFFECT OF INTERFERONS AND MEZEREIN, ALONE AND IN COMBINATION, ON GENE EXPRESSION IN HUMAN HO-1 MELANOMA CELLS

The ability to induce terminal differentiation in human melanoma cells by the appropriate treatment with IFN-β plus MEZ represents a valuable model system for the identification, molecular cloning, and characterization of gene(s) which mediate growth control and differentiation in human melanoma cells. Although the precise details have not been elucidated, studies employing many model systems suggest that differentiation is a multistage process, often consisting of an induction, reversible commitment to differentiate, and a terminal differentiation stage.[30,350,351] It is probable that the process of differentiation may involve many genes which must be expressed in a coordinated fashion for normal differentiation to occur. In addition, recent studies on myogenesis indicate that specific master-switch gene(s) exist which can directly induce cells to express a specific differentiation lineage, i.e., the myogenic phenotype.[47-52,352] At the present time, the identity and total number of genetic determinants of differentiation for the various tissue lineages, i.e., melanogenesis, adipogenesis, chondrogenesis, etc., or their temporal relationship of expression have not been determined. In the case of tumor cells, it is possible that the genes required for expression of a normal differentiated phenotype are (1) present, but not expressed as a consequence of transcriptional silencing (via changes in chromatin configuration, DNA methylation, or other mechanisms), (2) expressed, but the gene product(s) are not functional because of the presence of differentiation-inhibitory molecules, or (3) defective as a result of mutational changes occurring during development of the transformed state. The ability, by appropriate experimental manipulation, to induce terminal differentiation in tumor cells suggests that at least in some situations, the gene(s) which mediate differentiation may be present in a functional state in tumor cells, but they are not expressed (for review, see References 30, 97, 353, and 354). If this observation is a general phenomenon in tumor cells, then treatment of tumor cells with the appropriate agents may result in the activation of silent differentiation genes, resulting in the induction of terminal differentiation in the tumor cell and their loss of proliferative and tumorigenic capacity.

To begin to define the gene expression changes occurring in HO-1 cells following the induction of differentiation, cultures have been treated with IFN-β (2000 units per ml), MEZ (1 and 10 ng/ml), or IFN-β + MEZ (2000 units per ml + 10 ng/ml) for various time periods, including those encompassing the early commitment phase of differentiation, the late reversible commitment phase of differentiation, and the terminal phase of differentiation. For this purpose, the induction of two model genes (discussed previously), ISG-15 and ISG-54, in response to IFN-β and MEZ was investigated in HO-1 cells. Preliminary data (Figure 5) using Northern blot analysis indicates that ISG-15 is induced by IFN-β and by the combination of IFN-β and MEZ, but not by MEZ alone. The induction of ISG-15 occurs

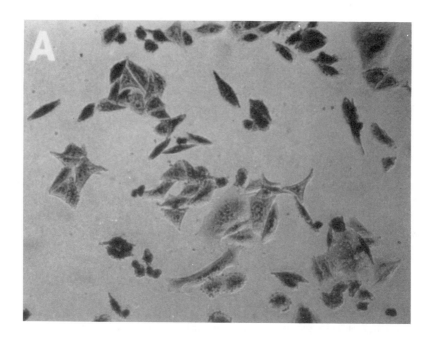

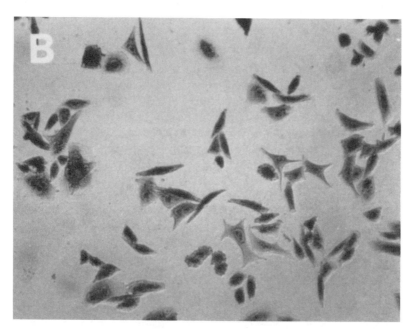

FIGURE 3.   Effect of recombinant human IFN-β and mezerein (MEZ), used alone and in combination, on the morphology of HO-1 human melanoma cells (Giemsa; × 150). (A) Control HO-1 cells 24 h postplating; (B) HO-1 cells exposed to 2000 units/ml IFN-β for 24 h; (C) HO-1 cells exposed to 10 ng/ml MEZ for 24 h; (D) HO-1 cells exposed to 2000 units/ml IFN-β and 10 ng/ml MEZ for 24 h. (From Fisher, P. B., Prignoli, D. R., Hermo, H., Jr., Weinstein, I. B., and Pestka, S., *J. Interferon Res.*, 5, 11, 1985. With permission.)

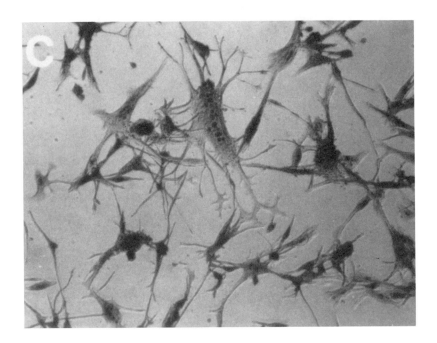

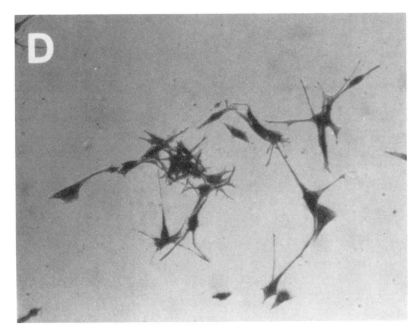

FIGURE 3C and D.

within 4 h and is sustained for 8 h, but by 16 h of continued exposure to IFN-β or IFNβ plus MEZ, ISG-15 mRNA has declined to undetectable levels. In contrast to ISG-15, the kinetics of the induction of ISG-54 are very different. Thus, ISG-54 mRNA, unlike ISG-15 mRNA, is undetectable after 4 h of exposure to drug treatment. ISG-54 mRNA appears after 8 h of exposure to IFN-β or IFN-β plus MEZ, but, unlike ISG-15 mRNA, the induction is maintained for at least 96 h. It therefore appears that in HO-1 cells there is a sequential induction of ISG-15 and ISG-54. This is in contrast to HeLa cells and Daudi cells, where

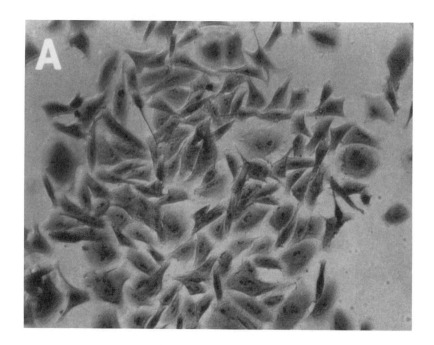

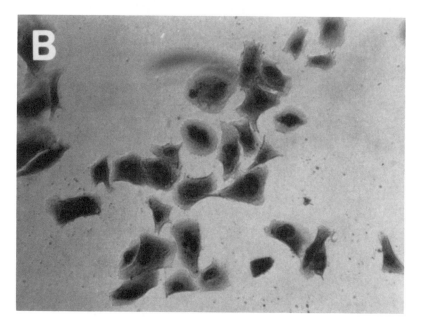

FIGURE 4.   Effect of recombinant human IFN-β and mezerein (MEZ), used alone and in combination, on the morphology of BO-2 human melanoma cells (Giemsa; × 150). (A) Control BO-2 cells 24 h postplating; (B) BO-2 cells exposed to 2000 units/ml IFN-β for 24 h; (C) BO-2 cells exposed to 10 ng/ml MEZ for 24 h; (D) BO-2 cells exposed to 2000 units/ml IFN-β and 10 ng/ml MEZ for 24 h. (From Fisher, P. B., Prignoli, D. R., Hermo, H., Jr., Weinstein, I. B., and Pestka, S., *J. Interferon Res.*, 5, 11, 1985. With permission.)

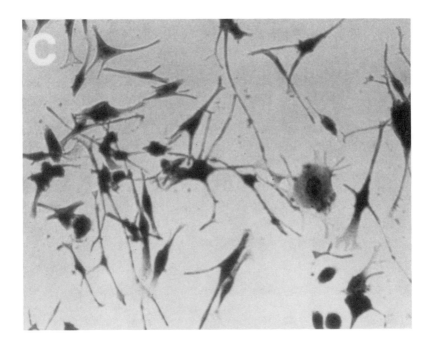

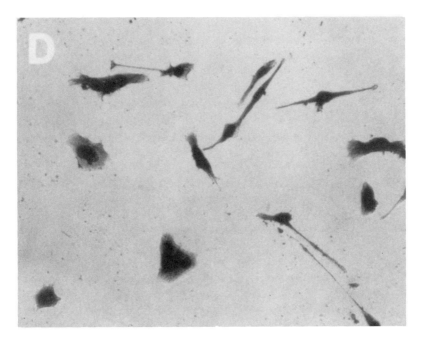

FIGURE 4C and D.

both ISG-15 and ISG-54 mRNAs are induced within 2 h and both mRNAs remain induced for at least 12 h of interferon treatment.[355] The relationship of these gene expression changes to the various stages of chemically induced differentiation in human melanoma cells are not presently known. By employing various molecular biological approaches, studies are currently in progress to identify and molecularly clone the gene(s) which regulate induction of terminal differentiation and loss of proliferative potential in human melanoma cells.

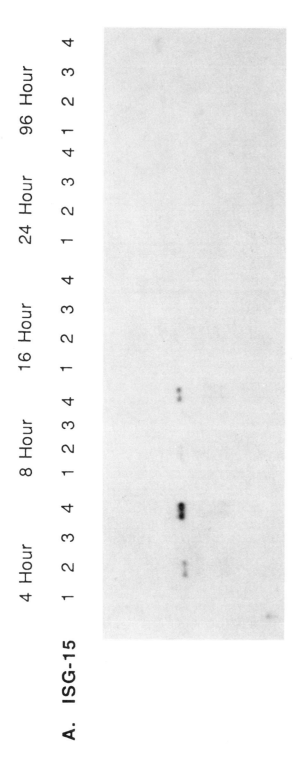

FIGURE 5. Kinetics of induction of ISG-15 and ISG-54 mRNAs in HO-1 cells. Human HO-1 melanoma cells were treated with IFN-β (2000 units per ml), mezerein (1 ng/ml), or a combination of the two compounds for 4, 8, 16, 24, or 96 h. Total cytoplasmic RNA was isolated and 20 μg aliquots were size fractionated in a 1.2% agarose-formaldehyde gel and transferred to Hybond-N nylon membranes. Northern blots were probed with ^{32}P-labeled nick-translated cloned ISG-15 or ISG-54 cDNAs, respectively. Lane 1, control; lane 2, IFN-β (2000 units/ml); lane 3, mezerein (1 ng/ml); and lane 4, IFN-β (2000 units/ml) plus mezerein (1 ng/ml).

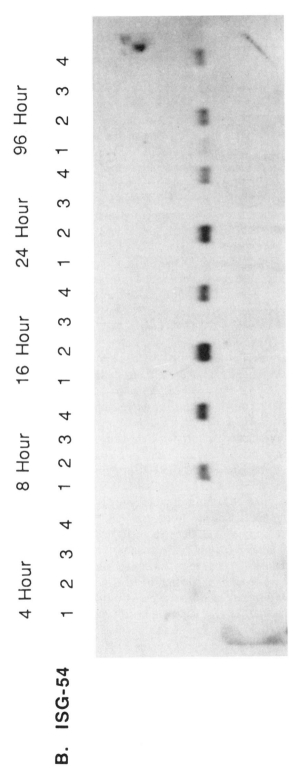

FIGURE 5 (continued).

B. EFFECT OF INTERFERONS AND MEZEREIN, ALONE AND IN COMBINATION, ON HISTOCOMPATIBILITY GENE AND PROTEIN EXPRESSION DURING THE INDUCTION OF DIFFERENTIATION IN HUMAN MELANOMA CELLS

It is well established that the immune system plays a critical role in determining both the initial survival of tumor cells and the evolution of neoplastic cells, as well as the elimination of tumor cells, in an organism. A potentially important component in the immunological network is expression by the tumor cell of major histocompatibility complex (MHC) molecules, including both HLA Class I and II antigens, on the surface of tumor cells. Several lines of evidence indicate that HLA Class I molecules function as restricting elements for recognition by T-cell.[356] For example, rodent cells transformed by different serotypes of human adenoviruses display different oncogenic phenotypes, i.e., cells transformed by type 2 or 5 adenovirus display low or negligible oncogenic potential, whereas cells transformed by type 12 adenovirus are highly oncogenic.[357-359] In a series of Ad2- and Ad12-transformed cells, a direct relationship was observed between expression of HLA Class I molecules and tumor rejection.[357] Rodent cells transformed by oncogenic Ad12 expressed negligible levels of HLA Class I antigens, whereas nononcogenic Ad2-transformed rodent cells expressed normal levels of HLA Class I proteins.[357] Induction of expression of HLA Class I proteins in Ad12-transformed cells either by transfection and expression of functional MHC genes[360] or by treatment with IFN-γ[361] resulted in a loss of the oncogenic phenotype. Although this relationship between expression of HLA Class I molecules and the lack of oncogenicity may not exist in all adenovirus-transformed rodent cells, i.e., some Ad12-transformed cells express high levels of HLA Class I molecules and are still oncogenic, and some Ad2- and Ad5-transformed cells express low levels of HLA Class I proteins but display low oncogenic potential, these results do suggest that under certain conditions, expression of HLA Class I molecules may serve as negative regulators of oncogenicity in specific transformed cells. Similarly, in other murine tumor systems, re-expression of HLA Class I molecules by transfection with functional MHC genes encoding these proteins can convert highly oncogenic or metastatic cells into nononcogenic or nonmetastatic cells.[362,363] These observations suggest that the expression of HLA Class I proteins in certain situations may be required for the appropriate immune recognition and subsequent elimination of transformed cells. In contrast, while the lack of HLA Class I expression in the tumor systems described above leads to escape of the tumor cells from destruction by cytotoxic cells, in other tumor systems it has been demonstrated that the absence of this antigen results in enhanced sensitivity of these cells to lysis by natural killer (NK) cells.[364,365]

The role of HLA Class I expression in the recognition of human tumors by the immune system is uncertain, although several human tumors, e.g., small cell lung carcinomas,[366] neuroblastoma,[367] and melanomas,[368] have reduced expression of HLA Class I molecules. Brocker et al.[369] have demonstrated that visceral metastases often have lower HLA Class I expression than do primary tumors. An analysis of a large number of human melanomas revealed a homogeneous expression of HLA Class I molecules in their primary lesions, while their metastases were generally negative.[370] In recent studies by Nistico et al.[371] and Giacomini et al.,[372] a series of early passage cell cultures derived from autologous human melanoma metastases from the same patients were found to display similar levels of HLA Class I expression and responsiveness to upregulation with recombinant human interferons. The mechanism by which human tumor cells display altered levels of HLA Class I expression is not presently known. A possible association between reduced HLA Class I expression and high expression of the c-*myc* and N-*myc* gene has recently been demonstrated.[373,374] The role of specific changes in oncogene expression and their subsequent effects on the status of proliferation, differentiation, and HLA Class I expression in tumor cells remains to be determined.

TABLE 7
Melanin Synthesis in Human Melanoma Cells Grown in the Presence of Recombinant Human Fibroblast Interferon, Mezerein, or Both Agents Combined

Cell line	Total melanin ($\mu g/10^6$ cells)						
	Control	IFN-β	$(\Delta)^a$	MEZ	$(\Delta)^a$	IFN-β + MEZ	$(\Delta)^a$
BO-2	9.55	16.17	(6.62)	13.68	(4.13)	26.25	(16.70)
DU-2	0.43	1.08	(0.65)	0.76	(0.33)	8.79	(8.36)
FO-1	1.70	1.43	(−0.27)	4.28	(2.58)	9.54	(7.84)
HO-1	3.38	5.12	(1.74)	4.98	(1.60)	8.32	(4.94)

Note: One million cells were seeded per 10 cm tissue culture plate and 4 h after plating the indicated compounds (2000 units per ml IFN-β, 10 ng/ml MEZ, or both) were added in fresh medium. Four (DU-2, FO-1, and HO-1) or 7 (BO-2) d after plating, the cell number was determined and the melanin extracted and quantitated by absorbance at 400 nm, as described in Fisher et al.[73]

[a] Designates difference between melanin content of treated and control cultures.

From Fisher, P. B., Prignoli, D. R., Hermo, H., Jr., Weinstein, I. B., and Pestka, S., *J. Interferon Res.,* 5, 11, 1985. With permission.

Interferons are potent enhancers of HLA Class I gene expression in human melanoma cells.[19,21,68,74,75,77,188,371,372,375] These biological response modulators also suppress growth to a variable degree and enhance the synthesis of melanin, a marker of melanoma cell differentiation, in specific human melanoma cells.[73] The compounds MEZ and TPA are also inducers of melanin synthesis in specific human melanoma cells.[72,73] When used in combination, i.e., IFN-β plus MEZ or TPA, growth suppression in human melanoma cells is potentiated and treated cells synthesize larger quantities of melanin (Table 7).[31,73] To determine if a relationship exists between the induction of differentiation and increased expression of HLA Class I expression in human melanoma cells, we have determined the effect of IFN-β, MEZ, and TPA, alone and in various combinations, on antigenic expression in a series of human melanoma cell lines.[76,77] If increased HLA Class I expression was correlated with the state of differentiation in human melanoma cells, it would be anticipated that the level of HLA Class I expression would be increased to the greatest extent when melanoma cells were exposed to IFN-β plus MEZ or TPA, which synergistically induce differentiation in human melanoma cells. However, our experimental results indicated that this is not the case, since the different compounds, alone or in combination, often exhibited divergent effects on the same melanoma culture.[77] The differential antigenic-modulatory response of the human melanoma cells to the differentiation-modulating agents was also not correlated with their innate growth-inhibitory response to IFN-β, MEZ, or TPA, used alone or in combination. These observations suggest that the enhanced expression of HLA Class I antigens by IFN-β in human melanoma cells is not related to changes in growth and/or differentiation. A lack of correlation between increased expression and enhanced HLA Class I expression in melanoma cells is also supported by the observation that cyclic AMP induces differentiation in specific human melanoma cells, and induction of differentiation is associated with a decreased expression of HLA Class I molecules.[77]

In addition to HLA Class I molecules, HLA Class II antigens are also major contributors to the immune response of an organism to environmental antigens.[376] It is believed that the unexpected, but frequent, expression of HLA Class II antigens on neoplastic cells of various histological origins might significantly influence the outcome of the interactions between the tumor and the host's immune system.[377,378] Several hypotheses have been proposed to explain the mechanism underlying the expression of HLA Class II antigens on specific tumor cells, although they are absent in normal tissue from which the tumor was derived. These include "freezing" of the tumor cell in a Class II-positive state (representing an early stage of differentiation), deregulation of MHC gene expression as a consequence of tumorigenesis, and/or secretion by tumor-infiltrating lymphocytes of Class II-inducing factors, such as IFN-γ.[379-381] In the case of human melanomas, a recently proposed schema to identify distinct stages in tumor progression suggests that the expression of HLA Class II molecules is correlated with higher states of malignancy.[382] However, as a consequence of the high degree of heterogeneity in the expression of these antigens in human melanomas, as well as in other tumors, a direct correlation betwen changes in HLA Class II antigens and states of malignancy is not always evident.[368,383-390] Indeed, up to 30% of cultured malignant human melanomas remain Class II negative.[386,389,390] Even within the large group of Class II-positive melanomas, a further level of heterogeneity is generated by the differential expression of the three gene products of the HLA-D region, namely DR, DP, and DQ.[386,389,390] This heterogeneity is further confirmed by recent studies indicating a differential response of different melanoma cell lines to the modulatory effects of IFN-γ on the genes of the HLA-D region.[386,391,392]

Interferons can induce or enhance, to varying degrees, the expression of Class II molecules on human melanoma cells *in vitro*.[19,21,68,74,75,77,371-373] Generally, IFN-γ is the most active of these agents in enhancing HLA Class II expression by human melanoma cells,[74,391-393] although such activity has also been observed in specific melanoma cultures treated with IFN-α and IFN-β.[19,74,371,372,392] In order to clarify the possible role of expression of HLA Class II antigens during the process of melanoma differentiation, we have conducted studies similar to those described above for the HLA Class I antigens.[77] We determined the effect of different interferons, employed alone or in combination with each other, MEZ, and the combination of MEZ and interferon on the expression of HLA Class II antigens on several human melanoma cell lines.[77] As was observed with HLA Class I molecules, the use of combinations of agents, which were synergistic in inducing either growth suppression and/or the induction of terminal differentiation in melanoma cells, failed to induce a potentiated or synergistic effect on the expression of HLA Class II molecules.[77] Similar results were obtained employing several different human melanoma cell lines and with different combinations of agents,[76] suggesting that the enhanced expression of Class II antigens in response to interferons is not directly correlated with the induction of human melanoma differentiation.

C. EFFECT OF INTERFERON ON THE EXPRESSION OF TUMOR-ASSOCIATED ANTIGENS AND OTHER ANTIGENS AS A FUNCTION OF CHEMICALLY INDUCED DIFFERENTIATION IN HUMAN MELANOMA CELLS

The advent of hybridoma technology has resulted in the successful preparation of monoclonal antibodies (MoAb) which react specifically with human melanoma associated antigens (MAA) and which are suitable reagents for the development of immunotherapeutic approaches for the treatment of melanoma.[394,395] A number of MAA are expressed by the transformed melanoma cells, but they are generally not expressed in their normal physiological counterpart, i.e., normal melanocytes and nevi. To date, a large number of MoAbs specific for MAA have been identified,[396] and substantial effort has been expended in trying to clarify the physiologic role of MAA. For example, a 96 kDa MAA that was originally identified following the *in vitro* exposure of human melanoma cells to IFN-γ[397] was eventually found to be identical to the intercellular adhesion molecule-1 (ICAM-1).[391] This molecule, which is an accessory molecule involved as a ligand for antigen-independent conjugates, antigen-specific cytotoxic T lymphocytes, and recognition and cytolysis of particular target cells,[398,399] has recently been shown to be expressed constitutively, or after incubation with

IFN-γ or tumor necrosis factor, on the surface of both human melanoma and central nervous system tumor cell cultures.[400,401] These observations suggest that ICAM-1 may play a role in the interactions between tumor cells and the host's immune system.

Since MAA are, by definition, present on malignant cells, but not on normal melanocytes, the hypothesis that these molecules may be associated with the process of differentiation of normal melanocytes has been proposed.[382] Indeed, a panel of MoAbs directed against eight distinct MAA has been reported to define at least five specific stages of tumor progression in malignant melanoma.[382] We have recently addressed this issue by determining the effect of chemically induced differentiation of human melanoma cells on the expression of a series of MAA.[76,77] We have analyzed three human melanoma cell lines, DU-2, FO-1, and HO-1, which exhibit a differential response to growth suppression and the induction of differentiation when exposed to IFN-β, MEZ, or TPA, alone and in combination.[73,76,77] When exposed to either agent alone, or in combination, the different cell lines displayed a differential response with respect to changes in the expression of specific MAA, including a HMW MAA, a 115-Kd MAA, a 100-Kd MAA, and ICAM-1.[76,77] These findings suggest that none of the MAA we have presently analyzed are markers of the process of reversible differentiation induced by treatment of human melanoma cells with IFN-β, TPA, MEZ, or various combinations of these agents.

D. USE OF INTERFERONS IN COMBINATION WITH OTHER AGENTS AS AN APPROACH TO THERAPY

Although the therapeutic use of interferon as a single agent for cancer treatment appears limited, combinations of interferon with other agents, which result in a more effective inhibition in the growth and an induction in the differentiation of neoplastic cells, may prove of greater therapeutic value (Table 4). In order to achieve this goal, a combination of agents must be found which is well tolerated by the patient (resulting in no or limited toxicity and effective at doses which do not cause abnormal physiologic changes) and effective in directly inhibiting the growth of the tumor cells or enhancing the ability of the tumor cell to be destroyed by the immune system. In the case of metastatic melanoma, which is refractory to the majority of primary chemotherapeutic agents and in most cases displays an even greater resistance to chemotherapy after failure to respond to the primary drug regimen, no consistently effective therapy is currently available.[402] These discouraging findings highlight the need for newer approaches to alter the negative clinical outcome of these patients. One approach is to utilize a combination of agents which are able to induce the expression of genes which mediate terminal differentiation in melanoma cells, resulting in an arrest of tumor cell proliferation. If successful, such an approach would also represent an important therapeutic procedure for other malignancies. As discussed in this review, we have recently demonstrated that IFN-β in combination with MEZ results in a synergistic suppression of the growth of human melanoma cells *in vitro*.[73,75,76] This inhibition of growth is associated with induction of differentiation, defined by both morphological criteria and an induction or increase in melanin synthesis.[73] This process of differentiation is still reversible after 72 h of exposure to specific concentrations of these compounds, but becomes irreversible after 7 d of continuous exposure to IFN-β plus MEZ.[73,75,76] These results indicate that the combination of IFN-β plus MEZ is capable of inducing the appropriate genes required for the reprogramming of a reversible commitment to differentiate as well as terminal differentiation and a loss of proliferative capacity in human melanoma cells. These findings support the hypothesis that under the appropriate conditions tumor cells can be reprogrammed to terminally differentiate and lose proliferative capacity, and they represent a rational for continuing to pursue this strategy for the ultimate use of a combinational approach for therapy in cancer patients. Another experimental procedure which may prove of value in cancer therapy is the use of combinations of agents to augment tumor-associated antigen expression by tumor cells, resulting in a more effective immunological response in patients and as a method for more effectively employing MoAbs for the therapy of specific human cancers.

V. SUMMARY AND FUTURE PERSPECTIVE

In the present review, we have attempted to present a current prospectus on interferon and its ability to modulate diverse programs of differentiation and to alter the expression of specific genes in target cells. Recent advances in cellular biology, immunology, and recombinant DNA technology offer exciting promise for defining the molecular basis of carcinogenesis and for developing powerful new weapons for the treatment of cancer. Although cancer therapy presently is still somewhat empirical in nature, the ability to produce large quantities of homogeneous reagents with immunomodulatory and direct antitumor activity should now permit a direct test of the efficacy of single and multiple modality treatment in cancer chemotherapy. As we become more aware of the fundamental mechanisms which are involved in tumor development and evolution, newer strategies and agents will undobtedly be developed which will result in a more effective early diagnosis and treatment for specific cancers. Interferons, in combination with other agents, may ultimately fill an important place among the arsenal of weapons used by clinicians to reverse the negative clinical prognosis associated with many currently unmanageable cancers afflicting humans.

ACKNOWLEDGMENTS

We gratefully acknowledge the contributions of our colleagues to the various studies described in this review. They include: Dr. A. Aguzzi, Dr. Lee E. Babiss, Dr. Kapil Bhalla, Dr. Arthur J. Bollon, Dr. Jeffrey N. Bruce, Dr. Subashree Datta, Marita C. Dietrich, Dr. Gregory J. Duigou, Dr. John G. Greiner, Gretchen M. Edwalds, Dr. Michael R. Fetell, Dr. Soldano Ferrone, Dr. Patrizio Giacomini, Dr. Harold, S. Ginsberg, Dr. Gabriel C. Goodman, Thomas Gorey, Dr. Michael M. Gottesman, Gary M. Graham, Dr. Steven Grant, Dr. Beena Gulwani, Dr. Edgar M. Housepian, Dr. Robert S. Kerbal, Dr. Mikihiro Kusama, Dr. Jerome A. Langer, Dr. Wen-shing Liaw, Dr. Michaele Maio, Dr. Maria D. Mileno, Dr. Armand F. Miranda, Dr. Thomas Moulton, Dr. R. Allan Mufson, Dr. P. G. Natali, Dr. Ira Pastan, Dr. Sidney Pestka, Diane R. Prignoli, Dr. Jeffrey Schlom, Wendy E. Solowey, Dr. Matsumo Temponi, Joseph R. Vita, Dr. I. Bernard Weinstein, and Dr. Stephen G. Zimmer. Support for the studies described in this review have come from the National Cancer Institute (CA35675 and CA43208), Hoffmann-LaRoche, Inc., Nutley, New Jersey, and Triton Biosciences, Inc., Alameda, California. Dr. Paul B. Fisher is a Chernow Research Scholar.

REFERENCES

1. **Stewart, W. E., II,** *The Interferon System,* Springer-Verlag, New York, 1979.
2. **Whitaker-Dowling, P. and Youngner, J. S.,** Antiviral effects of interferon in different virus-host cell systems, in *Mechanisms of Interferon Actions,* Vol. I, Pfeffer, L. M., Ed., CRC Press, Boca Raton, FL, 1987, 83.
3. **Faltynek, C. R. and Baglioni, C.,** Interferon induction of the antiviral state: early events, in *Mechanisms of Interferon Actions,* Vol. I, Pfeffer, L. M., Ed., CRC Press, Boca Raton, FL, 1987, 99.
4. **Branca, A. A. and Baglioni, C.,** Evidence that types I and II interferons have different receptors, *Nature,* 294, 768, 1981.
5. **Aguet, M. and Blanchard, B.,** High affinity binding of ^{125}I-labeled mouse interferon to a specific cell surface receptor, *Virology,* 115, 249, 1981.

6. **Aguet, M., Balardelli, F., Blanchard, B., Marcucci, F., and Gresser, I.,** Mouse γ interferon and cholera toxin do not compete for the receptor site of α/β interferon, *Virology,* 117, 541, 1982.
7. **Stewart, W. E., II, Blalock, J. E., Burke, D. C., Chany, C., Dunnick, J. K., Falcoff, E., Freidman, R. M., Galasso, G. J., Joklik, W. K., Vilcek, J. T., Youngner, J. S., and Zoon, K. C.,** Interferon nomenclature, *Nature,* 286, 110, 1980.
8. **Streuli, M., Nagata, S., and Weissmann, C.,** At least three human type α interferons, structure of α2, *Science,* 209, 1343, 1980.
9. **Staehelin, T., Hobbs, D. S., Kung, H.-F., Lai, C.-Y., and Pestka, S.,** Purification and characterization of recombinant human leukocyte interferon (IFLrα) with monoclonal antibodies, *J. Biol. Chem.,* 256, 9750, 1981.
10. **Goeddel, D. V., Leung, D. W., Dull, T. J., Gross, M., Lawn, R. M., McCandliss, R., Seeburg, P. H., Ullrich, A., Yelverton, E., and Gray, P. W.,** The structure of eight distinct cloned human leukocyte interferon cDNAs, *Nature,* 290, 20, 1981.
11. **Weissman, C., Nagata, S., Boll, W., Fountoulakis, M., Fujisawa, A., Fujisawa, J.-I., Haynes, J., Henco, K., Mantei, N., Ragg, H., Schein, C., Schmid, J., Shaw, G., Streuli, M., Taira, H., Todokora, K., and Wiedle, U.,** Structure and expression of human alpha interferon genes, in *Interferons: UCLA Symposium on Molecular Biology,* Vol. 25, Merigan, T. C. and Freidman, R. M., Eds., Academic Press, New York, 1982, 295.
12. **Langer, J. A. and Pestka, S.,** Structure of interferons, *Pharmacol. Therapeut.,* 27, 371, 1985.
13. **Goeddel, D. V., Sheppard, H. M., Yelverton, E., Leung, D., Crea, R., Sloma, A., and Pestka, S.,** Synthesis of human fibroblast interferon by E. coli, *Nucleic Acids Res.,* 8, 4057, 1980.
14. **Maeda, S., McCandliss, R., Gross, M., Sloma, A., Familletti, P. C., Tabor, J. M., Evinger, M., Levy, W. P., and Pestka, S.,** Construction and identification of bacterial plasmids containing nucleotide sequence for human leukocyte interferon, *Proc. Natl. Acad. Sci. U.S.A.,* 77, 7010, 1980; 78, 4648, 1981.
15. **Pestka, S., Langer, J. A., Zoon, K. C., and Samuel, C. E.,** Interferons and their actions, *Ann. Rev. Biochem.,* 56, 727, 1987.
16. **Gray, P. W., Leung, D. W., Pennica, D., Yelverton, E., Najarian, R., Simonsen, C. C., Derynck, R., Sherwood, P. J., Wallace, D. M., and Berger, S. L.,** Expression of human immune interferon cDNA in E. coli and monkey cells, *Nature,* 295, 503, 1982.
17. **Pan, Y.-C. E., Stern, A. S., Familletti, P. C., Khan, F. R., and Chizzonite, R.,** Structural characterization of human interferon γ. Heterogeneity of the carboxyl terminus, *Eur. J. Biochem.,* 166, 145, 1987.
18. **Fisher, P. B. and Grant, S.,** Effects of interferon on differentiation of normal and tumor cells, *Pharmacol. Therapeut.,* 27, 143, 1985.
19. **Greiner, J. W., Schlom, J., Pestka, S., Langer, J. A., Giacomini, P., Kusama, M., Ferrone, S., and Fisher, P. B.,** Modulation of tumor associated antigen expression and shedding by recombinant human leukocyte and fibroblast interferons, *Pharmacol. Therapeut.,* 31, 209, 1987.
20. **Pfeffer, L. M., Ed.,** *Mechanisms of Interferon Actions,* Vols. I and II, CRC Press, Boca Raton, FL, 1987.
21. **Guadagni, F., Kantor, J., Schlom, J., and Greiner, J. W.,** Regulation of tumor antigen expression by recombinant interferons, in *Mechanisms of Differentiation,* Vol. II, Fisher, P. B., Ed., CRC Press, Boca Raton, FL, 1990.
22. **Kirkwood, J. M. and Ernstoff, M. S.,** Interferons in the treatment on human cancer, *J. Clin. Oncol.,* 2, 336, 1984.
23. **Borden, E. C., and Krown, S. E.,** Clinical trails in neoplastic disease in the United States, in *Clinical Applications of Interferons and Their Inducers,* 2nd ed., Stringfellow, D., Ed., Marcel Dekker, New York, 1986, 101.
24. **Krown, S. E.,** Clinical trials of interferons in human malignancy, in *Mechanisms of Interferon Actions,* Vol. II, Pfeffer, L. M., Ed., CRC Press, Boca Raton, FL, 1987, 144.
25. **Goldstein, D. and Laszlo, J.,** Interferon therapy in cancer: from imagination to interferon, *Cancer Res.,* 46, 4315, 1986.
26. **Rossi, G. B.,** Interferons and differentiation, *Interferon,* 6, 31, 1985.
27. **Clemens, M. J. and McNurlan, M. A.,** Regulation of cell proliferation and differentiation by interferons, *Biochem. J.,* 226, 345, 1985.
28. **Moritz, T. and Kirchner, H.,** The effect of interferons on cellular differentiation, *Blut,* 53, 361, 1986.
29. **Tamm, I., Lin, S. L., Pfeffer, L. M., and Sehgal, P. B.,** Interferons alpha and beta as cellular regulatory molecules, *Interferon,* 9, 13, 1987.
30. **Fisher, P. B., Greiner, J. W., Mufson, R. A., and Scholm, J.,** Molecules for tumor diagnosis and therapy, in *Applied Genetic Engineering,* Bolivar, F. and Balbas, P., Eds., Marcel Dekker, New York, in press, 1990.
31. **Fisher, P. B., Hermo, H., Jr., Pestka, S., and Weinstein, I. B.,** Modulation of differentiation in murine and human melanoma cells by interferon and phorbol ester tumor promoters, in *Pigment Cell 1985: Biological, Molecular and Clinical Aspects of Pigmentation,* Bagnara, J., Klaus, S. N., Paul, E., and Schartl, M., Eds., University of Tokyo Press, 1985, 325.

32. **Johnson, H. M.,** Interferon-mediated modulation of the immune system, in *Mechanisms of Interferon Actions,* Vol. II, Pfeffer, L. M., Ed., CRC Press, Boca Raton, FL, 1987, 59.

33. **Miranda, A. F., DiMauro, S., and Somer, H.,** Isoenzymes as markers of differentiation, in *Muscle Regeneration,* Mauro, A., Ed., Raven Press, New York, 1979, 453.

34. **Miranda, A. F., Babiss, L. E., and Fisher, P. B.,** Transformation of human skeletal muscle cells by simian virus 40 (SV40), *Proc. Natl. Acad. Sci. U.S.A.,* 80, 6581, 1983.

35. **Blau, H. M. and Webster, C.,** Isolation and characterization of human muscle cells, *Proc. Natl. Acad. Sci. U.S.A.,* 78, 5623, 1981.

36. **Fisher, P. B., Miranda, A. F., Mufson, R. A., Weinstein, L. S., Fujiki, H., Sugimura, T., and Weinstein, I. B.,** Effects of teleocidin and the phorbol ester tumor promoters on cell transformation, differentiation and phospholipid metabolism, *Cancer Res.,* 42, 2829, 1982.

37. **Miranda, A. F., Babiss, L. E., and Fisher, P. B.,** Measurement of the effect of interferons on cellular differentiation in human skeletal muscle cells, *Methods Enzymol.,* 119, 619, 1986.

38. **Fisher, P. B., Miranda, A. F., Babiss, L. E., Pestka, S., and Weinstein, I. B.,** Opposing effects of interferon produced in bacteria and of tumor promoters on myogenesis in human myoblast cultures, *Proc. Natl. Acad. Sci. U.S.A.,* 80, 2961, 1983.

39. **Lough, J., Keay, S., Sabran J. L., and Grossberg, S. E.,** Inhibition of chicken myogenesis *in vitro* by partially purified interferon, *Biohem. Biophys. Res. Commun.,* 109, 92, 1982.

40. **Tomita, Y. and Hasegawa, S.,** Multiple effects of interferon on myogenesis in chicken myoblast cultures, *Biochim. Biophys. Acta,* 804, 370, 1984.

41. **Multhauf, C. and Lough, J.,** Interferon-mediated inhibition of differentiation in a murine myoblast cell line, *J. Cell. Physiol.,* 126, 211, 1986.

42. **Cohen, R., Pacifici, M., Rubinstein, N., Biehl, J., and Holtzer, H.,** Effect of a tumor promoter on myogenesis, *Nature,* 266, 538, 1977.

43. **Nishizuka, Y.,** The role of protein kinase C in cell surface signal transduction and tumor promotion, *Nature,* 308, 693, 1984.

44. **Nishizuka, Y.,** Studies and perspectives of protein kinase C, *Science,* 233, 305, 1986.

45. **Blumberg, P. M.,** Protein kinase C as the receptor for the phorbol ester tumor promoters, *Cancer Res.,* 48, 1, 1988.

46. **Fisher, P. B., Schachter, D., Mufson, R. A., and Huberman, E.,** The role of membrane lipid dynamics and translocation of protein kinase C in the induction of differentiation in human promyelocytic leukemic cells, in *Pharmacological Effects of Lipids III,* Kabara, J. J., Ed., Oil Chemists Society, Urbana, IL, 1989, 69.

47. **Davis, R. L., Weintraub, H., and Lassar, A. B.,** Expression of a single transfected cDNA converts fibroblasts to myoblasts, *Cell,* 51, 987, 1987.

48. **Tapscott, S. J., Davis, R. L., Thayer, M. J., Cheng, P.-F., Weintraub, H., and Lassar, A. B.,** MyoD1: a nuclear phosphoprotein requiring a *myc* homology region to convert fibroblasts to myoblasts, *Science,* 243, 405, 1988.

49. **Wright, W. E., Sassoon, D. A., and Lin, V. K.,** Myogenin, a factor regulating myogenesis, has a domain homologous to MyoD, *Cell,* 56, 607, 1989.

50. **Wright, W. E. and Lin, V. K.,** The cloning and characterization of myogenin, a factor regulating muscle cell differentiation, in *Mechanisms of Differentiation,* Fisher, P. B., Ed., Vol. I, CRC Press, Boca Raton, FL, 1990.

51. **Pinney, D. F., Pearson-White, S. H., Konieczny, S. F., Latham, K. E., and Emerson, C. P., Jr.,** Myogenic lineage determination and differentiation: evidence for a regulatory gene pathway, *Cell,* 53, 781, 1988.

52. **Braun, T., Bober, E., Buschhausen-Denker, G., Kohtz, S., Grzeschik, K. H., Arnold, H. H., and Kotz, S.,** Differential expression of myogenic determination genes in muscle cells: possible autoactivation by the Myf gene products, *EMBO J.,* 8, 4358, 1989.

53. **Kreider, J. and Schmoyer, M.,** Spontaneous maturation and differentiation of B-16 melanoma cells in culture, *J. Natl. Cancer Inst.,* 55, 641, 1975.

54. **Korner, A. and Pawelek, J.,** Mammalian tyrosinase catalyzes three reactions in the biosynthesis of melanin, *Science,* 217 1163, 1975.

55. **Mufson, R. A., Fisher, P. B., and Weinstein, I. B.,** Effect of phorbol ester tumor promoters on the expression of melanogenesis in B-16 melanoma cells, *Cancer Res.,* 39, 3915, 1979.

56. **Fisher, P. B., Mufson, R. A., and Weinstein, I. B.,** Interferon inhibits melanogenesis in B-16 mouse melanoma cells, *Biochem. Biophys. Res., Commun.,* 100, 823, 1981.

57. **Kwon, B. S., Haq, A. K., Pomerantz, S. H., and Halaban, R.,** Isolation and sequence of a cDNA clone for human tyrosinase that maps at the mouse c-albino locus, *Proc. Natl. Acad. Sci. U.S. A.,* 84, 7473, 1987.

58. **Kwon, B. S., Haq, A. S., Kim, G. S., Pomerantz, S. H., and Halaban, R.,** Cloning and characterization of a human tyrosinase cDNA, *Prog. Clin. Biol. Res.,* 256, 273, 1988.

59. **Kameyama, K., Tanaka, S., Ishida, Y., and Hearing, V. J.,** Interferons modulate the expression of hormone receptors on the surface of murine melanoma cells, *J. Clin. Invest.,* 83, 213, 1989.
60. **Streuli, M., Hall, A., Boll, W., Stewart, W. E., II, Shigekazu, N., and Weissmann, C.,** Target cell specificity of two species of human interferon-α produced in *Escherichia coli* and of hybrid molecules derived from them, *Proc. Natl. Acad. Sci. U.S.A.,* 78, 2848, 1981.
61. **Weck, P. K., Apperson, S., Stebbing, N., Gray, P. W., Leung, D., Shepard, H. M., and Goeddel, D. V.,** Antiviral activities of hybrids of two major human leukocyte interferons, *Nucleic Acids Res.,* 9, 6153, 1981.
62. **Rehberg, E., Kelder, B., Hoal, E. G., and Pestka, S.,** Specific molecular activities of recombinant and hybrid leukocyte interferons, *J. Biol. Chem.,* 257, 11497, 1982.
63. **Van Heuvel, M., Bosveld, I. J., Mooren, A. T. A., Trapman, J., and Zwarthoff, E. C.,** Properties of natural and hybrid murine alpha interferons, *J. Gen. Virol.,* 67, 2215, 1986.
64. **Raj, N. B. K., Kelley, K. A., Israeli, R., Kellum, M., and Pitna, P. M.,** Species specificity of the mouse human hybrid interferons produced in *E. coli,* in *Interferons as Cell Growth Inhibitors and Antitumor Factors,* Freidman, R. M., Merigan, T., and Sreevalsan, T., Eds., Alan R. Liss, New York, 1986, 121.
65. **Stebbing, N.,** Mechanisms of action of interferons: evidence from studies with recombinant DNA-derived subtypes and analogs, in *Mechanisms of Interferon Action,* Vol. II, Pfeffer, L. M., Ed., CRC Press, Boca Raton, FL, 1987, 79.
66. **Fisher, P. B., Hermo, H., Jr., Prignoli, D. R., Weinstein, I. B., and Pestka, S.,** Hybrid recombinant human leukocyte interferon inhibits differentiaition in murine B-16 melanoma cells, *Biochem. Biophys. Res. Commun.,* 119, 108, 1984.
67. **Balkwill, F. R., Proietti, E., Bodmer, J., Hart, I., and Ramani, P.,** Mechanisms of antitumor action of interferons on human tumor xenografts and in mouse metastases models, in *Interferons as Cell Growth Inhibitors and Antitumor Factors,* Friedman, R. M., Merigan, T., and Sreevalsan, T., Eds., Alan R. Liss, New York, 1986, 425.
68. **Giacomini, P., Aguzzi, A., Pestka, S., Fisher, P. B., and Ferrone, S.,** Modulation by recombinant DNA leukocyte (α) and fibroblast (β) interferons of the expression and shedding of HLA and tumor associated antigens by human melanoma cells, *J. Immunol.,* 133, 1649, 1984.
69. **Greiner, J. W., Fisher, P. B., Pestka, S., and Schlom, J.,** Differential effects of recombinant human leukocyte interferons on cell surface antigen expression, *Cancer Res.,* 46, 4894, 1986.
70. **Balkwill, F. R., Moodie, E. M., Mowshowitz, S., and Fantes, K. H.,** An animal model system for investigating the anti-tumor effect of human IFNs on human tumors, in *The Biology of the Interferon System,* Demaeyer, E. and Schellekens, H., Eds., Elsevier, New York, 1983, 443.
71. **Balkwill, F. R., Lee, A., Adam, G., Moodie, E., Thomas, J. A., Travernier, J., and Fiers, W.,** Human tumor xenografts treated with recombinant human tumor necrosis factor alone or in combiantion with interferons, *Cancer Res.,* 46, 3990, 1986.
72. **Huberman, E., Heckman, C., and Langenbach, R.,** Stimulation of differentiated functions in human mealnoma cells by tumor-promoting agents and dimethyl sulfoxide, *Cancer Res.,* 39, 2618, 1979.
73. **Fisher, P. B., Prignoli, D. R., Hermo, H., Jr., Weinstein, I. B., and Pestka, S.,** Effects of combined treatment with interferon and mezerein on melanogenesis and growth in human melanoma cells, *J. Interferon Res.,* 5, 11, 1985.
74. **Giacomini, P., Imberti, L., Aguzzi, A., Fisher, P. B., Trinchieri, G., and Ferrone, S.,** Immunochemical analysis of the modulation of human melanoma associated antigens by DNA recombinant immune interferon, *J. Immunol.,* 135, 2887, 1985.
75. **Fisher, P. B., Hermo, H., Jr., Solowey, W. E., Dietrich, M. C., Edwalds, G. M., Weinstein, I. B., Langer, J. A., Pestka, S., Giacomini, P., Kusama, M., and Ferrone, S.,** Effect of recombinant human fibroblast interferon and mezerein on growth, differentiation, immune interferon binding and tumor associated antigen expression in human melanoma cells, *Anticancer Res.,* 6, 765, 1986.
76. **Ahmed, M. A., Guarini, L., Ferrone, S., and Fisher, P. B.,** Induction of differentiation in human melanoma cells by the combination of different classes of interferons or interferon plus mezerein, *N.Y. Acad. Sci.,* 567, 328, 1989.
77. **Guarini, L., Temponi, M., Edwalds, G. M., Vita, J. R., Fisher, P. B., and Ferrone, S.,** In vitro differentiation and antigenic changes in human melanoma cell lines, *Cancer Immunol. Immunother.,* 30, 363, 1989.
78. **Yamasaki, H., Fibach, E., Nudel, U., Weinstein, I. B., Rifkind, R. A., and Marks, P. A.,** Tumor promoters inhibit spontaneous and induced differentiation of murine erythroleukemia cells in culture, *Proc. Natl. Acad. Sci. U.S.A,* 74, 3451, 1977.
79. **Rovera, G., O'Brien, T. A., and Diamond, L.,** Tumor promoters inhibit spontaneous differentiation of Friend erythroleukemia cells in culture, *Proc. Natl. Acad. Sci. U.S.A.,* 74, 2894, 1977.
80. **Ishii, D., Fibach, E., Yamasaki, H., and Weinstein, I. B.,** Tumor promoters inhibit morphological differentiation in cultured mouse neuroblastoma cells, *Science,* 200, 556, 1978.

81. **Weinstein, I. B., Lee, L. S., Fisher, P. B., Mufson, R. A., and Yamasaki, H.,** Action of phorbol esters in cell culture: mimicry of transformation, altered differentiation and effects on cell membranes, *J. Supramol. Struct.,* 12, 195, 1979.

82. **Fisher, P. B.,** Enhancement of viral transformation and expression of the transformed phenotype by tumor promoters, in *Tumor Promotion and Cocarcinogenesis In Vitro, Mechanisms of Tumor Promotion,* Slaga, T. J., Ed., CRC Press, Boca Raton, FL, 1984, 57.

83. **Green, H. and Meuth, M.,** An established preadipose cell line and its differentiation in culture, *Cell,* 3, 127, 1974.

84. **Green, H. and Kehinde, O.,** An established preadipose cell line and its differentiation in culture. II. Factors affecting adipose conversion, *Cell,* 5, 19, 1975.

85. **Diamond, L., O'Brien, T. G., and Rovera, G.,** Inhibition of adipose conversion of 3T3 fibroblasts by tumor promoters, *Nature,* 269, 247, 1977.

86. **Green, H.,** The adipose conversion of 3T3 cells, in *Differentiation and Development,* Ahmad, F., Ed., Tenth Miami Winter Symposium, Academic Press, New York, 1978, 13.

87. **O'Brien, T. G., Saladik, D., and Diamond, L.,** The tumor promoter 12-O-tetradecanoylphorbol-13-acetate stimulates lactate production in BALB/C 3T3 preadipose cells, *Biochem. Biophys. Res. Commun.,* 88, 103, 1979.

88. **Keay, S. and Grossberg, S. E.,** Interferon inhibits the conversion of 3T3-L1 mouse fibroblasts into adipocytes, *Proc. Natl. Acad. Sci. U.S.A.,* 77, 4099, 1980.

89. **Cioe, L., O'Brien, T. G., and Diamond, L.,** Inhibition of adipose conversion of BALB/c 3T3 cells by interferon and 12-O-tetradecanoyl-phorbol-13-acetate, *Cell Biol. Int. Rep.,* 4, 255, 1980.

90. **Taylor, J. L., Turo, K. A., McCann, P. P., and Grossberg, S. E.,** Inhibition of the differentiation of 3T3-L1 cells by interferon-beta and difluoromethyl ornithine, *J. Biol. Regul. Homeostatic Agents,* 2, 19, 1988.

91. **Spiegelman, B. M., Distel, R. J., Ro, H.-S., Rosen, B. S., and Satterberg, B.,** *Fos* protooncogene and the regulation of gene expression in adipocyte differentiation, *J. Cell Biol.,* 107, 829, 1988.

92. **Sassa, S.,** Sequential induction of heme pathway enzymes during erythroid differentiation of mouse Friend erythroleukemia cells in culture, *J. Exp. Med.,* 143, 305, 1976.

93. **Friend, C.,** The phenomenon of differentiation in murine erythroleukemia cells, *Harvey Lect.,* 72, 253, 1978.

94. **Marks, P. A. and Rifkind, R. A.,** Erythroleukemic differentiation, *Annu. Rev. Biochem.,* 47, 419, 1978.

95. **Reuben, R. C., Rifkind, R. A., and Marks, P. A.,** Chemically induced murine erythroleukemic differentiation, in *BBA Reviews on Cancer,* Vol. 605, Elsevier, Amsterdam, 1980, 325.

96. **Troxler, D. H., Ruscetti, S. K., and Scolnick, E. M.,** The molecular biology of Friend virus, in *BBA Reviews on Cancer,* Vol. 605, Elsevier, Amsterdam, 1980, 305.

97. **Rifkind, R. A., Breslow, R., and Marks, P. A.,** Induced differentiation of transformed cells with polar/apolar compounds and the reversibility of the transformed phenotype, in *Mechanisms of Differentiation,* Vol. II, Fisher, P. B., Ed., CRC Press, Boca Raton, FL, 1990.

98. **Ramoni, C., Rossi, G. B., Matarese, G. P., and Dolei, A.,** Production of Friend leukemia virus antigens in chronically-infected cells treated with interferon, *J. Gen. Virol.,* 37, 285, 1977.

99. **Luftig, R. B., Conscience, J. F., Skoultchi, A., McMillan, P., Revel, M., and Ruddle, F. H.,** Effect of interferon on dimethylsulfoxide-stimulated Friend erythroleukemic cells: ultrastructural and biochemical study, *J. Virol.,* 23, 799, 1977.

100. **Krieg, C. J., Ostertag, W., Klauss, U., Pragnell, J. B., Swetly, P., Roesler, G., and Weiman, B. J.,** Increase in intracisternal A-type particles in Friend cells during inhibition of Friend virus (SFFV) release by interferon or azidothymine, *Exp. Cell Res.,* 116, 21, 1978.

101. **Dolei, A., Colleta, G., Capobianchi, M. R., Rossi, G. B., and Vecchio, G.,** Interferon effects on Friend leukemia cells. I. Expression of virus and erythroid markers in untreated and dimethyl sulphoxide-treated cells, *J. Gen. Virol.,* 46, 227, 1980.

102. **Lieberman, D., Volloch, Z., Aviv, H., Nudel, V., and Revel, M.,** Effects of interferon on hemoglobin synthesis and leukemia production in Friend cells, *Mol. Biol. Rep.,* 1, 447, 1975.

103. **Cioe, L., Dolei, A., Rossi, G. B., Belardelli, F., Affabris, E., Gambari, E., and Fantoni, A.,** Potential for differentiation, virus production and tumorigenicity in murine erythroleukemic cells treated with interferon, in *In Vitro Aspects of Erythropoiesis,* Murphy, M. J., Ed., Springer-Verlag, New York, 1978, 159.

104. **Rossi, G. B., Matarese, G. P., Grappelli, C., and Belardelli, F.,** Interferon inhibits dimethyl sulphoxide-induced erythroid differentiation of Friend leukemia cells, *Nature,* 267, 50, 1977.

105. **Rossi, G. B., Dolei, A., Cioe, L., Benedetto, A., Matarese, G. P., and Belardelli, F.,** Inhibition of transcription and translation of globin messenger RNA in dimethyl sulfoxide-stimulated Friend erythroleukemia cells treated with interferon, *Proc. Natl. Acad. Sci. U.S.A.,* 74, 2036, 1977.

106. **Rossi, G. B., Dolei, A., Capobianchi, M. R., Peschle, C., and Affabris, E.,** Interactions of interferon with in vitro model systems involved in hematopoietic cell differentiation, *Ann. N.Y. Acad. Sci.,* 350, 279, 1980.

107. **Matarese, G. P. and Rossi, G. B.,** Effect of interferon on growth and division cycle of Friend erythro-leukemic murine cells in vitro, *J. Cell Biol.,* 75, 344, 1978.

108. **Affabris, E., Jemma, C., and Rossi, G. B.,** Isolation of interferon-resistant variants of Friend erythro-leukemia cells: effects of interferon and ouabain, *Virology,* 120, 441, 1982.

109. **Affabris, E., Romeo, G., Belardelli, F., Jemma, C., Mechti, N., Gresser, I., and Rossi, G. B.,** 2-5A synthetase activity does not increase in interferon-resistant Friend leukemia cell variants treated with α/β interferon despite the presence of high-affinity interferon receptor sites, *Virology,* 125, 508, 1983.

110. **Affabris, E., Romeo, G., Federico, M., Coccia, E. M., and Rossi, G. B.,** Friend leukemia cell clones resistant to murine IFNs, in *The Interferon System,* Vol. 24, Serono Symposia, Dianzani, F. and Rossi, G. B., Eds., Raven Press, Rome, 1985, 373.

111. **Romeo, G., Affabris, E., Federico, M., Mechti, N., Coccia, E. M., Jemma, C., and Rossi, G. B.,** Establishment of the antiviral state in α/β interferon-resistant Friend cells treated with γ interferon: induction of 67K protein kinase activity in absence of detectable 2-5A synthetase, *J. Biol. Chem.,* 260, 3833, 1985.

112. **Coccia, E. M., Federico, M., Romeo, G., Affabris, E., Cofano, F., and Rossi, G. B.,** Interferon-α/β and γ-resistant Friend cell variants exhibiting receptor sites for interferons but no induction of 2-5A synthetase and 67K protein kinase, *J. Interferon Res.,* 8, 113, 1988.

113. **Affabris, E., Federico, M., Romeo, G., Coccia, E. M., and Rossi, G. B.,** Opposite effects of murine interferons on erythroid differentiation of Friend cells, *Virology,* 167, 185, 1988.

114. **Locardi, C., Belardelli, F., Federico, M., Romeo, G., Affabris, E., and Gresser, I.,** Effect of mouse interferon alpha/beta on the expression of H-2 (Class I) antigens and on the levels of 2'-5' oligoadenylate synthetase activity in interferon-sensitive and interferon-resistant Friend leukemia cell tumors in mice, *J. Biol. Regul. Homeostatic Agents,* 1, 189, 1987.

115. **Gallagher, R., Collins, S., Trujill, J., McGredie, K., Ahern, M., Tsai, S., Metzgar, R., Aulak, G., Ting, R., Ruscetti, F., and Gallo, R.,** Characterization of the continuous differentiating myeloid cell line (HL-60) from a patient with acute promyelocytic leukemia, *Blood,* 54, 713, 1979.

116. **Palumbo, A., Minowada, J., Erikson, J., Croce, C. M., and Rovera, G.,** Lineage infidelity of a human leukemia cell line, *Blood,* 64, 1059.

117. **Sundstrom, C. and Vilsson, K.,** Establishment and characterization of a human histiocytic cell line (U937), *Int. J. Cancer,* 17, 565.

118. **Huberman, E. and Callaham, M. F.,** Induction of terminal differentiation in human promyelocytic leukemia cells by tumor-promoting agents, *Proc. Natl. Acad. Sci. U.S.A.,* 76, 1293, 1979.

119. **Rovera, G., Santoli, D., and Damsky, C.,** Human promyelocytic leukemia cells in culture differentiate into macrophage-like cells when treated with phorbol diesters, *Proc. Natl. Acad. Sci. U.S.A.,* 76, 2779, 1979.

120. **Miyawa, C., Abe, E., Kuribayashi, T., Tanaka, H., Kouno, K., Nishii, Y., and Suda, T.,** 1,25-dihydroxy vitamin D_3 induces differentiation of human myeloid cells, *Biochem. Biophys. Res. Commun.,* 102, 937, 1981.

121. **Elias, L., Wogenrich, F. J., Wallace, J. M., and Longmire, J.,** Altered pattern of differentiation and proliferation of HL-60 promyelocytic leukemia cells in the presence of leukocyte conditioned medium, *Leukemia Res.,* 4, 301, 1980.

122. **Collins, S. J., Ruscetti, M. W., Gallagher, R. E., and Gallo, R. C.,** Terminal differentiation of human promyelocytic leukemia cells induced by dimethyl sulfoxide and other polar compounds, *Proc. Natl. Acad. Sci. U.S.A.,* 75, 2458, 1978.

123. **Breitman, T. R., Scolnick, E. S., and Collins, S. J.,** Induction of differentiation of the human promye-locytic cell line (HL-60) by retinoic acid, *Proc. Natl. Acad. Sci. U.S.A.,* 77, 2936, 1980.

124. **Grant, S., Bhalla, K., Pestka, S., Weinstein, I. B., and Fisher, P. B.,** Differential effect of recombinant human leukocyte interferon on human leukemic and normal myeloid progenitor cells, *Biochem. Biophys. Res. Commun.,* 108, 1048, 1982.

125. **Tomida, M., Yamamoto, Y., and Hozumi, M.,** Stimulation by interferon of induction of differentiation of human promyelocytic leukemia cells, *Biochem. Biophys. Res. Commun.,* 104, 30, 1982.

126. **Dayton, E. T., Matsumoto-Kobayashi, M., Perrusia, B., and Trinchieri, G.,** Role of immune interferon in the monocytic differentiation of human promyelocytic cell lines induced by leukocyte conditioned-medium, *Blood,* 66, 583, 1985.

127. **Hemmi, H. and Breitman, T. R.,** Combinations of recombinant human interferons and retinoic acid synergistically induce differentiation of the human promyelocytic leukemia cell line HL-60, *Blood,* 67, 501, 1986.

128. **Lin, J. and Sartorelli, A. C.,** Stimulation by interferon of the differentiation of human promyelocytic leukemia (HL-60) cells produced by retinoic acid and actinomycin D, *J. Interferon Res.,* 7, 379, 1987.

129. **Kohlhepp, E. A., Condon, M. E., and Hamburger, A. W.,** Recombinant human interferon α enhancement of retinoic acid-induced differentiation of HL-60 cells, *Exp. Hematol.,* 15, 414, 1987.

130. **Ball, E. D., Guyre, P. M., Shen, L., Glynn, J. M., Maliszewski, C. R., Baker, P. E., and Fanger, M. W.,** Gamma inteferon induces monocytoid differentiation in the HL-60 cell line, *J. Clin. Invest.,* 73, 1072, 1984.

131. **Takei, M., Takeda, K., and Konno, K.,** The role of interferon-γ in induction of differentiation of human myeloid leukemia cell lines, ML-1 and HL-60, *Biochem. Biophys. Res. Commun.,* 124, 100, 1984.

132. **Harris, P. E., Ralph, P., Gabrilove, J., Welte, K., Karmali, R., and Moore, M. A. S.,** Distinct differentiation-inducing activities of γ-interferon and cytokine factors acting on the human promyelocytic leukemia cell line HL-60, *Cancer Res.,* 45, 3090, 1985.

133. **Hattori, T., Pack, M., Bougnoux, P., Chang, Z., and Hoffman, T.,** Interferon induced differentiation in U937 cells, *J. Clin. Invest.,* 72, 237, 1983.

134. **Gullberg, U., Nilsson, E., Einhorn, S., and Olsson, I.,** Combinations of interferon-γ and retinoic acid or 1α,25-dihydroxycholecalciferol induce differentiation of the human monoblast leukemia cell line U-937, *Exp. Hematol.,* 13, 675, 1985.

135. **Weinberg, J. B. and Larrick, J. W.,** Receptor-mediated monocytoid differentiation of human promyelocytic cells by tumor necrosis factor: synergistic actions with interferon-γ and 1,25-dihydroxyvitamin D_3, *Blood,* 70, 994, 1987.

136. **Testa, U., Ferbus, D., Gabbianelli, M., Pascucci, B., Boccoli, G., Louache, F., and Thang, M. N.,** Effect of endogenous interferons on the differentiation of human monocyte cell line U937, *Cancer Res.,* 48, 82, 1988.

137. **Trinchieri, G., Kobayashi, M., Rosen, M., Loudon, R., Murphy, M., and Perussia, B.,** Tumor necrosis factor and lymphotoxin induce differentiation of human myeloid cell lines in synergy with immune interferon, *J. Exp. Med.,* 164, 1206, 1986.

138. **Grant, S., Bhalla, K., Weinstein, I. B., Pestka, S., Mileno, M. D., and Fisher, P. B.,** Recombinant human interferon sensitizes resistant myeloid leukemic cells to induction of terminal differentiation, *Biochem. Biophys. Res. Commun.,* 130, 379, 1985.

139. **Murao, S.-I., Fukumoto, Y., Katayama, M., Maeda, H., Sugiyama, T., and Huberman, E.,** Recombinant gamma-interferon and lipopolysaccharide enhance 1,25-dihydroxyvitamin D_3-induced cell differentiation in human promyelocytic leukemia (HL-60) cells, *Jpn. J. Cancer Res.,* 76, 596, 1985.

140. **Weinberg, J. B., Misukonis, M. A., Hobbs, M. M., and Borowitz, M. J.,** Cooperative effects of gamma interferon and 1-alpha,25-dihydroxyvitamin D_3 in inducing differentiation of human promyeolocytic leukemia (HL-60) cells, *Exp. Hematol.,* 114, 138, 1986.

141. **Ball, E. D., Howell, A. L., and Shen, L.,** Gamma interferon and 1,25 dihydroxyvitamin D_3 cooperate in the induction of monocytoid differentiation but not in the functional activation of the HL-60 promyelocytic leukemia cell line, *Exp. Hematol.,* 14, 998, 1986.

142. **Gallagher, R. E., Lurie, K. J., Leavitt, R. D., and Wiernik, P. H.,** Effects of interferon and retinoic acid on the growth and differentiation of clongenic leukemic cells from acute myelogenous leukemia patients treated with recombinant-αA interferon, *Leukemia Res.,* 609, 1987.

143. **Riordin, J. R., Deuchars, K., Kartner, N., Alon, N., Trent, J., and Ling, V.,** Amplification of P-glycoprotein genes in multidrug-resistant mammalian cell lines, *Nature,* 316, 817, 1985.

144. **Fojo, A. T., Whang-Peng, J., Gottesman, M. M., and Pastan, I.,** Amplification of DNA sequences in human multidrug resistant KB carcinoma cells, *Proc. Natl. Acad. Sci. U.S.A.,* 82, 7661, 1985.

145. **Gros, P., Croop, J., Roninson, I. B., Varsharvsky, A., and Housman, D. E.,** Isolation and characterization of DNA sequences amplified in multidrug resistant hamster cells, *Proc. Natl. Acad. Sci. U.S.A.,* 83, 1986.

146. **Gros, P., Croop, J., and Housman, D. E.,** Mammalian multidrug resistance gene: complete cDNA sequence indicates strong homology to bacterial transport proteins, *Cell,* 47, 381, 1986.

147. **Bradley, G., Juranka, P. F., and Ling, V.,** Mechanism of multidrug resistance, *Biochim. Biophys. Acta,* 948, 87, 1988.

148. **Reddy P. G., Graham, G. M., Datta, S., Moulton, T. L., Guarini, L., Gottesman, M. M., and Fisher, P. B.,** Construction and characterization of human glioblastoma multiforme cells displaying a multidrug resistance phenotype, in preparation.

149. **Vita, J. R., Edwalds, G. M., Gorey, T., Housepian, E. M., Fetell, M. R., Guarini, L., Langer, J. A., and Fisher, P. B.,** Enhanced in vitro growth suppression of human glioblastoma cultures treated with the combination of recombinant fibroblast and immune interferons, *Anticancer Res.,* 8, 297, 1988.

150. **Fetell, M. R., Housepian, E. M., Oster, M. W., Cote, D. N., Sisti, M. B., Marcus, S. G., and Fisher, P. B.,** Intratumor administration of beta interferon in malignant gliomas: phase I clinical and laboratory study, *Cancer,* 65, 78, 1990.

151. **Guarini, L., Graham, G. M., Datta, S., Bollon, A. J., Ferrone, S., and Fisher, P. B.,** Differential immunomodulatory response of multidrug sensitive versus multidrug resistant human glioblastoma multiforme cells to interferons and tumor necrosis factor, in preparation.

152. **Michalevicz, R. and Revel, M.,** Interferons regulate the in vitro differentiation of multilineage lymphomyeloid stem cells in hairy cell leukemia, *Proc. Natl. Acad. Sci. U.S.A.,* 84, 2307, 1987.

153. **Improta, T., Salvatore, A. M., Di Luzio, A., Romeo, G., Coccia, E. M., and Calissano, P.,** IFN-γ facilitates NGF-induced neuronal differentiation in PC12 cells, *Exp. Cell Res.,* 179, 1, 1988.

154. **Andrews, P. W., Trinchieri, G., Perussia, B., and Baglioni, C.,** Induction of class I major histocompatibility complex antigens in human teratocarcinoma cells by interferon without induction of differentiation, growth inhibition, or resistance to viral infection, *Cancer Res.,* 47, 740, 1987.

155. **Yaar, M., Karassik, R. L., Schnipper, L. E., and Gilchrest, B. A.,** Effects of alpha and beta interferons on cultured human keratinocytes, *J. Invest. Dermatol.,* 85, 70, 1985.

156. **Stadler, R., Muller, R., and Orfanos, C. E.,** Effect of recombinant alpha A-interferon on DNA synthesis and differentiation of human keratinocytes in vitro, *Br. J. Dermatol.,* 114, 273, 1986.

157. **Chang, E. H., Ridge, J., Black, R., Zou, Z.-Q., Masnyk, T., Noguchi, P., and Harford, J. B.,** Interferon-γ induces altered oncogene expression and terminal differentiation in A431 cells, *Proc. Soc. Exp. Biol. Med.,* 186, 319, 1987.

158. **Rubin, B. Y. and Gupta, S. L.,** Interferon-induced proteins in human fibroblasts and development of the antiviral state, *J. Virol.,* 34, 446, 1980.

159. **Weil, J., Epstein, C. J., Epstein, L. B., Sedmak, J. J., Sabran, J. L., and Grossberg, S. E.,** A unique set of polypeptides is induced by γ interferon in addition to those induced in common with α and β interferons, *Nature,* 301, 437, 1983.

160. **Weil, J., Epstein, C. J., and Epstein, L. B.,** Cell growth regulation and growth regulatory substances, *Nat. Immunol. Cell Growth Regul.,* 3, 51, 1983/1984.

161. **Dron, M., Tovey, M. G., and Eid, P.,** Isolation of Daudi cells with reduced sensitivity to interferon. III. Interferon-induced proteins in relation to the phenotype of interferon resistance, *J. Gen. Virol.,* 66, 787, 1985.

162. **Knight, E., Fahey, D., and Blomstrom, D. C.,** Interferon-β enhances the synthesis of a 20,000 Dalton membrane protein: a correlation with cessation of cell growth, *J. Interferon Res.,* 5, 305, 1985.

163. **McMahon, M., Stark, G. R., and Kerr, I. M.,** Interferon-induced gene expression in wild-type and interferon resistant human lymphoblastoid (Daudi) cells, *J. Virol,* 57, 362, 1986.

164. **Horisberger, M. A. and Hochkeppel, H. K.,** IFN-α induced human 78kD protein: purification and homologies with the mouse Mx protein, production of monoclonal antibodies, and potentiation effect of IFN-γ, *J. Interferon Res.,* 7, 331, 1987.

165. **Horisberger, M., Staeheli, P., and Haller, O.,** Interferon induces a unique protein in mouse cells bearing a gene for resistance to influenza virus, *Proc. Natl. Acad. Sci. U.S.A.,* 80, 1910, 1983.

166. **Baglioni, C.,** Interferon-induced enzymatic activities and their role in the antiviral state, *Cell,* 17, 255, 1979.

167. **Lengyel, P.,** Biochemistry of interferons and their actions, *Annu. Rev. Biochem.,* 51, 251, 1982.

168. **Basham, T. Y., Bourgeade, M. F., Creasey, A. A., and Merigan, T. C.,** Interferon increases HLA synthesis in melanoma cells: interferon-resistant and -sensitive cell lines, *Proc. Natl. Acad. Sci. U.S.A.,* 79, 3265, 1982.

169. **Yoshie, O., Schmidt, H., Reddy, E. S. P., Weissman, S., and Lengyel, P.,** Mouse interferons enhance the accumulation of human HLA RNA and protein in transfected mouse and hamster cells, *J. Biol. Chem.,* 257, 13169, 1982.

170. **Rosa, F., Hatat, D., Abadie, A., Wallach, D., Revel, M., and Fellous, M.,** Differential regulation of HLA-DR mRNAs and cell surface antigens by interferons, *EMBO J.,* 9, 1585, 1983.

171. **Cheng, Y.-S. E., Colonno, R. J., and Yin, F. H.,** Interferon induction of fibroblast proteins with guanylate binding activity, *J. Biol. Chem.,* 258, 7746, 1983.

172. **Friedman, R. L. and Stark, G. R.,** α-Interferon-induced transcription of HLA and metallothionein genes containing homologous upstream sequences, *Nature,* 314, 637, 1985.

173. **Bersini, M. H., Lempert, M. J., and Epstein, L. B.,** Overlapping polypeptide induction in human fibroblasts in response to treatment with interferon-α, interferon-γ, interleukin 1a, interleukin 1b, and tumor necrosis factor, *J. Immunol.,* 140, 485, 1988.

174. **Karin, M., Haslinger, A., Holtgreve, H., Richards, I., Krauter, P., Westphal, H. M., and Beato, M.,** Characterization of DNA sequences through which cadmium and glucocorticoid hormones induce human metallothionein-II$_a$, *Nature,* 308, 513, 1984.

175. **Karin, M. and Herschman, H. R.,** Dexamethasone stimulation of metallothionein synthesis in HeLa cell culture, *Science,* 204, 176, 1979.

176. **Hager, L. J. and Palmiter, R. D.,** Transcriptional regulation of mouse liver metallothionein-I gene by glucocorticoids, *Nature,* 291, 340, 1981.

177. **Karin, M., Haslinger, A., Holtgreve, H., Cathala, G., Slater, E., and Baxter, J. D.,** Activation of a heterologous promoter in response to dexamethasone and cadmium by metallothionein gene 5′ flanking DNA, *Cell,* 36, 371, 1984.

178. **Karin, M., Imbra, R. J., Heguy, A., and Wong, G.,** Interleukin 1 regulates human metallothionein gene expression, *Mol. Cell. Biol.,* 5, 2866, 1985.

179. **Imbra, R. J. and Karin, M.,** Metallothionein gene expression is regulated by serum factors and activation of protein kinase C, *Mol. Cell. Biol.,* 7, 1358, 1987.

180. **Kelly, J. M., Gilbert, C. S., Stark, G. R., and Kerr, I. M.,** Differential regulation of interferon-induced mRNAs and c-myc mRNA by α- and γ-interferons, *Eur. J. Biochem.,* 153, 367, 1985.

181. **Kelly, J. M., Porter, A. C. G., Chernajovsky, Y., Gilbert, C. S., Stark, G. R., and Kerr, I. M.,** Characterization of a human gene inducible by α- and β-interferons and its expression in mouse cells, *EMBO J.,* 5, 1601, 1986.

182. **Pober, J. S., Collins, T., Gimbrone, M. A., Cotran, R. S., Gitlin, J. D., Fiers, W., Clayberger, C., Krensky, A. M., Burakoff, S J., and Reiss, C. S.,** Lymphocytes recognize human vascular endothelial and dermal I_a antigens induced by recombinant immune interferon, *Nature,* 305, 726, 1983.

183. **Wallach, D., Fellous, M., and Revel, M.,** Preferential effect of γ interferon on the synthesis of HLA antigens and their mRNAs in human cells, *Nature,* 299, 833, 1982.

184. **Sen, G. C. and Rubin, B. Y.,** Synthesis of interferon-inducible proteins is regulated differently by interferon-α and interferon-γ, *Virology,* 134, 483, 1984.

185. **Blasi, E., Herberman, R. B., and Varesio, L.,** Requirement for protein synthesis for induction of macrophage tumoricidal activity by IFN-β and IFN-β but not IFN-γ, *J. Immunol.,* 132, 3226, 1984.

186. **Dianzani, I., Zucca, M., Scupham, A., and Georgiades, J. A.,** Immune and virus induced interferons may activate cells by different derepressional mechanisms, *Nature,* 283, 400, 1980.

187. **Baglioni, C. and Maroney, P. A.,** Mechanism of action of human interferons, *J. Biol. Chem.,* 255, 8390, 1980.

188. **Giacomini, P., Tecce, R., Gambari, R., Sacchi, A., Fisher, P. B., and Natali, P. G.,** Recombinant human IFN-γ, but not IFN-α or IFN-β, enhances MHC- and Non-MHC-encoded glycoproteins by a protein synthesis-dependent mechanism, *J. Immunol.,* 140, 3073, 1988.

189. **Reich, N., Evans, B., Levy, D., Fahey, D., Knight, E., and Darnell, J. E.,** Interferon-induced transcription of a gene encoding a 15-kDa protein depends on an upstream enhancer element, *Proc. Natl. Acad. Sci. U.S.A.,* 84, 6394, 1987.

190. **Larner, A. C., Jonak, G., Cheng, Y.-S. E., Korant, B., Knight, E., and Darnell, J. E.,** Transcriptional induction of two genes in human cells by β interferon, *Proc. Natl. Acad. Sci. U.S.A.,* 81, 6733, 1984.

191. **Levy, D. E., Kessler, D. S., Reich, N., and Darnell, J. E.,** Interferon-induced nuclear factors that bind a shared promoter element correlate with positive and negative transcriptional control, *Genes Dev.,* 2, 383, 1988.

192. **Chebath, J., Merlin, G., Metz, R., Benech, P., and Revel, M.,** Interferon-induced 56,000 M_r protein and its mRNA in human cells: molecular cloning and partial sequence of the cDNA, *Nucleic Acids Res.,* 11, 1213, 1983.

193. **Larner, A. C., Chaudhuri, A., and Darnell, J. E.,** Transcriptional induction by interferon. New protein(s) determine the extent and length of the induction, *J. Biol. Chem.,* 261, 453, 11986.

194. **Kusari, J., Szabo, P., Grzeschik, K.-H., and Sen, G. C.,** Chromosomal localization of the interferon-inducible gene encoding mRNA 561, *J. Interferon Res.,* 7, 53, 1987.

195. **Friedman, R. L., Manly, S. P., McMahon, M., Kerr, I. M., and Stark, G. R.,** Transcriptional and posttranscriptional regulation of interferon-induced gene expression in human cells, *Cell,* 38, 745, 1984.

196. **Blanar, M. A., Boettger, E. C., and Flavell, R. A.,** Transcriptional activation of HLA-DRα by interferon γ requires a trans-acting protein, *Proc. Natl. Acad. Sci. U.S.A.,* 85, 4672, 1988.

197. **Staeheli, P., Haller, O., Boll, W., Lindenmann, J., and Weissmann, C.,** Mx protein: constitutive expression in 3T3 cells transformed with cloned Mx cDNA confers selective resistance to influenza virus, *Cell,* 44, 147, 1986.

198. **Merlin, G., Chebath, J., Benech, P., Metz, P., and Revel, M.,** Molecular cloning and sequence of partial cDNA for interferon-induced (2'-5') oligo (A) synthetase mRNA from human cells, *Proc. Natl. Acad. Sci. U.S.A.,* 80, 4904, 1983.

199. **Samanta, H., Engel, D. A., Chao, H. M., Thakui, A., Gracia-Blanco, M. A., and Lengyel, P.,** Interferons as gene activators, *J. Biol. Chem.,* 261, 11849, 1986.

200. **Engel, D. A., Snoddy, J., Toniato, E., and Lengyel, P.,** Interferons as gene activators: close linkage of two interferon-activatable murine genes, *Virology,* 166, 24, 1988.

201. **McGrath, J. P., Capon, D. J., Goeddel, D. V., and Levinson, A. D.,** Comparative biochemical properties of normal and activated human ras p21 protein, *Nature,* 310, 644, 1984.

202. **Downward, J., Yarden, Y., Mayes, E., Scrace, G., Totty, N., Stockwell, P., Ullrich, A., Schlessinger, J., and Waterfield, M. D.,** Close similarity of epidermal growth factor receptor and v-erb-B oncogene protein sequences, *Nature,* 307, 521, 1984.

203. **Waterfield, M. D., Scrace, G. T., Whittle, N., Stroobant, P., Johnsson, A., Wasteson, A., Westermark, B., Heloin, C.-H., Huang, J. S., and Deuel, T. F.,** Platelet derived growth factor is structurally related to the putative transforming protein p28^{sis} of Simian Sarcoma virus, *Nature,* 304, 35, 1983.

204. **Rapp, U. R., Storm, S. M., and Cleveland, J. L.,** Oncogenes and interferon, *Cancer Rev.,* 9, 34, 1987.

205. **Harris, D. E., Ralph, P., Gabrilove, J., Welte, K., Karmali, R., and Moore, M. A. S.,** Distinct differentiation inducing activities of γ-interferon and cytokine factors acting on the human promyelocytic leukemia cell line HL-60, *Cancer Res.,* 45, 3090, 1985.

206. **Ralph, P., Harris, P. E., Punjabi, C. J., Welte, K., Litcofsky, P. B., Ho, M. K., Rubin, B. Y., Moore, M. A. S., and Springer, T. A.,** Lymphokine inducing "terminal differentiation" of the human monoblast leukemia cell line U937: a role for γ interferon, *Blood,* 62, 1169, 1983.

207. **Moore, R. N., Pitruzzeilo, F. J., Deana, D. G., and Rouse, B. T.,** Endogenous regulation of macrophage proliferation and differentiation by E prostaglandins and interferon alpha/beta, *Lymphokine Res.,* 4, 43, 1985.

208. **Seghal, P. B., May, L. T., Tamm, I., and Vilcek, J.,** Human β_2 interferon and B cell differentiation factor BSF-2 are identical, *Science,* 235, 731, 1987.

209. **Haegman, G., Content, J., Volckaert, G., Derynck, R., Tavernier, J., and Fiers, W.,** Structural analysis of the sequence coding for an inducible 26 kDa protein in human fibroblasts, *Eur. J. Biochem.,* 159, 625, 1986.

210. **Zilberstein, A., Ruggieri, R., Korn, J. H., and Revel, M.,** Structure and expression of cDNA and genes for human interferon β-2, a distinct species inducible by growth stimulatory cytokines, *EMBO J.,* 5, 2529, 1986.

211. **May, L. T., Helfgott, D. C., and Seghal, P. B.,** Anti-β-interferon antibodies inhibit the increased expression of HLA-B7 mRNA in tumor necrosis factor treated human fibroblasts: structural studies of the β_2 interferon involved, *Proc. Natl. Acad. Sci. U.S.A.,* 83, 8957, 1986.

212. **Seghal, P. B., Zilberstein, A., Ruggieri, R. M., May, L. T., Ferguson-Smith, A., Slate, D. L., Revel, M., and Ruddle, F. H.,** Human chromosome 7 carries the β_2 interferon gene, *Proc. Natl. Acad. Sci. U.S.A.,* 83, 3663, 1986.

213. **Seghal, P. B. and May, L. T.,** Human interferon β_2, *J. Interferon Res.,* 7, 521, 1987.

214. **Seghal, P. B. and Sagar, A. D.,** Heterogeneity of poly (I):poly (C)-induced human fibroblast interferon mRNA species, *Nature,* 288, 95, 1980.

215. **Weissenbach, J., Chernajovsky, Y., Zeevi, M., Shulman, L., Soreq, H., Nir, U., Wallach, D., Perricaudet, M., Tiollais, P., and Revel, M.,** Two interferon mRNAs in human fibroblasts: in vitro translation and Escherichia coli cloning studies, *Proc. Natl. Acad. Sci. U.S.A.,* 77, 7152, 1980.

216. **Vilcek, J., Palombella, V. J., Henerikson-DeStefano, D., Swenson, C., Feinman, R., Hirai, M., and Tsujimoto, M.,** Fibroblast growth enhancing activity of tumor necrosis factor and its relationship to other polypeptide growth factors, *J. Exp. Med.,* 163, 632, 1986.

217. **Kohase, M., Henrikson-DeStefano, D., May, L. T., Vilcek, J., and Seghal, P. B.,** Induction of β_2 interferon by tumor necrosis factor: a homeostatic mechanism in control of cell proliferation, *Cell,* 45, 659, 1986.

218. **Chen, L., Mory, Y., Zilberstein, A., and Revel, M.,** Growth inhibition of human breast carcinoma and leukemia/lymphoma cell lines by recombinant interferon β_2, *Proc. Natl. Acad. Sci. U.S.A.,* 85, 8037, 1988.

219. **Kohase, M., May, L. T., Tamm, I., Vilcek, J., and Seghal, P. B.,** A cytokine network in human diploid fibroblasts: interaction of β-interferons, tumor necrosis factor, platelet derived growth factor and interleukin-1, *Mol. Cell. Biol.,* 7, 273, 1987.

220. **Content, J., De Wit, L., Poupart, P., Opdenakker, G., Van Damme, J., and Billiau, A.,** Production of a 26 kDa protein mRNA in human cells treated with an interleukin-1 related, lymphocyte drived factor, *Eur. J. Biochem.,* 152, 253, 1985.

221. **Bachwich, P. R., Chensur, W. W., Larrick, J. W.,and Kunkel, S. L.,** Tumor necrosis factor stimulates interleukin-1 and prostaglandin E_2 production in resting macrophages, *Biochem. Biophys. Res. Commun.,* 136, 94, 1986.

222. **Lin, S. L., Ts'O, P. O. P., and Hollenberg, M. D.,** Epidermal growth factor-urogastric action: induction of 2',5' oligoadenylate synthetase activity and enhancement of the mitogenic effect by anti-interferon antibody, *Life Sci.,* 32, 1479, 1983.

223. **Seghal, P. B., Walther, Z., and Tamm, I.,** Rapid enhancement of β_2 interferon/B cell differentiation factor BSF-2 gene expression in human fibroblasts by diacylglycerols and the calcium ionophore A23187, *Proc. Natl. Acad. Sci. U.S.A.,* 84, 3663, 1987.

224. **Poupart, P., DeWit, L., and Content, J.,** Induction and regulation of the 26 kDa protein in the absence of synthesis of β-IFN mRNA in human cells, *Eur. J. Biochem.,* 143, 15, 1984.

225. **Hirano, T., Yasuka, K., Harada, H., Taga, T., Watanabe, Y., Matsuda, T., Kashiwamura, S., Nakijima, K., Koyama, K., Iwamatsu, A., Tsunasawa, S., Sakiyama, F., Matsui, K., Takahara, Y., Taniguchi, T., and Kishimoto, T.,** Complementary DNA for a novel human interleukin (BSF-2) that induces B lymphocytes to produce immunoglobulin, *Nature,* 324, 73, 1986.

226. **Vaquero, C., Sanceau, J., Weissenbach, J., Beranger, F., and Falcoff, R.,** Regulation of human gamma interferon and beta interferon gene expression in PHA-activated lymphocytes, *J. Interferon Res.,* 6, 161, 1986.

227. **Roussel, M., Saules, S., Lagrou, C., Rommens, C., Beug, H., Graf, T., and Stehlin, D.,** Three new types of viral oncogenes of cellular origin specific for haematopoietic cell transformation, *Nature,* 282, 452, 1979.

228. **Adamson, E. D.,** Oncogenes in development, *Development,* 99, 449, 1987.

229. **Persson, H., Hennighausen, L., Taub, C., Degrado, W., and Leder, P.,** Antibodies to human c-myc oncogene product: evidence of an evolutionarily conserved protein induced during cell proliferation, *Science,* 225, 687, 1984.

230. **Hann, S. R., Thompson, C. B., and Eisenman, R. N.,** c-myc oncogene protein synthesis is independent of the cell cycle in human and avian cells, *Nature,* 314, 366, 1985.

231. **Muller, R., Bravo, R., Burckhardt, J., and Curran, T.,** Induction of c-fos gene and protein by growth factors precedes activation of c-myc, *Nature,* 312, 716, 1984.

232. **Kelly, K. and Sieberlist, U.,** The role of c-myc in the proliferation of normal and neoplastic cells, *J. Clin. Immunol.,* 5, 65, 1984.

233. **Dony, C., Kessel, M., and Gruss, P.,** Post-transcriptional control of myc and p53 expression during differentiation of the embryonal carcinoma cell line F9, *Nature,* 317, 636, 1985.

234. **Einat, M., Resnitzky, D., and Kimchi, A.,** Close link between reduction of c-myc expression by interferon and G_0/G_1 arrest, *Nature,* 313, 597, 1985.

235. **Einat, M., Resnitzky, D., and Kimchi, A.,** Inhibitory effects of interferon on the expression of genes regulated by platelet derived growth factor, *Proc. Natl. Acad. Sci. U.S.A.,* 82, 7608, 1985.

236. **Matsui, T., Takahashi, R., Mihara, K., Nakagawa, T., Koizumi, T., Nakao, Y., Sugiyama, T., and Fujita, T.,** Cooperative regulation of c-myc expression in differentiation of human promyelocytic leukemia cells induced by gamma interferon and 1,25-dihydroxyvitamin D3, *Cancer Res.,* 45, 4366, 1985.

237. **Dron, M., Modjtahedi, N., Brinson, O., and Tovey, M. G.,** Interferon modulation of c-myc expression in cloned Daudi cells relationship to the phenotype of interferon resistance, *Mol. Cell. Biol.,* 6, 1374, 1986.

238. **Jonak, G. J. and Knight, E., Jr.,** Selective reduction of c-myc mRNA in Daudi cells by human β interferon, *Proc. Natl. Acad. Sci. U.S.A.,* 81, 1747, 1984.

239. **Dani, C., Mechti, N., Piechaczyk, M., Lebleu, B., Jeanteur, P., and Blanchard, J. M.,** Increased rate of degradation of c-myc m-RNA in interferon treated Daudi cells, *Proc. Natl. Acad. Sci. U.S.A.,* 82, 4896, 1985.

240. **Knight, E., Jr., Friedland, B. K., Anton, E. D., Fahey, D., and Jonak, G. J.,** The inibition of cell growth and regulation of the c-myc gene expression by interferon β, in *The Interferon System,* Dianzani, F. and Rossi, G. B., Eds., Raven Press, Rome, 1985, 405.

241. **Jonak, G. K. and Knight, E., Jr.,** Interferons and the regulation of oncogenes, in *Interferon 7,* Gresser, I., Ed., Academic Press, New York, 1986, 167.

242. **Jalava, A. M., Heikkila, J. E., and Akerman, K. E. O.,** Decline in c-myc mRNA expression but not the induction of c-fos mRNA expression is associated with differentiation of SH-SY5Y human neuroblastoma cells, *Exp. Cell Res.,* 179, 10, 1988.

243. **Morin, M. J., Kreutter, D., Rasmussen, H. I., and Sartorelli, A. C.,** Disparate effects of activators of protein kinase C on HL-60 promyelocytic leukemia cell differentiation, *J. Biol. Chem.,* 262, 11758, 1987.

244. **Kraft, A. S., Smith, J. B., and Berkow, R. L.,** Bryostatin, an activator of the calcium phospholipid-dependent protein kinase, blocks phorbol ester induced differentiation of human promyelocytic leukemia cells HL-60, *Proc. Natl. Acad. Sci. U.S.A.,* 83, 1334, 1986.

245. **Muller, R., Curran, T., Muller, D., and Guilbert, L.,** Induction of c-fos during myelomonocytic differentiation and macrophage proliferation, *Nature, (London)* 314, 546, 1985.

246. **Bravo, R., MacDonald-Bravo, H., Muller, R., Hubsch, D., and Almendral, J. M.,** Bombesin induces c-fos and c-myc expression in quiescent Swiss 3T3 cells, *Exp. Cell. Res.,* 170, 103, 1987.

247. **Kelly, K., Cochran, B. H., Stiles, C. D., and Leder, P.,** Cell specific regulation of the c-myc gene by lymphocyte mitogens and platelet derived growth factor, *Cell,* 35, 603, 1983.

248. **Dony, C., Kessel, M., and Gruss, P.,** Post-transcriptional control of myc and p-53 expression during differentiation of the embryonal carcinoma cell line F9, *Nature, (London),* 317, 636, 1985.

249. **Endo, T. and Nadal-Ginard, B.,** Transcriptional and posttranscriptional control of c-myc during myogenesis: its mRNA remains inducible in differentiated cells and does not suppress the differentiated phenotype, *Mol. Cell Biol.,* 6, 1412, 1986.

250. **Gonda, T. and Metcalf, D.,** Expression of myb, myc, and fos proto-oncogenes during the differentiation of a murine myeloid leukemia, *Nature,* 310, 249, 1984.

251. **Lachman, H. and Skoultchi, A.,** Expression of c-myc changes during differentiation of mouse erythroleukemia cells, *Nature,* 310, 592, 1984.

252. **Westin, L., Wong-Staal, F., Gelman, E., Dalla-Favera, R., Papas, T., Lautenberger, J., Eva, A., Reddy, E., Tronick, S., Aaronson, S., and Gallo, R.,** Expression of cellular homologues of retroviral onc genes in human hematopoietic cells, *Proc. Natl. Acad. Sci. U.S.A.,* 79, 2490, 1982.

253. **Reitsma, P., Rothberg, P., Astrin, S., Trial, J., Bar-Shavit, A., Teitalbaum, S., and Kahn, A.,** Regulation of myc gene expression in HL-60 leukemia cells by a vitamin D metabolite, *Nature,* 306, 492, 1983.

254. **Yokoyama, K. and Imamoto, F.,** Transcriptional control of the endogenous myc protooncogene by antisense RNA, *Proc. Natl. Acad. Sci. U.S.A.,* 84, 7363, 1987.

255. **Freytag, S. O.,** Enforced expression of the c-myc oncogene inhibits cell differentiation by precluding entry into a distinct predifferentiation state in G_0/G_1, *Mol. Cell. Biol.,* 8, 1614, 1988.

256. **Curran, T., Peters, G., Piechaczyk, M., Lebleu, B., Jeanteur, P., and Blanchard, J. M.,** FBJ murine osteosarcoma virus: identification and molecular cloning of biologically active proviral DNA, *J. Virol.,* 44, 674, 1982.

257. **Muller, R. and Verma, I. M.,** Expression of cellular oncogenes, *Curr. Top. Microbiol. Immunol.,* 112, 74, 1984.

258. **Miller, A. D., Curran, T., and Verma, I. M.,** c-fos protein can induce cellular transformation: a novel mechanism of activation of a cellular oncogene, *Cell,* 36, 51, 1984.

259. **Mitchell, R. C., Zokas, L., Schreiber, R. D., and Verma, I. M.,** Rapid induction of the expression of protooncogene fos during human monocyte differentiation, *Cell,* 40, 209, 1985.

260. **Muller, R., Muller, D., and Guilbert, L.,** Differential expression of c-fos in hematopoietic cells: correlation with differentiation of monomyelocytic cells *in vitro, EMBO. J.,* 3, 1887, 1984.

261. **Wan, Y. J. Y., Levi, B. Z., and Ozato, K,** Induction of c-fos gene expression by interferons, *J. Interferon Res.,* 8, 105, 1988.

262. **Bravo, R., Burckhardt, J., Curran, T., and Muller, R.,** Stimulation and inhibition of growth by EGF in different A431 cell clones is accompanied by the rapid induction of c-fos and c-myc protooncogenes, *EMBO J.,* 4, 1193, 1985.

263. **Bravo, R., Neuberg, M., Burckhardt, J., Almendral, J., Wallich, R., and Muller, R.,** Involvement of common cell type specific pathways in c-fos gene control: stable induction by cAMP in macrophages, *Cell,* 48, 251, 1987.

264. **Wan, Y. J. Y., Orrison, B. M., Lieberman, R., Lazarovici, P., and Ozato, K.,** Induction of major histocompatibility class I antigens by interferon in undifferentiated F9 cells, *J. Cell. Physiol.,* 130, 276, 1987.

265. **Mitchell, R. L., Hennings-Chubb, C., Huberman, E., and Verma, I. M.,** c-fos expression is neither sufficient nor obligatory for differentiation of monocytes to macrophages, *Cell,* 45, 497, 1986.

266. **Calabretta, B.,** Dissociation of c-fos induction from macrophage differentiation in human myeloid leukemic cell lines, *Mol. Cell. Biol.,* 7, 769, 1987.

267. **Kruijer, W., Schubert, D., and Verma, I. M.,** Induction of the proto-oncogene fos by nerve growth factor, *Proc. Natl. Acad. Sci. U.S.A.,* 82, 7330, 1985.

268. **Ruther, U., Wagner, E. F., and Muller, R.,** Analysis of the differentiation-promoting potential of inducible c-fos genes introduced into embryonal carcinoma cells, *EMBO J.,* 4, 1775, 1985.

269. **Sidhu, R.,** Two dimensional electrophoretic analysis of proteins synthesized during differentiation of 3T3-L1 preadipocytes, *J. Biol. Chem.,* 254, 11111, 1979.

270. **Spielgelman, B. and Green, H.,** Control of specific protein biosynthesis during the adipose conversion of 3T3 cells, *J. Biol. Chem.,* 25, 8811, 1980.

271. **Bernlohr, D. A., Angus, C. W., Lane, M. D., Bolanowski, A., and Kelly, T. J., Jr.,** Expression of specific mRNA during adipose differentiation: Identification of an mRNA encoding a homologue of myelin P2 protein, *Proc. Natl. Acad. Sci. U.S.A.,* 81, 5468, 1984.

272. **Cook, K. S., Hunt, C. R., and Spielgelman, B.,** Developmentally regulated mRNA in 3T3-adipocytes: analysis of transcriptional control, *J. Cell Biol.,* 100, 514, 1985.

273. **Hunt, C. R., Ro, H.-S., Dobson, D. E., Min, H. Y., and Spielgelman, B. M.,** Adipocyte P2 gene: developmental expression and homology of 5′ flanking sequences among fat cell specific genes, *Proc. Natl. Acad. Sci. U.S.A.,* 83, 3786, 1986.

274. **Phillips, M., Djian, P., and Green, H.,** The nucleotide sequence of three genes participating in the adipose differentiation of 3T3 cell, *J. Biol. Chem.,* 261, 10821, 1986.

275. **Franza, B. R., Jr., Rauscher, F. J., III, Josephs, S. F., and Curran, T.,** The fos complex and fos related antigens recognize sequence elements of the AP-1 binding sites, *Science,* 239, 1150, 1988.

276. **Rauscher, F. J., III, Sambucetti, L., Curran, T., Distel, R. J., and Spielgelman, B. M.,** Common DNA binding sites for fos protein complex-transcription factor AP-1, *Cell,* 52, 471, 1988.

277. **Lee, W. P., Mithchell, P., and Tjian, R.,** Purified transcription factor AP-1 interacts with TPA inducible enhancer elements, *Cell,* 49, 741, 1987.

278. **Struhl, K.,** The DNA binding domains of jun oncoprotein and yeast GCN4 transcriptional activator proteins are functionally homologous, *Cell,* 50, 841, 1987.

279. **Bohman, D., Bos, T. J., Admon, A., Nihimura, T., Vogt, P. K., and Tjian, R.,** Human proto-oncogene c-jun encodes a DNA binding protein with structural and functional properties of transcriptional factor AP-1, *Science,* 238, 1386, 1987.

280. **Chiu, R., Boyle, W. J., Meek, J., Smeal, T., Hunter, T., and Karin, M.,** The c-fos protein interacts with c-jun/AP-1 to stimulate transcription of AP-1 responsive genes, *Cell,* 54, 541, 1988.

281. **Rauscher, F. J., III, Cohen, D. R., Curran, T., Bos, T. J., Vogt, P. K., Bohman, D., Tjian, R., and Franza, B. R.,** fos associated p39 is the product of the jun protooncogene, *Science,* 240, 1010, 1988.

282. **Sassone-Corsi, P., Ransone, L. J., Lamph, W. W., and Verma, I. M.,** Direct interaction between fos and jun nuclear oncoproteins: role of the 'leucine zipper' domain, *Nature (London),* 336, 662, 1988.

283. **Spielgelman, B. M., Distel, R. J., Ro, H.-S., Rosen, B. S., and Satterberg, B.,** fos-protooncogene and the regulation of gene expression in adipocyte differentiation, *J. Cell Biol.,* 107, 829, 1988.

284. **Barbacid, M.,** ras genes, *Annu. Rev. Biochem.,* 56, 779, 1987.

285. **Marshall, C. J.,** Human oncogenes, in *RNA Tumor Viruses,* 2nd ed., R. Weis, R., Teich, N., Varmus, H., and Coffin, J., Eds., Cold Spring Harbor laboratory, Cold Spring Harbor, NY, 1985, 487.

286. **Pulciani, S., Santos, E., Lauver, A. V., Long, L. L., Aaronson, S. A., and Barbacid, M.,** Oncogenes in solid human tumors, *Nature (London),* 300, 539, 1982.

287. **Shimizu, K. D., Birnbaum, D., Ruley, M. A., Fasano, O., Suard, Y., Edlund, L., Taparowsky, E., Goldfarb, M., and Wigler, M.,** Structure of Ki-ras gene of the human lung carcinoma cell line Calu-1, *Nature (London),* 304, 497, 1983.

288. **Hall, A., Marshall, C. J., Spurr, N., and Weiss, R. A.,** The transforming gene in two human sarcoma lines is a new member of the ras family located on chromosome one, *Nature,* 304, 135, 1983.

289. **Goldfarb, M., Shimizu, K., Perucho, M., and Wigler, M.,** Isolation and preliminary characterization of a human transforming gene from T24 bladder carcinoma cells, *Nature (London),* 296, 404, 1982.

290. **Buss, J. E. and Sefton, B. M.,** Direct identification of palmitic acid as a lipid attached to p21 ras, *Mol. Cell. Biol.,* 6, 116, 1986.

291. **Scolnick, E. M., Papageorge, A. G., and Shih, T. Y.,** Guanine nucleotide binding activity as an assay for the src protein of rat derived murine sarcoma viruses, *Proc. Natl. Acad. Sci. U.S.A.,* 76, 5335, 1979.

292. **Shih, T. Y., Papageorge, A. G., Strokes, P. E., Weeks, M. O., and Scolnick, E. M.,** Guanine nucleotide binding and autophosphorylation activities associated with p21 src protein of Harvey murine sarcoma virus, *Nature,* 287, 689, 1980.

293. **Gibbs, J. B., Siegal, I. S., Poe, M., and Scolnick, E. M.,** Intrinsic GTPase activity distinguishes normal and oncogenic ras p21 molecules, *Proc. Natl. Acad. Sci. U.S.A.,* 81, 5704, 1984.

294. **McGrath, J. P., Capon, D. J., Goeddel, D. V., and Levinson, A. D.,** Comparative biochemical properties of normal and activated human ras p21 protein, *Nature,* 310, 644, 1984.

295. **Manne, V., Bekesi, E., and Kung, H. F.,** Ha-ras proteins exhibit GTPase activity: point mutations that activate H-ras gene product result in decreased GTPase activity, *Proc. Natl. Acad. Sci. U.S.A.,* 82, 376, 1985.

296. **Gilman, A. G.,** G proteins and dual control of adenylate cyclase, *Cell,* 36, 577, 1984.

297. **Fleishman, L. F., Chahwala, S. B., and Cantley, L.,** Ras transformed cells: altered levels of phosphotidylinositol-4,5-biphosphate and catabolites, *Science,* 231, 407, 1986.

298. **Wakelam, M. J. D., Davies, S. A., Houelay, M. D., McKay, I., Marshall, C. J., and Hall, A.,** Normal p21 N-ras couples bombesin and other growth factor receptors to inositol phosphate production, *Nature,* 323, 173, 1986.

299. **Bos, J. L.,** The ras gene family and human carcinogenesis, *Mutat. Res.,* 195, 255, 1988.

300. **Feramisco, J. R., Gross, M., Kamata, T, Rosenberg, M., and Sweet, R. W.,** Microinjection of the oncogene form of the human H-ras (T-24) protein results in a rapid proliferation of quiescent cells, *Cell,* 38, 109, 1984.

301. **Stacey, D. W. and Kung, H. F.,** Transformation of NIH 3T3 cells by microinjection of Ha-ras p21 protein, *Nature,* 310, 508, 1984.

302. **Noda, M., Ko, M., Ogura, A., Liu, D., Amano, T., Takano, T., and Ikawa, Y.,** Sarcoma viruses carrying ras oncogenes induce differentiation associated properties in neuronal cell line, *Nature,* 318, 73, 1985.

303. **Alima, S., Casabore, P., Agostini, E., and Tato, F.,** Differentiation of PC12 phaeochromocytoma cells induced by v-src oncogene, *Nature,* 316, 557, 1985.

304. **Bar-Sagi, D. and Feramisco, J.,** Microinjection of the ras oncogene protein into PC12 cells induces morphological differentiation, *Cell,* 42, 841, 1985.

305. **Hagag, N., Halegoua, S., and Viola, M.,** Inhibition of growth factor induced differentiation of PC 12 cells by microinjection of antibody to ras p21, *Nature,* 319, 680, 1982.

306. **Guerrero, I., Wong, H., Pellicer, A., and Burstein, D. E.,** Activated N-ras gene induces neuronal differentiation of PC 12 rat phaeochromocytoma cells, *J. Cell Physiol.,* 129, 71, 1986.

307. **Soslau, G., Bogucki, A., Gillespie, D., and Hubbel, H. R.,** Phosphoproteins altered by antiproliferative dose of human interferon β in a human bladder carcinoma cell line, *Biochem. Biophys. Res. Commun.,* 119, 941, 1984.

308. **Giacomini, P., Gambari, R., Barbieri, R., Nastruzzi, C., Fraioli, R., Spandidos, D., Fisher, P. B., and Natali, P. G.,** Recombinant immune interferon downregulates the Ha-ras-1 protooncogene in a human melanoma cell line, *Anticancer Res.,* in press, 1990.

309. **Samid, D., Chang, E. H., and Friedman, R. M.,** Biochemical correlates of phenotypic reversion in interferon treated mouse cells transformed by a human oncogene, *Biochem. Biophys. Res. Commun.,* 119, 21, 1984.

310. **Samid, D., Flessate, D., and Friedman, R. M.,** Interferon-induced revertants of ras transformed cells: resistance to transformation by specific oncogenes and retransformation by 5-azacytidine, *Mol. Cell. Biol.,* 7, 2196, 1987.

311. **Shimizu, K., Goldfarb, M., Suard, Y., Peruchio, M., Li, Y., Kamata, T., Feramisco, J., Stavenzere, E., Fogh, J., and Wigler, M.,** Three human transforming genes are related to the viral ras oncogenes, *Proc. Natl. Acad. Sci. U.S.A.,* 80, 2112, 1983.

312. **Taparowsky, E., Shimizu, K., Goldfarb, M., and Wigler, M.,** Structure and activation of the human N-ras gene, *Cell,* 34, 581, 1983.

313. **Jonak, G. J., Friedland, B. K., Anton, E. D., and Knight, E., Jr.,** Regulation of c-myc RNA and its proteins in Daudi cells by interferon β., *J. Interferon Res.,* 7, 41, 1987.

314. **Cotton, P. C. and Brugge, J. S.,** Neural tissues express high levels of the cellular src gene product pp60^{c-src}, *Mol. Cell Biol.,* 3, 1157, 1983.

315. **Fults, W., Towle, A. C., Lander, J. M., and Maness, P. F.,** Viral and cellular fos proteins are complexed with a 39000 dalton cellular protein, *Mol. Cell. Biol.,* 5, 167, 1985.

316. **Sorge, L. K., Levy, B. T., and Maness, P. F.,** pp60^{c-src} is developmentally regulated in the neural retina, *Cell,* 36, 249, 1984.

317. **Brugge, J., Cotton, P. C., Queral, A. E., Barrett, J. N., Nouner, D., and Keane, R. W.,** Neurones express high levels of a structurally modified, activated form of pp60^{c-src}, *Nature,* 316, 554, 1985.

318. **Simon, M. A., Drees, B., Kornberg, T., and Bishop, J. M.,** The nucleotide sequence and the tissue specific expression of Drosophila c-src, *Cell,* 42, 831, 1985.

319. **Lin, S. L., Garber, E. A., Wang, E., Caliguiri, L. A., Schellekens, H., Goldberg, A. R., and Tamm, I.,** Reduced synthesis of pp60^{src} and expression of the transformation related phenotype in interferon treated Rous sarcoma virus transformed rat cells, *Mol. Cell. Biol.,* 3, 1656, 1983.

320. **Sherr, C. J., Rettenmeier, G. W., Sala, R., Roussel, M. F., Look, A. T., and Stanley, E. R.,** The c-fms protooncogene product is related to the receptor for the mononuclear phagocyte growth factor, CSF-1, *Cell,* 40, 665, 1985.

321. **Rettenmeier, C. W., Chen, J. H., Roussel, M. F., and Sherr, C. J.,** The product of the c-fms protooncogene: a glycoprotein with associated tyrosine kinase activity, *Science,* 228, 320, 1985.

322. **Coussens, L., Van Beveren, C., Smith, D., Chen, E., Mitchell, R. L., Isacke, C. M., Verma, I., and Ullrich, A.,** Structural alterations of viral homologue of receptor protooncogene fms at carboxy terminus, *Nature (London),* 320, 277, 1986.

323. **Saruban, E., Mitchell, T., and Kuff, D.,** Expression of the c-fms proto-oncogene during human monocytic differentiation, *Nature,* 311, 438, 1985.

324. **Muller, R.,** Proto-oncogenes and differentiation, *TIBS,* 11, 129, 1986.

325. **Krueger, L., Bresser, J., Andryuk, P. J., and Gillespie, D. H.,** Interferon induced oncogene regulation: the gene bank hypothesis, *J. Cell. Biochem. Suppl.,* 8A, 85, 1984.

326. **Oishi, M. and Watanabe, T.,** The early reactions and factors involved in in vitro erythroid differentiation of mouse erythroleukemia (MEL) cells, in *Mechanisms of Differentiation,,* Vol. II, Fisher, P. B., Ed., CRC Press, Boca Raton, FL, 1990.

327. **Giroldi, L., Hollstein, M., and Yamasaki, H.,** Cellular-oncogene expression in Friend erythroleukemia cells: ralationship to differentiation, commitment and TPA effects, *Carcinogenesis,* 9, 817, 1988.

328. **Robert-Lezenes, J., Meneceur, P., Ray, D., and Moreau-Gachelin, T.,** Protooncogene expression in normal preleukemic, and leukemic murine erythroid cells and its relationship to differentiation and proliferation, *Cancer Res.,* 48, 3972, 1988.

329. **Ramsey, R. G., Ikeda, K., Rifkind, R. A., and Marks, P. A.,** Changes in gene expression associated with induced differentiation of erythroleukemia: protooncogenes, globin genes and cell division, *Proc. Natl. Acad. Sci. U.S.A.,* 83, 6849, 1986.

330. **Ishikura, H., Honma, Y., Honma, C., Hozumi, M., Black, J. D., Kiebermans, T., and Bloch, A.,** Inhibition of messenger RNA transcriptional activity in ML-1 human myeloblastic leukemia cell nuclei by antiserum to a c-myb specific peptide, *Cancer Res.,* 47, 1052, 1987.

331. **Thompson, C. B., Challoner, P. B., Neiman, P. E., and Groudine, M.,** Expression of the c-myb protooncogene during cellular proliferation, *Nature,* 319, 374, 1986.

332. **Orchansky, P., Novick, D., Fischer, D. G., and Rubinstein, M.,** Type I and type II interferon receptors, *J. Interferon Res.,* 4, 275, 1984.

333. **Rashidbaigi, A., Kung, H. F., and Pestka, S.,** Characterization of receptors for immune interferon, *J. Biol. Chem.,* 260, 8514, 1986.

334. **O'Rourke, E. C., Drummond, R. J., and Creasey, A. A,** Binding of ^{125}I-labeled recombinant beta interferon (IFN-βser$_{17}$) to human cells, *Mol. Cell. Biol.* 4, 2745, 1984.

335. **Merlin, G., Falcoff, E., and Aguet, M.,** ^{125}I-labeled human interferon alpha, beta and gamma: comparative receptor-binding data, *J. Gen. Virol.,* 66, 1149, 1985.

336. **Thompson, M. R., Zhang, Z., Fournier, A., and Tan, Y. H.,** Characterization of human beta interferon binding sites on human cells, *J. Biol. Chem.,* 260, 563, 1985.

337. **Orchansky, P., Rubinstein, M., and Fischer, D. G.,** The interferon gamma receptor in human monocytes is different from the one in nonhematopoietic cells, *J. Immunol.,* 136, 169, 1988.

338. **Fischer, D. G., Novick, D., and Rubinstein, M.,** Two molecular forms of the IFN-gamma receptor: ligand binding, internalization and down regulation, *J. Interferon Res.,* 7, 765, 1987.

339. **Gordon, I., Stevenson, D., Tumas, V., and Natham, C. R.,** Interferon receptors: a difference in their response to α and β interferons, *J. Gen. Virol.,* 641, 2777, 1983.
340. **Branca, A. A.,** Interferon receptors, *In Vitro Cell. Dev. Biol.,* 24, 155, 1988.
341. **Cohen, B., Peretz, D., Vaiman, D., Benech, P., and Chebath, J.,** Enhancer-like interferon responsive sequences of the human and murine (2'-5') oligoadenylate synthetase gene promoter, *EMBO J.,* 7, 1411, 1988.
342. **Porter, A. G. G., Chernajovski, Y., Dale, T. C., Gilber, C. S., Stark, G. R., and Kerr, I. M.,** Interferon response element of the human gene 6/16, *EMBO J.,* 7, 85, 1988.
343. **Kusari, J., Tiwari, R. K., Kumar, R., and Sen, G. C.,** Expression of interferon-inducible genes in RD-114 cells, *J. Virol.,* 61, 1524, 1987.
344. **Faltynek, C. R., McCandless, S., Chebath, J., and Baglioni, C.,** Different mechanisms for activation of gene transcription by interferons α and γ, *Virology,* 144, 173, 1985.
345. **Kusari, J. and Sen, G. C.,** Regulation of synthesis and turnover of an interferon-inducible mRNA, *Mol. Cell. Biol.,* 6, 2062, 1986.
346. **Branca, A. A., Faltynek, C. R., D'Alessandro, S. B., and Baglioni, C.,** Interaction of interferon with cellular receptors, *J. Biol. Chem.,* 257, 1329, 1982.
347. **Weber, J. M. and Stewart, R. B.,** Cyclic AMP potentiation of interferon antiviral activity and effect of interferon on cellular cyclic AMP levels, *J. Gen. Virol.,* 28, 363, 1975.
348. **Meldolesi, M. F., Friedman, R. M., and Kohn, L. D.,** An interferon-induced increase in cyclic AMP levels precedes the establishment of the antiviral state, *Biochem. Biophys. Res. Commun.,* 79, 239, 1977.
349. **Tovey, M. G.,** Interferon and cyclic nucleotides, *Interferon,* 4, 23, 1982.
350. **Waxman, S., Rossi, G. B., and Takaku, F., Eds.,** *The Status of Differentiation Therapy,* Serono Symposia, Raven Press, New York, 1988.
351. **Davila, D. G., Minoo, P., Estervig, D. N., Kasperbauer, J. L., Tzen, C.-Y., and Scott, R. E.,** Linkages in the control of differentiation and proliferation in murine mesenchymal stem cells and human keratinocyte progenitor cells: the effects of carcinogenesis, in *Mechanisms of Differentiation,* Vol. I, Fisher, P. B., Ed., CRC Press, Boca Raton, FL, 1990.
352. **Lassar, A. B., Buskin, J. N., Lockshon, D., Davis, R. L., Apone, S., Hauschka, S. D., and Weintraub, H.,** MyoD is a sequence-specific DNA binding protein requiring a region of myc homology to bind to the muscle creatine kinase enhancer, *Cell,* 58, 823, 1989.
353. **Yamamoto, K. and Koeffler, H. P.,** Induction of terminal differentiation of human leukemic cells by chemotherapeutic and cytotoxic drugs, in *Mechanisms of Differentiation,* Vol. II, Fisher, P. B., Ed., CRC Press, Boca Raton, FL, 1990.
354. **Jetten, A. M.,** Regulation of gene expression by retinoic acid. Embryonal carcinoma cell differentiation, in *Mechanisms of Differentiation,* Vol. I, Fisher, P. B., Ed., CRC Press, Boca Raton, FL, 1990.
355. **Kessler, D. S., Pine, R., Pfeffer, L. M., Levy, D. E., and Darnell, E., Jr.,** Cells resistant to interferon are defective in activation of a promoter-binding factor, *EMBO J.,* 7, 3778, 1988.
356. **Zinkernagel, R. M. and Doherty, P. C.,** MHC-restricted cytotoxic T cells: studies on the biological role of polymorphic major transplantation antigens determining T-cell restriction-specificity, *Adv. Immunol.,* 27, 51, 1979.
357. **Schrier, P. I., Bernards, R., Vaessen, R. T. M. J., Houweling, A., and van der Eb, A. J.,** Expression of class I major histocompatibility antigens switched off by highly oncogenic adenovirus 12 in transformed rat cells, *Nature,* 305, 771, 1983.
358. **Babiss, L. E., Liaw, W.-S., Zimmer, S. G., Godman, G. C., Ginsberg, H. S., and Fisher, P. B.,** Mutations in the E1a gene of adenovirus type 5 alter the tumorigenic properties of transformed cloned rat embryo fibroblast cells, *Proc. Natl. Acad. Sci. U.S.A.,* 83, 2167, 1986.
359. **Duigou, G. J., Babiss, L. E., Liaw, W.-S., Zimmer, S. G., Ginsberg, H. S., and Fisher, P. B.,** Mutations in the E1a gene of type 5 adenovirus result in oncogenic transformation of Fischer rat embryo cells, *J. Cell. Biochem.,* 33, 117, 1987.
360. **Tanaka, K., Isselbacher, K. J., Khoury, G., and Jay, G.,** Reversal of oncogenesis by the expression of a major histocompatibility complex class I gene, *Science,* 228, 26, 1985.
361. **Hayashi, H., Tanaka, K., Jay, F., Khoury, G., and Jay, G.,** Modulation of tumorigenicity of human adenovirus-12-transformed cells by interferon, *Cell,* 43, 263, 1985.
362. **Wallich, R., Bulbuc, N., Hammerling, G. J., Katzav, S., Segal, S., and Feldman, M.,** Abrogation of metastatic properties of tumor cells by de novo expression of H-2K antigens following H-2 gene transformation, *Nature,* 315, 301, 1985.
363. **Hui, K. M., Grosveld, F., and Festenstein, H.,** Rejection of transplantable AKR leukemia cells following MHC DNA-mediated cell transformation, *Nature,* 311, 750, 1984.
364. **Taniguchi, K., Karre, K., and Klein, G.,** Lung colonization and metastatization by disseminated B16 melanoma cells: H-2 associated control at the level of the host and the tumor cell, *Int. J. Cancer,* 36, 503, 1985.
365. **Karre, K., Ljunggren, H. G., Pointek, G., and Kiessling, R.,** Selective rejection of H-2 deficient lymphoma variants suggests alternative immune defense strategy, *Nature,* 319, 675, 1986.

366. **Doyle, A., Martin, W. J., Funa, K., Gazdar, A., Carney, D., Martin, S., Linoila, I., Cuttitta, F., Mulshine, J., Brunn, P., and Minna, J.,** Markedly decreased expression of class I histocompatibility antigens, protein, and mRNA in human small-cell lung cancer, *J. Exp. Med.,* 161, 1135, 1985.

367. **Lampson, L. A., Fisher, C. A., and Whelan, J. P.,** Striking paucity of HLA-A, B, C and beta 2-microglobulin on human neuroblastoma cell lines, *J. Immunol.,* 130, 2471, 1983.

368. **Ruiter, D. J., Bergman, J. W., Welvaart, K., Sheffer, E., Van Vloten, W. A., Russo, C., and Ferrone, S.,** Immunohistochemical analysis of malignant melanomas and nevocellular nevi with monoclonal antibodies to distinct monomorphic determinants of HLA antigens, *Cancer Res.,* 44, 3930, 1984.

369. **Brocker, E. B., Suter, L., Bruggen, J., Ruiter, D. J., Macher, C., and Sorg, C.,** Phenotypic dynamics of tumor progression in human malignant melanoma, *Int. J. Cancer,* 36, 29, 1985.

370. **Ruiz-Cabello, F., Lopez, G., Nevot, M. A., Gutierrez, J., Oliva, M. R., Romero, C., Ferron, A., Huelin, C., Piris, M. A., Rivas, C., and Garrido, F.,** Phenotypic expression of histocompatibility antigens in human primary tumors and metastasis, *Clin. Exp. Metastasis,* 7, 213, 1989.

371. **Nistico, P., Tecce, R., Giacomini, P., De Filippo, F., Fisher, P. B., and Natali, P. G.,** Effect of recombinant human leukocyte, fibroblast and immune interferons on expression of Class I and II MHC and Ii chain in early passage human melanoma cells, *Cancer Res.,* submitted.

372. **Giacomini, P., Fraioli, R., Nistico, P., Tecce, R., Nicotra, M. R., Fisher, P. B., and Natali, P. G.,** Modulation of the antigenic phenotype of early passage human melanoma cells derived from multiple autologous metastases by recombinant human leukocyte, fibroblast and immune interferons, *Int. J. Cancer,* submitted.

373. **Versteeg, R., Noordermeer, I. A., Kruse-Wolters, M., Ruiter, D. J., and Schrier, P. I.,** C-myc down-regulates Class I HLA expression in human melanomas, *EMBO J.,* 7, 1023, 1988.

374. **Versteeg, R., Kruse-Wolters, K. M., Plomp, A. C., van Leeuwen, A., Stam, N. J., Ploegh, H. L., Ruiter, D. J., and Schrier, P. I.,** Suppression of class I human histocompatibility leukocyte antigen by c-myc is locus specific, *J. Exp. Med.,* 170, 621, 1989.

375. **Dolei, A., Capobianchi, R., and Ameglio, F.,** Human interferon γ enhances the expression of class I and class II major histocompatibility complex products in neoplastic cells more effectively than interferon α and interferon β, *Infect. Immun.,* 40, 172, 1983.

376. **Giles, R. C. and Capra, J. D.,** Structure, function and genetics of human Class II molecules, *Adv. Immunol.,* 37, 1, 1985.

377. **Doherty, P. C., Knowles, B. B., and Wettstein, P. J.,** Immunological surveillance of tumors in the context of major histocompatibility complex restriction of T cell function, *Adv. Cancer Res.,* 42, 1, 1985.

378. **Fossati, G. A., Anichini, A., Taramelli, D., Balsari, A., Gambacorti-Passerini, C., Kirkwood, J. M., and Parmiani, G.,** Immune response to autologous human melanoma: implication of class I and class II MHC products, *Biochem. Biophys. Acta,* 865, 1, 1986.

379. **Natali, P. G., De Martino, V., Quaranta, V., Bigotti, M., Pellegrino, M. A., and Ferrone, S.,** Changes in Ia-like expression of malignant human cells, *Immunogenetics,* 12, 409, 1981.

380. **Houghton, A. N., Eisinger, M. A., Albino, P., Cairncross, J. G., and Old, L. J.,** Surface antigens of melanocytes and melanomas. Markers of melanocyte differentiation and melanoma subsets, *J. Exp. Med.,* 156, 1755, 1982.

381. **Real, F. X., Houghton, A. N., Albino, A. P., Cordon-Cardo, C., Melamed, M. R., Oettegen, H. F., and Old, L. J.,** Surface antigens of melanomas and melanocytes defined by mouse monoclonal antibodies: specificity analysis and comparison of antigen expression in cultured cells and tissue, *Cancer Res.,* 45, 4401, 1985.

382. **Holzman, B., Brocker, E. B., Lehman, J. M., Ruiter, D. J., Sorg, C., Riethmuller, G., and Johnson, J. P.,** Tumor progression in human malignant melanoma: five stages defined by their antigenic phenotypes, *Int. J. Cancer,* 39, 466, 1987.

383. **Yeh, M. Y., Hellstrom, I., and Hellstrom, K. E.,** Clonal variation in expression of a human melanoma antigen defined by a monoclonal antibody, *J. Immunol.,* 126, 1312, 1981.

384. **Suter, L., Brocker, E. B., Bruggen, J., Ruiter, D. J., and Sorg, C.,** Heterogeneity of primary and metastatic human malignant melanoma as detected with monoclonal antibodies in cryostat sections of biopsies, *Cancer Immunol. Immunother.,* 16, 53, 1983.

385. **Cillo, C., Mach, J. P., Schreyer, M., and Carrell, S.,** Antigenic heterogeneity of clones and subclones from human melanoma cell lines demonstrated by a panel of monoclonal antibodies and flow microfluorometry analysis, *Int. J. Cancer,* 34, 11, 1984.

386. **Pollack, M. S., Chin-Louie, J., and Moshief, R. D.,** Functional characteristics and differential expression of class II DR, DS, and SB antigens of human melanoma cell lines, *Hum. Immunol.,* 9, 75, 1984.

387. **De Muralt, B., De Tribolet, N., Diserens, A. C., Stravov, D., Mach, J. P., and Carrel, S.,** Phenotyping of 60 cultured human gliomas and 34 other neuroectodermal tumors by means of monoclonal antibodies against glioma, melanoma and HLA-DR, *Eur. J. Cancer Clin. Oncol.,* 21, 207, 1985.

388. **Anichini, A., Mortarini, R., Fossati, G., and Parmiani, G.,** Phenotypic profile of clones from early cultures of human metastatic melanomas and its modulation by recombinant interferon γ, *Int. J. Cancer,* 38, 505, 1986.

389. **Ghosh, A. K., Moore, M., Street, A. J., Howat, J. M. T., and Scholfield, P. P.,** Expression of HLA-D subregion products on human colorectal carcinoma, *Int. J. Cancer,* 38, 459, 1986.

390. **Natali, P. G., Bigotti, A., Cavalieri, R., Nicotra, M. R., Tecce, R., Manfred, D., Chen, Y. X., Nadler, L. M., and Ferrone, S.,** Gene products of the HLA-D region in normal and malignant tissue of nonlymphoid origin, *Hum. Immunol.,* 15, 220, 1986.

391. **Maio, M., Gulwani, B., Langer, J. A., Kerbel, R. S., Duigou, G. J., Fisher, P. B., and Ferrone, S.,** Modulation by interferons of HLA antigen, high molecular weight melanoma-associated antigen, and intercellular adhesion molecule 1 expression by cultured melanoma cells with different metastatic potential, *Cancer Res.,* 49, 2980, 1989.

392. **Maio, M., Gulwaini, B., Tombesi, S., Langer, J. A., Duigou, G. J., Kerbal, R. S., Fisher, P. B., and Ferrone, S.,** Differential induction by immune interferon of the gene products of the HLA-D region on the melanoma cell line MeWo and its metastatic variant MeM 50-10, *J. Immunol.,* 141, 913, 1988.

393. **Bashman, T. Y. and Merigan, T.,** Recombinant interferon-γ increases HLA-DR synthesis and expression, *J. Immunol.,* 130, 1492, 1983.

394. **Lloyd, K. O.,** Human tumor antigens: detection and characterization with monoclonal antibodies, in *Basic and Clinical Tumor Immunology,* Herbermann, R. B., Ed., Martinus Nijhoff, The Hague, The Netherlands, 1983, 159.

395. **Natali, P. G., Aguzzi, A., Vegli, F., Imai, K., Burlage, R. S., Giacomini, P., and Ferrone, S.,** The impact of monoclonal antibodies on the study of human malignant melanoma, *J. Cutan, Pathol.,* 10, 514, 1983.

396. **Ferrone, S., Temponi, M., Gargiulo, D., Scassellati, G. A., Cavaliere, R., and Natali, P. G.,** Selection and utilization of monoclonal antibody defined melanoma associated antigens for immunoscintography in patients with melanoma, in *Radiolabeled Monoclonal Antibodies for Imaging and Therapy — Potential, Problems, and Prospects,* Srivatsava, S., Ed., Plenum Press, New York, 1988, 55.

397. **Matsui, M., Temponi, M., and Ferrone, S.,** Characterization of a monoclonal antibody-defined human melanoma-associated antigen susceptible to induction by immune interferon, *J. Immunol.,* 139, 2088, 1987.

398. **Dustin, M. L., Rothlein, R., Bhan, A. K., Dinarello, C. A., and Springer, T. A.,** Induction by IL 1 and IFN-γ: tissue distribution, biochemistry, and function of a natural adherence molecule (ICAM-1), *J. Immunol.,* 137, 245, 1986.

399. **Makgoba, M. W., Sanders, M. E., Lucey, G. E. G., Gugel, E. A., Dustin, M. L., Springer, T. A., and Shaw, S.,** Functional evidence that intercellular adhesion molecule-1 (ICAM-1) is a ligand for LFA-1 dependent adhesion in T cell-mediated cytotoxicity, *Eur. J. Immunol.,* 18, 637, 1988.

400. **Temponi, M., Romano, G., D'Urso, C. M., Wang, Z., Kekish, U., and Ferrone, S.,** Profile of intercellular adhesion molecule-1 (ICAM-1) synthesized by human melanoma cell lines, *Semin. Oncol.,* 15, 595, 1988.

401. **Guarini, L., Temponi, M., Bruce, J. N., Bollon, A. P., Ferrone, S., and Fisher, P. B.,** Expression of the intercellular adhesion molecule and its modulation by recombinant immune interferon and tumor necrosis factor in human central nervous system tumor cell cultures, submitted, *Int. J. Cancer,* 1990.

402. **Kirkwood, J. M. and Ernstoff, M.,** Melanoma: therapeutic options with recombinant interferons, *Semin. Oncol.,* 12,7, 1985.

Chapter 2

REGULATION OF TUMOR ANTIGEN EXPRESSION BY RECOMBINANT INTERFERONS

F. Guadagni, J. Kantor, J. Schlom, and J. W. Greiner

TABLE OF CONTENTS

I. INTRODUCTION

Tumor-associated antigens that are expressed on the surface of human carcinoma cells have been extensively studied as (1) markers for the transformed phenotype of human epithelial cells, (2) markers of the biology of the human tumor cell population, particularly as related to the state of cellular differentiation, and (3) targets for the *in vivo* localization of monoclonal antibodies (MAb) to occult human carcinoma lesions. It has been recognized that an important consideration for the delivery of an effective amount of conjugated MAb for tumor diagnosis and/or therapy is an understanding of the biology of the human tumor cell population. It has been widely documented that most, if not all, human tumor cell populations express a certain degree of antigenic heterogeneity.[1-4] It is not uncommon to demonstrate heterogeneous staining of a human tumor cell population using MAbs in established immunohistochemical procedures.[4,5] Most often, antigenic heterogeneity is manifested as focal MAb staining adjacent to a region of tumor that is antigen negative. Clearly, those tumor cells that either do not express the tumor antigen or express it in quantitatively low levels will escape detection and, possibly, therapy due to their failure to bind the MAb. Therefore, it has been our interest to study those factors that regulate the level of human tumor antigen expression. In particular, we are interested in the identification of those agents that can selectively enhance the level of tumor antigen expression. It is postulated that combining such a strategy may result in an enhanced level of binding of a conjugated MAb, thereby increasing the effectiveness of the MAb in immunoscintigraphy and/or immunotherapy.

Our initial studies demonstrated that human tumor cells can, in fact, intrinsically regulate the level of expression of certain surface tumor antigens. Such cellular factors as cell cycle kinetics, clonal expansion of subpopulations of tumor cells, and the cell-to-cell communication permitted when cells are grown in a three-dimensional matrix all contribute to the ability of a human tumor cell population to alter its antigen phenotype.[4,6,7] These and other studies demonstrate that the antigen phenotype of a human tumor cell population is a result of dynamic interactions of cellular factors, preferential expansion of subpopulations of cells, influence from stromal elements, and cell-to-cell communication, which is a culmination of the diverse genotype of the human tumor cell population. The result is a highly diverse, heterogenous human tumor cell population that also has the ability to constantly alter its antigenic phenotype. Our efforts have been directed toward the identification of agents that, when administered extrinsically, can override the cellular factors modulating tumor antigen expression, resulting in an augmentation in the level of expression of the MAb-defined surface antigens.

Our initial attention was focused on the identification of those differentiation agents that can also upregulate the level of expression of cell-surface tumor antigens. The list of such agents included several organic solvents, cyclic AMP (cAMP) and related analogues, vitamin A and D, DNA intercalating agents, tumor necrosis factors, transforming growth factors (TGF), interleukins, several butyrate compounds, and the human interferons (Table 1).[8-16] Some of these, particularly cyclic AMP and selected analogs as well as sodium butyrate, were reported to enhance the level of carcinoembryonic antigen (CEA) expression in human colorectal tumor cells.[8-14] However, most of these induced changes require the addition of cytotoxic doses of the agents, which makes it difficult to determine if the changes in antigen expression were a result of selection of a subpopulation of cells that constitutively expressed high antigen levels or preferential antigen enhancement. The human interferons activated at least 15 to 20 cellular genes that mediate various biological actions of this cytokine family.[16] Such proteins as $(2'-5')$oligo(A) synthetase are believed to play a role in the antiviral actions of the interferons.[17] Others, such as the mouse Mx protein that inhibits the replication of, and thus confers resistance to, the influenza virus, are induced by the human interferons.[18] In contrast, other biological activities of the interferons, such as natural killer activity, have

TABLE 1
List of Differentiation Agents Analyzed for Their Ability to Regulate the Expression of Tumor Antigens on the Surface of Human Carcinoma Cells

Group	Compound	Effect on tumor antigen expression	Ref.
Organic solvents	DMSO	No change	LTIB[a], 9
	NMF	Decrease CEA	LTIB
	DMF	Increase CEA	8
Butyrates	Butyric acid	Increase CEA	9
	Sodium butyrate	Increase CEA	10—12
cAMP & analogues	cAMP	Increase CEA	13
	Dibutyryl-cAMP	Increase CEA	13
	8-Cl-cAMP	Increase CEA	LTIB
DNA intercalating agents	5-Azacytidine	No change	LTIB
Vitamins	Retinoic acid,	Decrease/increase CEA	LTIB, 9, 10
	Vitamin D, and derivatives	No change	LTIB
Interleukins	IL-2	No change	LTIB
Tumor necrosis factors	α	Increase/decrease	LTIB
	β	Not done	
TGF	α	Not done	
	β	Increase shed CEA	14
Interferons	α	Increase CEA, TAG-72, 90 kD	LTIB, 15, 80
	β_{ser}	Increase CEA, TAG-72, 90 kD	LTIB
	γ	Increase CEA, TAG-72, 90 kD	LTIB

[a] Our laboratory (Laboratory of Tumor Immunology and Biology, National Cancer Institute, National Institutes of Health).

not been associated with the activation of any single or group of genes. Our interest in the ability of the interferons to regulate tumor antigen expression stems from the considerable amount of experimental evidence linking both types of interferon to the regulation of class I and II major histocompatibility (MHC) antigens.[19-23] Both type I (i.e., α and β) and II (i.e., γ) interferons can augment or *de novo* induce the expression of class I and II HLA. The level of expression of class I HLA can be increased with treatment by either type I or type II interferon. For example, interferon (IFN)-γ preferentially increases the transcription of genes encoding for HLA-A, -B, -C, and β_2-microglobulin at concentrations 100 times lower than those needed for the antiviral or (2'-5')oligo(A) synthetase activity.[19] IFN-γ treatment can induce the *de novo* expression of class II HLA on a variety of normal and tumor cell types.[23,24] Those alterations in class I and II HLA expression are believed to be involved in the recognition of virus-infected cells by cytotoxic lymphocytes and in antigen presentation to the helper T-cell receptor, thereby initiating an immune response.[25,26] In contrast, little was known of the ability of the human interferons to regulate the wide variety of non-MHC cell-surface tumor antigens.

In this chapter, we will outline several different studies performed in our laboratory and with the collaboration of scientists at the National Cancer Institute, as well as other research institutions. The initial study will demonstrate the ability of the human interferons to regulate the constitutive expression of CEA on several different human colorectal tumor cell lines at different stages of cellular differentiation. In addition, the data will also show that IFN-γ can increase the steady-state level of CEA mRNA in some moderately differentiated human colorectal cell lines. Next, we will investigate whether interferon treatment can enhance the localization of a radionuclide-conjugated MAb in athymic mice bearing human carcinomas.

TABLE 2
Comparison of CEA Expression in Different Established Human Colorectal Cell Lines

Cell line	CEA content[a] (ng/ml)	Cell surface CEA expression[b] (% CEA-positive cells)		Degree of differentiation
		COL-1	COL-12	
DLD-1	Neg	1.3	0.7	Undifferentiated
MIP	Neg	0.7	1.0	Undifferentiated
HT-29	12.5	58.2	48.7	Moderately
WiDr	140	42.7	42.9	Moderately
LS174T	711	72.3	56.6	Well
CBS	821	95.3	72.5	Well
GEO	1813	94.9	84.9	Well
MRC-5	Neg	1.1	0.2	N/A
Granulocytes	Neg	4.3	3.9	N/A

[a]   CEA content was measured in extracts prepared from the human colorectal cells MRC-5 and fresh granulocytes, as listed, using a commercially available CEA-RIA monoclonal *in vitro* test kit (Abbott Laboratories, Inc., Chicago, IL). Extracts of the different cell types were prepared,[4] and the protein concentration was measured. All extracts were diluted in buffer to a final protein concentration of 1.0 mg/ml. The amount of CEA was measured using a 0.1-ml aliquot of the extract. Neg = <2.5 ng CEA per ml.

[b]   The complete details of the flow cytometric analyses have been previously described.[58] The cells were harvested, washed twice with Dulbecco's PBS (Ca^{2+}/Mg^{2+} free; pH 7.2), and incubated for 1 h at 4°C in the presence of 2 μg/10^6 cells for each MAb. The cells were subsequently washed with Dulbecco's PBS (Ca^{2+}/Mg^{2+} free) and incubated in the presence of fluorescent goat anti-mouse IgG (1:50 dilution) (Cooper Biomedical, Malvern, PA) for 1 h at 4°C. The cells were then washed, resuspended in 0.3 ml of Dulbecco's PBS containing 1 g glucose per l, and fixed by adding 0.7 ml of cold 1% paraformaldehyde. The analysis was done using an Ortho Cytofluorograf system 50H equipped with a blue laser excitation of 200 mW at 488 nm.

Finally, data from a recently completed study will demonstrate the ability of human interferons to augment the level of expression of several different human carcinoma cells isolated from patients' serous effusions. It is hoped that the data presented from these studies will provide the framework from which to design clinical trials to assess whether interferon treatment could be used as an adjuvant for MAb targeting of human carcinoma lesions.

II. *IN VITRO* REGULATION OF CEA BY HUMAN INTERFERONS

CEA is a 180-kD glycoprotein that was first described by Gold and Freedman[27] and has been most prominently used as a clinical marker for malignancy of the gastrointestinal tract.[28-32] CEA is highly glycosylated and is a member of a family of structurally related genes[33-38] whose complexity of expression varies in a tissue-specific manner. The normal function of CEA, its role in carcinogenesis, and the molecular basis for its differential expression in human tumors is not known. Recently, several laboratories have reported the nucleotide sequence of several cDNA clones for the family of CEA-related genes.[33-35]

In this study, seven different human colorectal tumor cell lines were used that vary in both their constitutive level of CEA expression and their stage of differentiation (Table 2). The study was designed to determine whether (1) alterations in CEA expression mediated by IFN-γ were dependent on the stage of colorectal tumor cell differentiation, (2) IFN-γ treatment can alter the steady-state levels of CEA mRNA in moderately and well-differentiated human colorectal cells, and (3) some insight could be developed into those human tumor cell populations that may be chosen for antigen augmentation. Table 2 summarizes the constitutive level of CEA expression in the seven different human colorectal cell lines used in that study. For comparison, a single human fibroblast line, MRC-5, and fresh human granulocytes obtained from buffy coats were also included. The use of a commercially available CEA kit for the quantitation of the antigen in cell extracts revealed that five of the seven lines had constitutive CEA expression. DLD-1 and MIP, two undifferentiated colon tumor cell lines, were CEA negative using the 2.5 ng/ml for a positive CEA value. There is an approximate 150-fold difference in the CEA content of the five cell lines that constitutively express the antigen. Clearly, the GEO tumor cell line expressed the highest level of CEA; in fact, CEA comprised approximately 0.2% of the total cellular protein for that cell line. The level of CEA cell-surface expression on those seven human colorectal cell lines also differed greatly, as illustrated by the difference in the percent of CEA positive cells using either anti-CEA MAb COL-1 or COL-12. These MAbs are 2 of a series of 15 developed in our laboratory[39] and are shown to preferentially react with human colon carcinomas. Previous results have shown that COL-1 preferentially reacts with human colorectal tumor cells and does not recognize any CEA-related gene product expressed by human granulocytes.[39,40] Using either anti-CEA MAb COL-1 or -12, the MIP and DLD-1 cells were found not to express any cell-surface CEA. Using MAb COL-1, the percent of cells positive for surface CEA on the five remaining colorectal cell lines ranged from approximately 42% for the WiDr cells to >90% for the highly differentiated CBS and GEO cell lines. The range of cells positive for COL-12 binding was 32 to 48%, suggesting that this epitope is not as well expressed on those cell lines as is the COL-1 CEA epitope. Those cell lines that contained the highest CEA content in their extracts were the same cell lines with the highest percent of CEA positive cells. Interestingly, HT-29 and WiDr have cell-surface antigen phenotypes for CEA that are similar (i.e., 40 to 60% CEA positive), yet there is over a tenfold difference in the CEA content in the extract of these cells. All colorectal tumor cells were grown as subcutaneous tumors in athymic mice. The histological grade of each colorectal tumor cell line correlated with the CEA content of the whole-cell extract; that is, the more differentiated the colorectal tumor, the higher the level of CEA expression.

Dose- and time-dependent changes in CEA expression following the addition of IFN-γ were analyzed using several cell lines. In Figure 1A, WiDr cells were treated with 1000 antiviral units of Hu-IFN-γ, and the time-dependent alterations in CEA expression (i.e., COL-1 binding) and class II HLA were measured. Significant increase in either COL-1 or the anti-HLA-DR MAb binding were evident within 24 h of the addition of IFN-γ. Maximum induction for the class II HLA occurred after 24 to 48 h, whereas that for CEA occurred after 48 to 72 h. Figure 1B summarizes the dose-dependent changes in cell-surface CEA following treatment of the DLD-1, LS174T, and WiDr cells for 72 h with different antiviral titers of IFN-γ. The LS174T cell line is unresponsive to the ability of IFN-γ to alter the level of CEA expression. In contrast, CEA expression as measured by the binding of MAb COL-1 was increased in a dose-dependent manner by IFN-γ using WiDr or DLD-1 cells. The maximum level of CEA increase for both of these cell lines occurred after the addition of 1000 antiviral units of IFN-γ for 72 h. The level of CEA expression on the WiDr cells increased approximately threefold, whereas that for the DLD-1 rose from CEA negative to binding approximately 7000 cpm in a live-cell radioimmunoassay (RIA).

The mechanism(s) involved in the interferon-mediated increase of surface CEA expression was further investigated using flow cytometry (Figure 2). Briefly, the moderately differentiated human colorectal cell line, WiDr, was treated with 800 antiviral units per ml of IFN-γ for 48 h. The cells were harvested, washed with Ca^{2+}/Mg^{2+}-free Dulbecco's phosphate-buffered saline, and incubated for 1 to 2 h in the presence of 2 μg of the anti-CEA COL-1 MAb. Panel A represents the background fluorescence of untreated and IFN-γ-treated WiDr cells. The addition of the anti-CEA MAb COL-1 to the untreated WiDr cell population resulted in approximately 40% of the cells expressing constitutive CEA (Figure 2B). The WiDr cell-surface antigen phenotype clearly shows an increase in CEA expression as a result

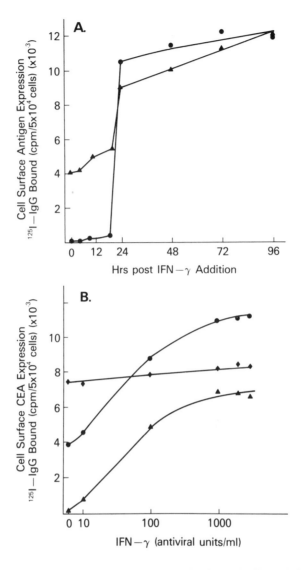

FIGURE 1.   Temporal (A) and dose-dependent (B) changes in CEA and class II
HLA expression following the addition of Hu-IFN-γ. A. WiDr cells were seeded
in 96-well microtiter plates at densities that ranged from 1 to 4 × 10⁴ cells/well.
Hu-IFN-γ (1000 units/ml) was added to the appropriate wells at the indicated time
intervals, and a live-cell RIA was performed 96 h later. Wells that contained 1000
antiviral units Hu-IFN-γ/ml for 72 and 96 h had approximately 15% fewer cells
than the untreated wells. Therefore, all measurements of MAb binding were
adjusted to cpm per 5 × 10⁴ cells. COL-1 (▲—▲) and anti-HLA-DR (●—●)
binding are the mean of two separate experiments. B. DLD-1 (▲—▲), WiDr
(●—●), and LS174T (♦—♦) were plated in 96-well microtiter plates and treated
with 10 to 3000 antiviral units Hu-IFN-γ/ml for 72 h. Untreated cells received
an equal aliquot of complete medium. An antiviral titer of 1000 units/ml or greater
resulted in a 15 to 20% decrease in cell growth; therefore, the amount of bound
COL-4 was adjusted for 5 × 10⁴ cells. Results shown are the mean of two to
four separate experiments.

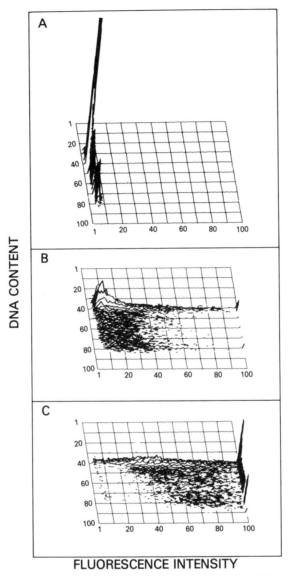

FIGURE 2.   Fluorescence-activated cell sorter (FACS) analysis of MAb COL-1 binding to the surface of human colon carcinoma cells: effect of IFN-γ. The flow cytometric analysis was done using an Ortho Cytofluorograf system 50H with blue laser excitation of 200 mW at 488 nm. WiDr cells were incubated for 48 h in growth medium with or without 500 antiviral units IFN-γ per ml, harvested, and incubated in culture medium containing 2 μg MAb COL-1/10⁶ cells for 30 min at 4°C. The cells were then washed with Ca^{2+}/Mg^{2+}-free Dulbecco's PBS and then treated with fluoresceinated goat anti-mouse IgG (heavy and light chains; 1:30 dilution) for 30 min at 4°C. The cells were washed, excess antibody was removed, and the cells were resuspended at a concentration of 10⁶ cells/ml and stained with propidium iodide (18 μg/ml) and RNAse (2000 units/ml) for 4 h at room temperature. The stained cells were analyzed and, under these conditions, propidium iodide bound to nuclear DNA fluoresces red, while surface immunofluorescence bound to CEA fluoresces green. Data from 25,000 cells were stored on an Ortho Model 2150 computer system and used to generate each three-dimensional isometric display of DNA content (Y axis), fluorescence intensity, i.e., cell-surface expression of CEA (X axis), and the number of cells (Z axis). (A) WiDr cells stained for DNA content but no MAb; (B) WiDr cells stained for nuclear DNA and COL-1 binding; (C) WiDr cells treated with 800 antiviral units IFN-γ for 48 h and stained as in B.

TABLE 3

Comparison of the Effects of IFN-γ on the Expression of Tumor-Associated and Normal Antigens in Extracts of Human Colorectal Tumor Cells and Fibroblasts

Cell line	IFN-γ[b]	COL-4	COL-12	W6/32	Anti-cytokeratins	Anti-HLA-DR	MOPC-21
MIP	−	Neg[c]	Neg	Neg	ND	Neg	Neg
	+	Neg	Neg	Neg	ND	Neg	Neg
DLD-1	−	Neg	Neg	Neg	4850	Neg	ND
	+	8290[d]	1080	Neg	4990(− 3%)	870	ND
WiDr	−	8120	1940	5415	27480	Neg	Neg
	+	17765(119%)	4060(109%)	21710(301%)	26090(− 5%)	4060	Neg
HT-29	−	4160	960	2390	ND	Neg	Neg
	+	10190(145%)	1545(61%)	11680(389%)	ND	9895	Neg
LS174T	−	11330	2880	2840	22850	Neg	Neg
	+	12055(6%)	3725(29%)	2610(− 8%)	19670(− 14%)	Neg	Neg
CBS	−	20620	5165	2150	24990	Neg	Neg
	+	20810(1%)	7835(52%)	12660(489%)	24465(− 2%)	7610	Neg
GEO	−	22155	8995	4150	19400	Neg	Neg
	+	22740(2%)	9175(2%)	9375(126%)	17410(− 10%)	4200	Neg
MRC-5	−	Neg	Neg	6800	Neg	Neg	Neg
	+	Neg	Neg	12220(80%)	Neg	7270	Neg

MAb binding to the cell extract[a] (cpm/5 μg protein)

[a] Cells were scraped from T-150 flasks and pelleted by centrifugation at 1000 × g for 10 to 15 min. The pellet was suspended in extraction buffer and the extracts prepared as previously described.[4]

[b] IFN-γ was added to subconfluent T-150 flasks of cells at antiviral titers of 800 to 1500 units/ml and treated for 48 to 72 h.

[c] Neg = <1000 cpm/5 μg protein.

[d] Values represent the mean of triplicate determinations. The numbers in parentheses are the percentage increases of the MAb bound to the tumor cell extract above that measured in the untreated cells. The formula used was

$$\% \text{ increase MAb binding} = \frac{\text{cpm bound IFN-}\gamma\text{-treated cells } - \text{cpm bound untreated cells}}{\text{cpm bound untreated cells}} \times 100$$

of IFN-γ treatment (Figure 2C). The percent CEA-positive WiDr cells rose to >85%, and the mean fluorescence intensity increased approximately fourfold as a result of IFN-γ treatment. Furthermore, the change in CEA expression was not accompanied by any substantial alterations in cell proliferation. These data indicate a selective nature in IFN-γ regulation of CEA expression and not a preferential expansion of a subpopulation of cells expressing high CEA levels. The increase measured in the mean fluorescence intensity suggests that, in addition to an increase in the number of WiDr cells that are CEA positive, IFN-γ treatment is also increasing the amount of antigen expressed per cell. In any case, IFN-γ treatment is capable of "recruiting" previously antigen-negative cells to express the cell-surface tumor antigen. The overall effect is a human tumor cell population that is more homogeneous for the expression of CEA.

The effects of IFN-γ treatment on the expression of CEA as well as other normal cellular antigens (e.g., MHC, cytokeratins, etc.) were also analyzed in extracts prepared from different human colorectal tumor cell lines (Table 3). Using both the changes of CEA expression and induction of the HLA-DR antigens as evidence of responsiveness to IFN-γ, two cell lines, the LS174T and MIP, emerged as being unresponsive to the ability of IFN-γ to alter their antigen phenotype. However, treatment of either of these cell types with increasing antiviral titers of IFN-γ results in a dose-dependent increase in (2′-5′)oligo(A) synthetase activity.[81] In addition, LS174T as well as the MIP cells can be growth inhibited by the addition of IFN-γ to their respective culture media. The findings provide a strong argument for the presence of a viable membrane interferon receptor on both the LS174T and MIP cells. Thus, the association of IFN-γ with its receptor can elicit the biochemical pathway(s) that mediates antiviral and antiproliferative activities. However, little is known of the intracellular triggers that mediate the ability of interferon to regulate the level of expression of surface HLA and/or human tumor antigen. Recent evidence by Zuckerman et al.[41] suggests that a *trans*-activating factor is one possible mediator required for *de novo* class II HLA induction by IFN-γ. Lack of such a factor could explain the inability of IFN-γ to alter class II antigen, and possibly CEA, expression in the LS174T and MIP cells. The remaining five human colorectal cell lines (DLD-1, WiDr, HT-29, CBS, and GEO) are shown to express *de novo* HLA-DR expression as well as elevated class I HLA levels after IFN-γ treatment (Table 3). The HT-29 and WiDr cell lines exhibit the most substantial increase in both COL-4 (COL-1 and COL-4 react with the same CEA epitope[40]) and COL-12 binding as a result of IFN-γ treatment. CBS and GEO cells express high constitutive levels of CEA, and only the level of COL-12 binding to the CBS cell extract was enhanced after IFN-γ treatment. The level of cytokeratin expression does not seem to be altered with IFN-γ treatment. The observed changes in the binding of the anti-CEA MAbs in extracts of the human colorectal cell lines were further corroborated by flow cytometric analysis.[81]

The most dramatic change in CEA expression after IFN-γ treatment was measured in the HT-29 and WiDr cell lines, two moderately differentiated colorectal cell types. Following IFN-γ treatment, the level of CEA expression in the whole cell extracts prepared from either cell type rose by >100%; the percentage of CEA-positive cells as analyzed by flow cytometry increased from 30 to 45% to >80%. Until recently, studies of the regulation of tumor antigens by human interferons have been confined to analysis of antigen levels through the use of MAbs. The data suggested that an increased synthesis of the tumor antigen may be one component of that process. Recently, molecular probes of CEA have become available that have permitted the study of the regulation of this antigen at the transcriptional level. Additional studies were performed to investigate the effect of IFN-γ on the steady-state levels of CEA mRNA. The CEA cDNA probe has been described elsewhere by Zimmermann et al.[34] Briefly, it is a 550-bp *Pst*I fragment isolated from the vector pCEA1, containing domain II and part of domain III of the glycopeptide, subcloned into the *Pst*I site of pGEM-1. The *Pst*I fragment was isolated, and the insert was nick translated with a nick translation kit according to the manufacturer's instructions (Bethesda Research Laboratories, Gaithersburg, MD), yielding a specific activity of 1×10^8 cpm/μg of DNA.[42]

Northern blot analysis of CEA expression with poly(A)+-selected RNA from HT-29, WiDr, and LS174T colon carcinoma cells and WI38 and MRC-5 fibroblast cell lines before and 72 h after treatment with 2000 antiviral units per ml of Hu-IFN-γ is shown in Figure 3. The CEA cDNA probe identified three predominant species of CEA transcripts, having sizes of 4.2, 3.5, and 2.8 kb, respectively, in the human colorectal carcinoma cells. The three species were found in varying amounts in untreated LS174T (Figure 3, lane 1), WiDr (Figure 3, lane 3), and HT-29 (Figure 3, lane 5) cells. After treatment with Hu-IFN-γ, all three CEA transcripts were amplified in both the WiDr and HT-29 cells (Figure 3, lanes 4 and 6, respectively); their levels in LS174T cells were relatively unchanged (Figure 3, lane 2). The HT-29 cell line also had low constitutive levels of a 1.7-kb CEA transcript; this transcript was also enhanced several fold after Hu-IFN-γ treatment. The 1.7-kb transcript was more pronounced under less stringent washing conditions and did not appear in any of the other colon carcinoma cell lines or in the normal fibroblast cell lines. This 1.7-kb transcript may thus represent the gene product of another member of the CEA-related family.[36,38] CEA transcripts were not detected in the RNA from either fibroblast cell line, nor could these cells be induced by Hu-IFN-γ to express any of the CEA transcripts (Figure 3, lanes 7 through 10). Probing these blots with an actin probe disclosed similar amounts of RNA loaded into each lane for all five cell lines (Figure 3B).

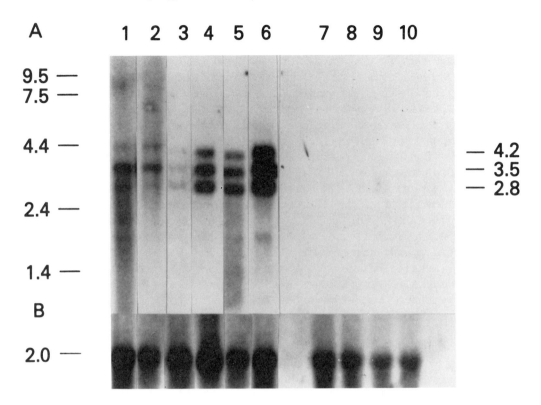

FIGURE 3.   Effect of Hu-IFN-γ on the levels of CEA poly(A)$^+$ RNA from human colon tumor cell lines and normal fibroblast cell lines. Poly(A)$^+$ RNA from colon tumor cell lines (lanes 1 through 6) and normal fibroblasts (lanes 7 through 10) were extracted, blotted, and hybridized to a CEA cDNA probe as described and then exposed for 3 d. RNA markers are 9.5, 7.5, 4.4, 2.4, 1.4, and 0.3 kb (Bethesda Research Laboratories). Lanes 1 and 2, LS174T; lanes 3 and 4, WiDr; lanes 5 and 6, HT-29; lanes 7 and 8, WI38; lanes 9 and 10, MRC-5. The odd-numbered lanes contain samples prepared from untreated cells, and the even-numbered lanes contain samples (i.e., extracts) from cells treated with 2000 units/ml Hu-IFN-γ.

The analysis of the data derived from the densitometric profiles of the Northern auto-radiograms is shown in Figure 4. The addition of 2000 antiviral units per ml for 72 h increased the steady-state levels of the 2.8-, 3.5-, and 4.2-kb transcripts by 9-, 11-, and 13-fold, respectively, in the WiDr cells (Figure 4A) and 6-, 8-, and 11-fold, respectively, in HT-29 cells (Figure 4B). In contrast, the steady-state levels of the 4.2-kb CEA transcripts were unaffected in the LS174T cell line, whereas the 2.8- and 3.5-kb transcripts decreased approximately twofold (Figure 4C).

The results demonstrate that the addition of IFN-γ to the growth media of two moderately differentiated human colorectal cell lines can significantly increase the steady-state level of CEA mRNA. The highly differentiated LS174T cell line does not exhibit any enhanced mRNA levels for CEA or *de novo* induction of mRNA-related transcripts for class II HLA. The CEA and MHC genes belong to the immunoglobulin superfamily,[43,44] and both are modulated by IFN-γ, suggesting that they may share some regulatory sequences and/or factors.[41,45-49] Thus, the failure of LS174T cells to respond to Hu-IFN-γ may result from the absence or defective function of a common regulatory element involved in modulation of the CEA and HLA-DR genes. In this context, Hu-IFN-γ treatment of responsive and nonresponsive colon carcinoma cells could provide valuable information for the identification and purification of those inducible factor(s) that may regulate transcription of CEA and MHC genes.

The cell line (i.e., HT-29 and WiDr) in which IFN-γ treatment enhanced the level of

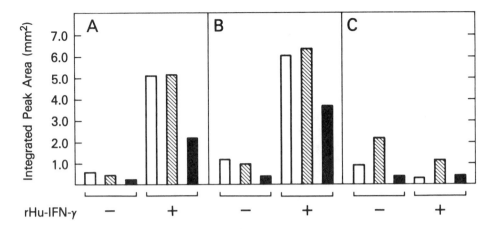

FIGURE 4. Densitometric analysis of the experiments shown in Figure 3. The autoradiograms containing fractionated poly(A)$^+$ mRNA from colon tumor cell lines WiDr (A), HT-29 (B), and LS174T (C), nontreated (−) and treated (+) with Hu-IFN-γ, were scanned with an LKB ultrascan densitometer. The transcripts scanned are 2.8 kb (□); 3.5 kb (▨); and 4.2 kb (■).

binding of the anti-CEA MAb COL-1 were also those in which interferon treatment increased the steady-state levels of the three CEA-related transcripts. Generally, there appears to be a correlation between the increase in COL-4 binding and the increase in the steady-state levels of the 3.5-kb transcript. Kamarck et al.[38] have shown that a 3.6-kb CEA mRNA comprises 75% of the three mRNA bands observed in the colon carcinoma cell line, LoVo, which expresses high levels of CEA glycoprotein. In studies recently completed, Zimmermann et al.[50] have determined that the 4.2- and 3.5-kb transcripts are encoded by the CEA gene, and the nonspecific cross-reacting antigen (NCA) gene encodes for the 2.8-kb transcript. Our findings thus demonstrate that IFN-γ can also upregulate NCA transcripts. Since NCA is a constituent of normal human granulocytes and its function is thus far unknown, it would be of interest to determine in future studies the role of IFN-γ-mediated upregulation of NCA in granulocytes in correlation studies with the granulocyte function. The appearance of the 4.2- and 3.5-kb transcripts may also indicate differential splicing patterns[36] or alternative poly(A)$^+$ addition sites[33] from one CEA gene. In addition, these multiple mRNAs may arise from the transcription of different members of the same gene family. There has been an estimate of eight to ten genes in the CEA family,[34-38,44] and these genes are believed to be 80 to 85% homologous. In any case, it is unclear at this time whether this increase in CEA mRNA accumulation resulted from an increased rate of transcription of the CEA gene(s) or from increased stability of the mRNAs produced.[51] Further studies are clearly required to define the mechanism(s) of the biological action of Hu-IFN-γ on the tumor-associated antigen CEA in colon carcinoma cells. As a result of findings reported here, we can conclude that the interaction of Hu-IFN-γ with its surface receptor triggers the induction of internal signals to an actively transcribing CEA gene, resulting in an increase in the levels of steady-state mRNA.

III. EFFECT OF INTERFERON ON THE *IN VIVO* LOCALIZATION OF MAb

Recently, investigators have designed experimental models to determine whether the administration of a human interferon can alter the antigenic phenotype of a human tumor cell population *in vivo*. Specifically, Rowlinson et al.[52] reported that IFN-γ treatment of athymic mice bearing human breast carcinoma can induce the expression of class II HLA in the tumors. In addition, they reported that IFN-γ treatment also increased the accumulation of a radiolabeled anti-HLA-DR MAb TAL1B5 in the malignant tissues. Other reports have shown that IFN-γ, when administered to patients, can increase the constitutive level of class II HLA on circulating human lymphoid cells.[53] A few reports have focused on the *in vivo* regulation of human tumor antigens by interferon. Matsui et al.[54] reported that a 96-kDa antigen recognized by MAb CL207 can be induced on human carcinoma and melanoma cells by IFN-γ. Moreover, a combination of IFN-γ treatment and the administration of MAb CL207 conjugated to daunomycin resulted in a synergistic cytotoxic effect on human tumor cells. We performed a series of experiments designed to determine whether type I interferon, IFN-αA, was capable of altering the level of expression of MAb B6.2 on human breast and colorectal cells grown as subcutaneous tumors in athymic mice.[55,56] MAb B6.2 recognizes a 90-kDa tumor antigen that is expressed in approximately 75% of the breast carcinomas and >90% of the colorectal tumors. Extensive *in vitro* studies have shown that the constitutive level of expression of this antigen can be substantially increased following the administration of IFN-αA or IFN-γ.[57] In addition, MAb B6.2 has served as an important experimental model in our laboratory. Prior studies have utilized B6.2 to generate F(ab) and F(ab')$_2$ fragments as well as the development of protocols for the conjugation of radionuclides.[58] Other studies evaluated the effect of whole IgG and their fragments of B6.2 on the localization to the tumor site *in vivo* and their relative whole-body clearance. B6.2 was also used to produce a series of human-mouse chimeric antibodies.[59]

Table 4 lists the different human tumor cell types that were grown as subcutaneous tumors in athymic mice and subsequently treated with IFN-αA. Clouser is a transplantable human breast carcinoma, WiDr and LS174T cells are moderately and well-differentiated human colorectal cells, and the A375 are human melanoma cells. Each group of tumor-bearing mice was subsequently divided into two or three groups: untreated and treated with either 50,000 or 250,000 antiviral units per d for 5 to 7 d. The interferon was routinely given as two daily intramuscular (i.m.) injections. After a 5- to 7-d treatment, the tumors were removed, extracts were prepared,[4] and direct binding of ^{125}I-labeled B6.2 was measured in a solid-phase RIA. The control antibody, MOPC-21, was also included in all assays and shown not to react with any tumor extract either before or after IFN-αA treatment (Table 4). In addition, MOPC-21 does not react with a variety of normal mouse tissues (e.g., liver, spleen, kidney, etc.), indicating that interferon treatment of athymic mice does not induce expression of Fc receptors on normal and tumor tissues. IFN-αA treatment of mice bearing Clouser or WiDr tumors resulted in a two- to threefold increase in the binding of ^{125}I-B6.2 to tumor extracts (Table 4). In addition, the degree to which the B6.2-reactive 90-kDa antigen is increased is somewhat related to the circulating plasma level of IFN-αA. For example, WiDr-bearing athymic mice administered either 50,000 or 250,000 antiviral units per d for 5 d had circulating interferon plasma titers of 75 and 310 units per ml, respectively. Furthermore, the level of ^{125}I-B6.2 binding was increased by approximately 45 and 240%, respectively, above that measured in untreated mice, suggesting that the increase in the level of the B6.2-reactive antigen is somewhat correlated with the plasma interferon levels. The LS174T cells grown as subcutaneous tumors in athymic mice constitutively express the B6.2-reactive 90-kDa antigen. However, consistent with the *in vitro* findings, treatment with IFN-αA was not able to increase the level of antigen expression in the LS174T tumor extracts. Therefore, the unresponsiveness demonstrated by that colorectal cell line *in vitro* to IFN-αA is a stable phenotype of the tumors *in vivo*. The human melanoma cell type A375 does not bind MAb B6.2 either constitutively or after interferon administration *in vivo*.

The initial *in vivo* studies utilized IFN-αA as a twice daily i.m. administration. The plasma clearance of the recombinant interferon after a single i.m. injection of 5×10^5 antiviral units to WiDr-bearing athymic mice was approximately 3 to 4 h.[56] Therefore, approximately 8 h after the single injection, >90% of the IFN-αA had been removed from the plasma. Due to this rapid clearance, it was decided that perhaps a more sustained delivery

TABLE 4
Effect of IFN-αA Administration on the Binding of ^{125}I-B6.2 to Extracts of Human Tumors Prepared from Athymic Mice

Tumor-bearing athymic mice[a]	IFN-αA treatment[b] (units/d)	Plasma IFN levels[c] (antiviral units/ml)	^{125}I-IgG Bound[d] (cpm/10 μg tumor extract) B6.2	MOPC-21
Clouser	None	<30	3,480 (2,610—4,940)	Neg
	50,000	90 (60—120)	7,910 (6,770—12,410)	Neg
WiDr	None	<30	5,460 (3,680—7,240)	Neg
	50,000	75 (<30—90)	7,890 (5,810—12,890)	Neg
	250,000	310 (120—940)	18,460 (7,890—24,140)	Neg
LS174T	None	<30	7,840 (5,880—9,120)	Neg
	250,000	340 (160—860)	8,690 (5,410—10,540)	Neg
A375	None	<30	Neg	Neg
	250,000	460 (310—940)	Neg	Neg

[a] The WiDr, LS174T, Clouser, and A375 cells were grown as s.c. tumors in athymic mice on a Balb/c background. Mice were inoculated s.c. with 0.1 to 1.0 × 10^7 WiDr, LS174T, or A375 cells. Female athymic mice were also inoculated with a 1- to 2-mm^3 piece of the Clouser transplantable breast carcinoma. In all studies, the average diameter of each tumor ranged from 0.7 to 1.2 cm.

[b] IFN-αA, kindly supplied by Hoffman-LaRoche (Nutley, NJ), was received in ampules containing 50 × 10^6 antiviral units with a specific activity of 5.3 × 10^8 units/mg protein. The interferon-treated group received twice daily i.m. injections, and the total units/d is shown.

[c] Plasma IFN-αA levels were measured using MDBK cells infected with vesicular stomatitis virus. The values in parentheses represent the range of antiviral units per milliliter for each group. Each group contained 5 to 12 mice.

[d] The procedure for iodination of MAb B6.2 and MOPC-21 was outlined in a prior study.[56] The tumors were removed, extracts were prepared, and 10 μg of each tumor extract was dried to each well of a 96-well microtiter plate. ^{125}I-B6.2 and ^{125}I-MOPC-21 was added beginning at 200,000 cpm/well and with subsequent 1:2 dilution. The values in parentheses represent the range of ^{125}I-B6.2 bound to tumor extracts from 5 to 10 individual mice per group. The single value is the mean for each group. Neg = <1000 cpm/10 μg tumor tissue.

of IFN-αA could result in higher, more constant plasma levels. The WiDr tumor-bearing mice were implanted subcutaneously with a miniosmotic pump (Alzet pump, model 2001, Alza Corp.) containing 2 to 3 × 10^6 units of IFN-αA. The constant delivery of IFN-αA via the osmotic pump maintained plasma IFN-αA levels of 300 to 800 antiviral units per ml for 6 to 7 d.[56] Upon removal of the pump, the clearance rate of the IFN-αA was similar to that reported for the single injection.

It was demonstrated that the tumor:blood ratio of ^{125}I-B6.2-F(ab')$_2$ localized to human tumors in athymic mice increased with time post-injection of the antibody.[56] Indeed, as shown in Table 5, the tumor:blood ratios for the localization of ^{125}I-B6.2-F(ab')$_2$ to WiDr tumors were 2.5, 4.0, and 6.9 at 24, 48, and 72 h post-MAb administration, respectively. It is believed that the increase in tumor:blood ratios, as well as the concomitant increase in tumor:liver and tumor:spleen ratios, with time post-MAb injection is due to the clearance of the ^{125}I-labeled antibody from nontumor sites.[58] This same phenomenon was evident in the WiDr-bearing athymic mice implanted with the IFN-αA-containing osmotic pump. In addition, at 24, 48, and 96 h post-^{125}I-MAb administration, the ratio of antibody localized

TABLE 5

Effect of IFN-αA Treatment on the Localization of ^{125}I-B6.2-F(ab')$_2$ to WiDr Tumors Grown in Athymic Mice

Ratio	+/− IFN-αA	^{125}I-B6.2-F(ab')$_2$ localization at various times post-i.v. administration		
		24 h	48 h	72 h
Tumor:blood	−	2.5 ± 0.2	4.0 ± 0.4	6.9 ± 0.3
	+	6.1 ± 0.8	7.1 ± 0.9	12.3 ± 1.4
Tumor:spleen	−	4.4 ± 0.2	8.7 ± 0.4	8.8 ± 1.0
	+	10.8 ± 1.2	14.8 ± 1.4	16.9 ± 1.5
Tumor:liver	−	4.0 ± 0.2	6.1 ± 0.4	7.9 ± 0.3
	+	7.8 ± 0.4	12.4 ± 1.1	11.9 ± 0.8
% injected	−	3.3 ± 0.6	2.1 ± 0.3	1.2 ± 0.2
	+	8.9 ± 1.1	6.8 ± 0.7	6.0 ± 0.5

Note: WiDr tumor-bearing athymic mice were administered IFN-αA in a constant infusion osmotic pump. Five d after installation of the pump containing 2 to 3 × 10^6 antiviral units IFN-αA, the mice received a single i.v. injection of ^{125}I-B6.2-F(ab')$_2$ (2 × 10^6 cpm/mouse). The IFN-αA-treated mice and appropriate control-treated mice (i.e., osmotic pump filled with 0.9% NaCl) were sacrificed 24, 48, and 72 h post-MAb administration. The levels of ^{125}I-B6.2-F(ab')$_2$ localized to the tumor relative to that of the normal organs were determined as previously described.[56] Values represent the mean ± standard error of four different experiments with each group containing 6 to 10 mice.

and measured on the basis of tumor-to-normal tissues (i.e., blood, liver, and spleen) was approximately twofold greater than those measured at these same time intervals in the untreated mice (Table 5). It should also be noted that the percent injected dose of MAb per gram tumor was increased fivefold at 72 h post-MAb administration as a result of interferon treatment. These results suggest that the increase in MAb localization was primarily due to the increase in the level of expression of the 90-kDa antigen and not to alterations in the blood flow to the tumor site.

The effects of constant infusion of IFN-αA for 9 and 13 d on the localization of ^{125}I-B6.2-F(ab')$_2$ were also assessed on individual mice (Table 6). In those experiments, the osmotic pump containing IFN-αA was replaced after 6 d to ensure the maintenance of high plasma levels of the recombinant interferon (Table 6). The results indicate a temporal dependence on the length of time the mice are treated with IFN-αA, resulting in an increase in the percent injected dose per gram of tumor localized by the WiDr tumors. For example, after a 9-d treatment with IFN-αA, the mean percent injected dose per gram of tumor tissue was increased 148% above control levels. Likewise, after a 13-d treatment, the percent injected dose of ^{125}I-B6.2-F(ab')$_2$ was increased approximately 250%, from 3.1 for the untreated mice to 10.9 for the interferon-treated group. In another experiment, a group of WiDr tumor-bearing athymic mice were implanted with a miniosmotic pump containing 0.5 to 2.5 × 10^6 antiviral units of IFN-αA. After a 6-d treatment, the mice were given a single bolus injection of ^{125}I-B6.2-F(ab')$_2$ and sacrificed 24 h later. The amount of radioactivity localized to the WiDr tumors per tumor wet weight, as well as the circulating plasma IFN-αA levels (i.e., antiviral units per ml), were determined. The results shown in Figure 5 indicate a relationship between plasma interferon levels and the level of localization of the antibody to the tumor site. The results also indicate that a considerable difference exists among the individual mice as to their ability to localize the ^{125}I-MAb. This is particularly evident in the groups of mice treated with IFN-αA and exhibiting circulatory plasma IFN-αA levels of 150 antiviral units or greater (Figure 5). This could be explained by a single or combination of several factors, including the existence of a wide range of sensitivity among the different WiDr tumors in the athymic mice for the antigen augmentation by IFN-

TABLE 6

Temporal-Dependent Increase in the Localization of ^{125}I-B6.2-F(ab')$_2$ to WiDr Tumors After a 9- and 13-D Treatment with IFN-αA

WiDr tumor-bearing nude mice		^{125}I-B6.2-F(ab')$_2$ localization (% injected dose per g tissue)			
		Tumor	Blood	Liver	Spleen
Control	1	2.7	0.5	0.5	0.3
	2	2.2	0.8	0.4	0.3
	3	4.2	1.0	0.7	0.5
	4	4.4	0.8	0.7	0.5
	5	2.6	1.0	1.0	0.7
	6	3.0	0.7	0.5	0.4
	7	2.6	0.9	0.8	0.4
	8	3.2	0.8	0.6	0.3
	9	3.1	0.7	0.3	0.2
	10	2.8	0.7	0.7	0.4
		$\bar{x} = 3.1 \pm 0.2$			
+IFN-αA (9 d)	1	7.2	0.7	0.8	0.4
	2	3.5	1.0	0.7	0.3
	3	6.3	0.9	0.9	1.2
	4	8.6	0.4	0.5	0.4
	5	8.5	0.9	0.8	0.4
	6	6.7	0.7	1.0	0.6
	7	9.2	0.9	0.8	0.4
	8	4.3	0.4	0.9	0.4
	9	6.2	0.6	0.7	0.4
	10	8.0	0.4	0.5	0.7
	11	11.9	0.8	0.4	0.3
	12	4.4	0.6	0.7	0.5
	13	8.2	1.0	0.7	0.4
	14	15.3	0.9	0.5	0.4
	15	7.1	0.8	1.0	0.7
		$\bar{x} = 7.7 \pm 0.7$[a]			
+IFN-αA (13 d)	1	10.5	1.1	1.4	0.5
	2	6.5	0.9	1.0	0.7
	3	13.9	1.0	0.6	0.5
	4	10.1	1.1	1.0	0.6
	5	4.6	1.1	0.8	0.5
	6	18.6	1.0	0.9	0.6
	7	12.6	1.2	1.0	0.6
	8	9.4	0.9	1.2	0.8
	9	14.8	1.0	0.6	0.4
	10	11.0	0.9	1.0	0.7
	11	5.3	1.0	0.7	0.4
	12	7.5	0.8	1.0	0.6
	13	19.3	1.1	1.5	1.0
	14	12.9	1.3	1.2	0.7
	15	6.8	0.7	0.6	0.5
		$\bar{x} = 10.9 \pm 1.1$[b]			

Note: The WiDr-bearing athymic mice were divided into three groups: the 13-d-treated group received an osmotic pump containing 2 to 3 × 10^6 antiviral units of IFN-αA; 4 d later, the 9-d-treated group received the same implant. Each group was given new osmotic pumps containing IFN-αA and sacrificed 24 h after injection of ^{125}I-B6.2-F(ab')$_2$. The percent injected dose per g of the ^{125}I-MAb localized to the tumor as well as to the appropriate normal tissue was determined as previously outlined.[56] Values represent the mean ± standard error.

[a] $P < 0.05$ (vs. untreated group).   [b] $P < 0.05$ (vs. the 9-d interferon-treated group).

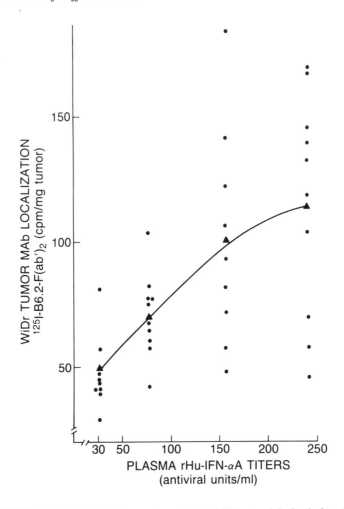

FIGURE 5.   Relationship between plasma levels of IFN-αA and the level of *in vivo* localization of ^{125}I-B6.2-F(ab')$_2$ to WiDr tumors in athymic mice. WiDr tumor-bearing athymic mice were implanted with miniosmotic pumps containing 0.5 to 2.5 × 10^6 antiviral units of IFN-αA. Control mice received pumps filled with 0.9% NaCl. After 6 d of interferon treatment, all mice received ^{125}I-B6.2-F(ab')$_2$ as previously described. Twenty-four hours later, all mice were sacrificed, and the localization of ^{125}I-B6.2-F(ab')$_2$ to the human colon carcinoma xenograft was measured as total counts per minute per milligram tumor wet weight and plasma IFN-αA antiviral titers as previously described. Each data point on the graph represents a single mouse.

αA, differences in tumor vascularization, or difference among the mice in metabolizing the interferon.

The results demonstrate that the *in vivo* administration of a type I human leukocyte interferon can increase the expression of a 90-kDa tumor antigen in human breast and colorectal tumors grown in athymic mice. The administration of IFNαA to carcinoma-bearing athymic mice increases the level of expression of the B6.2-reactive 90-kDa antigen in tumor extracts. In addition, interferon treatment also enhances the level of cell-surface expression of the 90-kDa antigen on human colorectal tumors. This results in a three- to fourfold increase in the amount of ^{125}I-B6.2-F(ab')$_2$ localized at the tumor site. The increase in MAb localization was also shown to be related to the circulating plasma levels of the interferon. Furthermore, the ability to deliver IFN-αA via a constant-flow osmotic pump maintains high circulating plasma levels, which results in a temporal-dependent increase in antibody localization.

TABLE 7
Properties of the Tumor-Associated Glycoprotein-72 (TAG-72)

Characteristics	Ref.
High-molecular-weight glycoprotein ($>10^6$ Da)	61
Expressed in:	60,62
50% breast carcinomas	
85% colorectal carcinomas	
95% ovarian carcinomas	
95% adenocarcinomas and adenosquamous carcinomas of the lung	
95% gastric carcinomas	
95% endometrial carcinomas	
Not expressed in:	62
Leukemias, lymphomas, and sarcomas, with only trace amounts found in benign and normal adult tissues	
Exception: high TAG-72 levels found in endometrium (secretory stage)	63
Density — 1.45 g/ml in cesium chloride	61
Shed and found in the plasma of cancer patients	66
Contains blood group-related oligosaccharides	61
Contains sialic acid	61
Contains the B72.3-reactive epitope sialosyn Tn	65
Amino acid composition rich in Thr, Ser, Pro, Gly	64
Poorly expressed in established human tumor cell lines MCF-7 and LS174T	6,7

Additional studies are needed to determine if interferon treatment might also alter the blood flow (i.e., MAb clearance) as well as other factors contributing to the availability of the antibody to the tumor site. Nevertheless, it may be possible that with the selection of a MAb that has a limited amount of reactivity with normal tissues (i.e., MAb B72.3), the administration of IFN-αA may partially overcome the antigen heterogeneity generally associated with human tumor cell populations and result in better discrimination between tumor and nontumor tissue. The results reported here may also be of value for a more accurate diagnosis of occult lesions in immunohistochemical analyses of biopsy specimens or malignant effusions.

IV. INTERFERON-INDUCED ENHANCEMENT OF B72.3-REACTIVE TAG-72 EXPRESSION ON HUMAN ADENOCARCINOMA CELLS ISOLATED FROM SEROUS EFFUSIONS

We have attempted to demonstrate that recombinant human interferons can regulate the state of differentiation of human adenocarcinoma cells as defined by the expression of MAb-defined cell-surface tumor antigens. IFN-γ treatment of human colorectal tumor cells can substantially increase CEA mRNA transcripts as well as cell-surface CEA indicative of an alteration in the state of cellular differentiation. In addition, *in vivo* administration of a type I human interferon can increase the localization of a radionuclide-conjugated MAb by enhancing the cell-surface expression of another tumor antigen. A third human tumor antigen, termed TAG-72, has been extensively studied by our laboratory.[60-64,66-70] A summary of some of the characteristics of the TAG-72 antigen is found in Table 7. TAG-72 is found in a wide variety of human carcinomas, including breast, colon, ovary, pancreas, stomach, endometrial, and non-small cell lung carcinoma.[63] It is a high-molecular-weight mucin (i.e., $>10^6$ Da) that is found in high quantities in highly differentiated mucinous colorectal tumors.[62,65,67] A TAG-72-reactive MAb, B72.3, has been successfully used in clinical trials for the localization of metastatic colorectal and ovarian carcinoma lesions.[67,68] The study of the regulation of the TAG-72 antigen has been hampered by the fact that few human tumor

TABLE 8

**Expression of TAG-72 and Class I HLA on the Surface of Human Tumor
Cells Isolated from Serous Effusions of Patients with Adenocarcinomas
and Various Nonepithelial Malignant Neoplasms**

| | | No. of antigen-positive cases[a] | |
Diagnosis	No. of cases	TAG-72 (%)	Class I HLA(%)
Adenocarcinomas			
Breast	10	8/10 (80)	9/9 (100)
Lung	7	6/7 (86)	6/6 (100)
Ovary	16	12/16 (75)	8/8 (100)
Uterus	3	3/3 (100)	2/2 (100)
Pancreas	3	3/3 (100)	3/3 (100)
Unknown	4	3/4 (75)	3/3 (100)
Total	43	35/43 (81)	31/31 (100)
Malignant nonepithelial neoplasms[b]	10	0/10 (0)	5/5 (100)
Benign effusions[c]	8	0/8 (0)	8/8 (100)

[a] Expression of each antigen was considered positive when binding in a suspension RIA exceeded 500 cpm/10^5 cells above background.

[b] Nonepithelial malignancy samples included melanoma, lymphoma, Wilms' tumor, carcinoid tumor, etc.

[c] Patients from whom effusions were isolated and diagnosed as either benign or reactive mesothelium.

cell lines constitutively express this antigen.[6] In fact, the only adherent human cell lines to constitutively express TAG-72 are the human breast MCF-7 cell line and the highly differentiated LS174T colorectal cell line.[7] Previous studies have shown that *in vitro* treatment of the MCF-7 cells with recombinant interferon can enhance TAG-72 expression.[57] However, as previously shown, the LS174T cell is unresponsive to the ability of either type I or type II interferons to regulate normal or tumor antigen expression. Recent studies[69,70] have shown that a high percentage of human tumor cells isolated from malignant serous effusions (i.e., pleural effusions and ascites) constitutively express TAG-72. Thus, human adenocarcinoma cells isolated from serous effusions of patients diagnosed with adenocarcinoma present a unique opportunity to investigate the ability of recombinant interferon to regulate TAG-72 expression.

Table 8 lists a total of 61 effusions collected from patients diagnosed with adenocarcinomas, malignant nonepithelial neoplasms, and benign diseases. The cytological diagnosis of each effusion was based on accepted, standard cytologic criteria as outlined in a previous study.[71] In most cases, the cytological diagnosis was confirmed by a subsequent surgical diagnosis. The isolation of malignant as well as nonmalignant tumor cells was performed using an established cell isolation procedure. The procedure includes density-gradient centrifugation, enzymatic tissue dissociation, and removal of infiltrating lymphoid cells using immunomagnetic beads. Table 8 summarizes the percentage of each group of serous effusions expressing constitutive levels of the tumor-associated and class I HLA. As previously reported using immunohistochemical techniques,[62] the range of distribution of the high-molecular-weight glycoprotein antigen, TAG-72, is that of a pancarcinoma antigen in that approximately 81% of the malignant effusions were found to constitutively express this antigen. None of ten serous effusions that were isolated from patients with nonepithelial malignancies and none of eight benign effusions (i.e., reactive mesothelium) were found to contain cells that reacted with B72.3, thus confirming the previous immunohistochemical findings.[69,70]

Figure 6 summarizes the results of incubating fresh isolated human malignant and

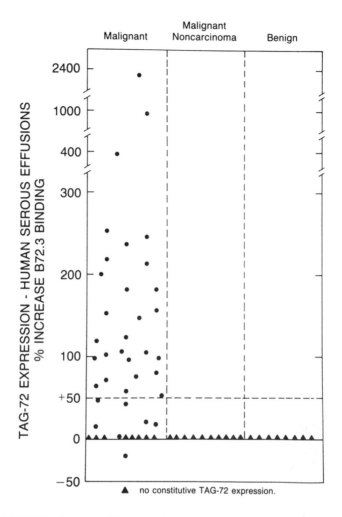

FIGURE 6. Summary of the percent increase in constitutive TAG-72 expression following treatment of freshly isolated human tumor cells with IFN-αA, IFN-β$_{ser}$, or IFN-γ. Human carcinoma, malignant noncarcinoma, or benign (i.e., reactive mesothelium) cells were isolated from patients' serous effusions. The cells were incubated in the presence of 500 to 1000 antiviral units of IFN-αA, IFN-β$_{ser}$, and/ or IFN-γ for 48 to 72 h. The level of expression of TAG-72 was measured by the binding of MAb B72.3. Each circle or triangle represents individual pleural or peritoneal effusions. The isolated human malignant cell populations that constitutively expressed TAG-72 are identified by a circle (●). Those cell populations isolated from patients diagnosed with malignant, malignant noncarcinoma, and benign effusions and did not constitutively express TAG-72 are represented by a triangle (▲). The percent increase in TAG-72 was determined by the equation shown in Table 3.

nonmalignant cells in the presence of 500 to 1000 antiviral units per ml of either type I (i.e., IFN-αA and IFN-β$_{ser}$) or type II (i.e., IFN-γ) interferons on the cell-surface expression of the B72.3-reactive TAG-72 tumor antigen. None of the malignant noncarcinoma or benign human cell populations isolated from their respective serous effusions expressed TAG-72 either before or after interferon treatment. In fact, on selected benign cell populations, antiviral titers of IFN-γ or IFN-αA were increased to >5,000 units per ml, resulting in no *de novo* induction of TAG-72 expression. Of the 43 effusions isolated from patients with confirmed metastatic adenocarcinoma, 35 constitutively expressed TAG-72. Furthermore,

TABLE 9
A Partial List of the Human Tumor Antigens Whose Level of Expression
Can Be Enhanced by Recombinant Interferons

Cell type	Antigen	Interferon	Ref.
Colon	CEA	Lymphoblastoid	15
		Recombinant α(A)	7, 57, 71
		Recombinant γ	42, 71
	90 kDa	Recombinant α(A)	55
	M111	Recombinant γ	77
Ovarian	TAG-72	Recombinant α, β, γ	71
Breast	90 kDa	Recombinant α-2b	80
		Recombinant α(A)	57, 71
	CEA	Recombinant α(A)	71
		Recombinant γ	79
	TAG-72	Recombinant α(A), β, γ	71
	M111	Recombinant γ	77
Melanoma	80 kDa and Ganglioside	Recombinant γ	73
	Cytoplasmic MAA	Recombinant γ	74
	32, 38, 48, and 50 kDa	Recombinant γ	75
	M111	Recombinant γ	77
	96 kDa	Recombinant γ	43
	96.5 kDa (Patients)	α	78
Neuroblastoma	4—6 Antigens	Recombinant γ	76

treatment with recombinantly derived human interferon increased TAG-72 expression by at least 50% in 27 of the 35 (77.1%) antigen-positive cell populations. From a diagnostic viewpoint, the ability to enhance the expression of tumor antigens on cells isolated from patients serous effusions may be helpful in the accurate discrimination between malignant and benign effusions. In addition, analysis of serous effusions might be useful for the selection of the proper MAb and the determination of whether interferon treatment might increase MAb localization to the tumor site.

V. SUMMARY

Human interferons regulate the expression of a variety of different cellular genes. Some of these genes have been shown to be directly involved in mediating some interferon-induced biological actions, including antiviral activity, antiproliferation, and several immunomodulatory actions. Among the latter, we can include the ability of the interferons to regulate natural killer activity, the expression of class I and II MHC antigen, and, as presented here, the expression of certain human tumor antigens. Table 9 summarizes the studies that have shown which human tumor antigen can be upregulated by the administration of human interferons. Because some of the human tumor antigens, in particular CEA, have also functioned as cellular markers of differentiation, one can add the control of differentiation to the growing list of biological actions of the interferons. From a practical standpoint, the ability to alter the antigen phenotype of a human tumor cell population may have important implications in the utilization of MAbs as immunodiagnostic or immunotherapeutic agents. The evolving MAb technologies have or will produce new procedures for the conjugation of drugs or radionuclides, the production of second-generation MAb to purified tumor antigens, and the development of heteroantibodies and recombinant/chimeric antibodies. Concomitant with the evolution of these technologies must be a better understanding of the factors involved with cellular differentiation. It is believed that interferon or any other biological response modifier capable of increasing tumor antigen expression will serve as an important adjuvant in the future use of MAbs in cancer management.

REFERENCES

1. **Foulds, L.,** The experimental study of tumor progression: a review, *Cancer Res.,* 14, 327, 1954.
2. **Prehn, R. T.,** Analysis of antigenic heterogeneity within individual 3-methylcholanthrene-induced mouse sarcomas, *J. Natl. Cancer Inst.,* 45, 1039, 1970.
3. **Miller, F. R. and Heppner, G. H.,** Immunologic heterogeneity of tumor cell subpopulations from a single mouse mammary tumor, *J. Natl. Cancer Inst.,* 63, 1457, 1979.
4. **Horan-Hand, P., Nuti, M., Colcher, D., and Schlom, J.,** Definition of antigenic heterogeneity and modulation among human mammary carcinoma cell populations using monoclonal antibodies to tumor-associated antigen, *Cancer Res.,* 43, 728, 1983.
5. **Stramignoni, D., Bowen, R., Atkinson, B., and Schlom, J.,** Differential reactivity of monoclonal antibodies with human colon adenocarcinomas and adenomas, *Int. J. Cancer,* 31, 543, 1982.
6. **Horan-Hand, P., Colcher, D., Salomon, D., Ridge, J., Noguchi, P., and Schlom, J.,** Spatial configuration of carcinoma cell populations influences the expression of a tumor-associated glycoprotein, *Cancer Res.,* 45, 833, 1985.
7. **Greiner, J. W., Horan-Hand, P., Colcher, D., Weeks, M., Thor, A., Noguchi, P., Pestka, S., and Schlom, J.,** Modulation of human tumor antigen expression, *J. Lab. Clin. Med.,* 109, 244, 1987.
8. **Hager, J. C., Gold, D. V., Barbosa, J. A., Fligiel, Z., Miller, F., and Dexter, D. L.,** N,N-dimethylformamide-induced modulation of organ- and tumor-associated markers in cultured human colon carcinoma cells, *J. Natl. Cancer Inst.,* 64, 439, 1980.
9. **Friedman, J., Seger, M., Levinsky, H., and Allolouf, D.,** Modulation of carcinoembryonic antigen release by HT-29 colon carcinoma line in the presence of different agents, *Experientia,* 43, 1121, 1987.
10. **Niles, R. M., Wilhelm, S. A., Thomas, P., and Zamcheck, N.,** The effect of sodium butyrate and retinoic acid on growth and CEA production in a series of human colorectal tumor cell lines representing different states of differentiation, *Cancer Invest.,* 6, 39, 1988.
11. **Abe, M. and Kufe, D. W.,** Effects of sodium butyrate on human breast carcinoma (MCF-7) cellular proliferation, morphology and CEA production, *Breast Cancer Res. Treat.,* 4, 269, 1984.
12. **Tsao, D., Shi, Z.-R., Wong, A., and Kim, Y. S.,** Effect of sodium butyrate on carcinoembryonic antigen production by human colonic adenocarcinoma cells in culture, *Cancer Res.,* 43, 1217, 1983.
13. **Hwang, W. I., Sack, T. L., and Kim, Y. S.,** Effects of cyclic adenosine 3':5'-monophosphate upon glycoprotein and carcinoembryonic antigen synthesis and release by human colon cancer cells, *Cancer Res.,* 46, 3371, 1986.
14. **Chakrabarty, S., Tobon, A., Varani, J., and Brattain, M. G.,** Induction of carcinoembryonic antigen secretion and modulation of protein/expression and fibronectin/laminin expression in human colon carcinoma cells by transforming growth factor-β, *Cancer Res.,* 48, 4059, 1988.
15. **Attallah, A. M., Neely, C. F., Noguchi, P. D., and Elisberg, B. L.,** Enhancement of carcinoembryonic expression by interferon, *Int. J. Cancer,* 24, 49, 1979.
16. **Revel, M. and Chebath, J.,** Interferon-activated genes, *Trends Biochem. Sci.,* 11, 166, 1986.
17. **St. Laurent, G., Yoshie, O., Floyd-Smith, G., Samanta, H., and Sehgel, P. B.,** Interferon action: two $(2'-5')$ $(A)_n$ synthetases specified by distinct mRNAs in Ehrlich ascites tumor cells treated with interferon, *Cell,* 33, 95, 1983.
18. **Horisberger, M. A., Staehelin, P., and Haller, O.,** Interferon induces a unique protein in mouse cell bearing a gene for resistance to influenza virus, *Proc. Natl. Acad. Sci. U.S.A.,* 80, 1910, 1983.
19. **Wallach, D., Fellous, M., and Revel, M.,** Preferential effect of γ interferon on the synthesis of HLA antigens and their mRNAs in human cells, *Nature,* 299, 833, 1982.
20. **Fellous, M., Nir, U., Wallach, D., Merlin, G., Rubinstein, M., and Revel, M.,** Interferon-dependent induction of mRNA for the major histocompatibility antigens in human fibroblasts and lymphoblastoid cells, *Proc. Natl. Acad. Sci. U.S.A.,* 79, 3082, 1982.
21. **Collins, T., Kormon, A. J., Wuke, C. T., Bass, J. M., Kappas, D. J., Fiers, W., Ault, K. A., Gimbrone, M. A., Strominger, J. L., and Pobe, J. S.,** Immune interferon activates multiple class II major histocompatibility complex genes and the associated invariant chain gene in human endothelial cells and dermal fibroblasts, *Proc. Natl. Acad. Sci. U.S.A.,* 81, 4917, 1984.
22. **Friedman, R. L., Manly, S. P., McMahon, M., Kerr, I. M., and Stark, G. R.,** Transcriptional and posttranscriptional regulation of interferon-induced gene expression in human cells, *Cell,* 38, 745, 1984.
23. **Basham, T. Y., Bourgeade, M. F., Creasey, A. A., and Merigan, T. C.,** Interferon increase HLA synthesis in melanoma cells: interferon-resistant and -sensitive cell lines, *Proc. Natl. Acad. Sci. U.S.A.,* 79, 3265, 1982.
24. **Rosa, F. M. and Fellous, M.,** Regulation of HLA-DR gene by IFN-γ. Transcriptional and post-transcriptional control, *J. Immunol.,* 140, 1660, 1988.
25. **Schwartz, R. H.,** T-lymphocyte recognition of antigen in association with gene products of the major histocompatibility complex, *Annu. Rev. Immunol.,* 3, 237, 1985.

26. **Todd, J. A., Acha-Orbea, H., Bell, J. I., Chao, N., Fronek, Z., Jacob, C. D., McDermott, M., Sinha, A. A., Timmerman, L., Steinman, L., and McDevitt, H. D.,** A molecuar basis for MHC class II-associated autoimmunity, *Science,* 246, 1003, 1988.

27. **Gold, P. and Freedman, S. O.,** Demonstration of tumor specific antigens in human colonic carcinomata by immunologic tolerance and absorption techniques, *J. Exp. Med.,* 121, 439, 1965.

28. **Booth, S. N., King, J. P. G., Leonard, J. C., and Dyres, P. W.,** Serum carcinoembryonic antigen in clinical disorders, *Gut,* 14, 794, 1973.

29. **Zamcheck, N.,** The present status of CEA in diagnosis, prognosis, and evaluation of therapy, *Cancer,* 36, 2460, 1975.

30. **Wiggers, T., Arends, J. W., Vesstijnen, C., Moerkerk, P. M., and Bosman, F. T.,** Prognostic significance of CEA immunoreactivity patterns in large bowel carcinoma tissues, *Br. J. Cancer,* 54, 409, 1986.

31. **Sikorska, H., Shuster, J., and Gold, P.,** Clinical applications of carcinoembryonic antigen, *Cancer Detect. Prev.,* 12, 321, 1988.

32. **Leibovitz, A., Stinson, J. C., McCombs, W. B., III, Mayer, K. C., and Mabry, N. D.,** Classification of human colorectal adenocarcinoma cell lines, *Cancer Res.,* 36, 4562, 1976.

33. **Oikawa, S., Nakazato, H., and Kosaki, G.,** Primary structure of human carcinoembryonic antigen (CEA) induced from cDNA sequence, *Biochem. Biophys. Res. Commun.,* 142, 511, 1987.

34. **Zimmermann, W., Ortlieb, B., Friedrich, R., and von Kleist, S.,** Isolation and characterization of cDNA clones encoding the human carcinoembryonic antigen revealed a highly conserved repeating structure, *Proc. Natl. Acad. Sci. U.S.A.,* 84, 2960, 1987.

35. **Beauchemin, N., Benchimol, S., Cournoyer, D., Fuks, A., and Stanners, C.,** Isolation and characterization of full length functional cDNA clones for human carcinoembryonic antigen, *Mol. Cell. Biol.,* 7, 3221, 1987.

36. **Thompson, J. A., Pande, H., Paxton, R. J., Shively, L., Padma, A., Simmer, R. L., Todd, C. W., Riggs, A. D., and Shively, J. E.,** Molecular cloning of a gene belonging to the carcinoembryonic antigen gene family and discussion of a domain model, *Proc. Natl. Acad. Sci. U.S.A.,* 84, 2965, 1987.

37. **Oikawa, S., Imajo, S., Noguchi, T., Kosaki, G., and Nakazato, H.,** The carcinoembryonic antigen (CEA) contains multiple immunoglobulin like domains, *Biochem. Biophys. Res. Commun.,* 144, 634, 1987.

38. **Kamarck, M. E., Elting, J. J., Hart, J. T., Goebel, S. J., Rae, P. M. M., Nothdurft, M. A., Nedwin, J. J., and Barnett, T. R.,** The carcinoembryonic antigen family: expression in an L-cell transfectant and characterization of a partial cDNA in lambda gtll, *Proc. Natl. Acad. Sci. U.S.A.,* 84, 5350, 1987.

39. **Muraro, R., Wunderlich, D., Thor, A., Lundy, J., Noguchi, P., Cunningham, R., and Schlom, J.,** Definition of monoclonal antibodies of a repertoire of epitopes on carcinoembryonic antigen differentially expressed in human colon carcinomas versus normal adult tissues, *Cancer Res.,* 45, 5769, 1985.

40. **Kuroki, M., Greiner, J. W., Simpson, J. F., Primus, F. J., Guadagni, F., and Schlom, J.,** Serologic mapping and biochemical characterization of the carcinoembryonic antigen epitopes using fourteen distinct monoclonal antibodies, *Int. J. Cancer,* 44, 208, 1989.

41. **Zuckerman, S. H., Schreiber, R. D., and Marder, P.,** Reactivation of class II antigen expression in murine macrophage hybrids, *J. Immunol.,* 140, 978, 1987.

42. **Kantor, J., Tran, R., Greiner, J. W., Pestka, S., Fisher, P. B., Shively, J., and Schlom, J.,** Modulation of carcinoembryonic antigen mRNA levels in human colon carcinoma cells by recombinant human gamma interferon, *Cancer Res.,* 49, 2651, 1989.

43. **Larhammar, D., Gustafsson, K., Claesson, L., Bill, P., Wiman, K., Schenning, L., Sundelin, J., Widmark, E., Peterson, P., and Rask, L.,** Alpha chain of HLA-DR transplantation antigens is a member of the same protein superfamily as the immunoglobulins, *Cell,* 30, 153, 1982.

44. **Paxton, R. J., Mooser, G., Pande, H., Lee, T. D., and Shively, J. E.,** Sequence analysis of carcinoembryonic antigen: identification of glycosylation sites and homology with the immunoglobulin supergene family, *Proc. Natl. Acad. Sci. U.S.A.,* 84, 920, 1987.

45. **Friedman, R. L. and Stark, G. R.,** Alpha-interferon induced transcription of HLA and metallothionein genes containing homologous upstream sequences, *Nature,* 314, 637,1985.

46. **Larner, A. C., Chaudhuri, A., and Darnell, J. E., Jr.,** Transcriptional induction by interferon: new protein(s) determine the extent and length of induction, *J. Biol. Chem.,* 261, 453, 1986.

47. **Reich, N., Evans, B., Levy, D., Fahey, D., Knight, E., Jr., and Darnell, J. E., Jr.,** Interferon-induced transcription of a gene encoding a 15 kDa protein depends on an upstream enhancer element, *Proc. Natl. Acad. Sci. U.S.A.,* 84, 6394, 1987.

48. **Levy, D., Larner, A., Chaudhuri, A., Babiss, L. E., and Darnell, J. E., Jr.,** Interferon-stimulated transcription: isolation of an inducible gene and identification of its regulatory region, *Proc. Natl. Acad. Sci. U.S.A.,* 83, 8929, 1986.

49. **Luster, A. D. and Ravetch, J. V.,** Genomic characterization of a gamma-interferon inducible hypersensitive site, *Mol. Cell. Biol.,* 7, 3723, 1987.

50. **Zimmermann, W., Weber, B., Ortlieb, B., Rudert, F., Schempp, W., Fiebig, H.-H., Shively, J. E., von Kleist, S., and Thompson, J. A.,** Chromosomal localization of the carcinoembryonic antigen gene family and differential expression in various tumors, *Cancer Res.,* 48, 2550, 1988.

51. **Darnell, J. E., Jr.,** Variety in the level of gene control in eukaryotic cells, *Nature,* 297, 365, 1982.

52. **Rowlinson, G., Balkwill, F., Snook, D., Hooker, G., and Epenetos, A. A.,** Enhancement by γ-interferon of *in vivo* tumor radiolocalization by a monoclonal antibody against HLA-DR antigen, *Cancer Res.,* 46, 6413, 1986.

53. **Spear, G. T., Paulnock, D. M., Jordan, R., Hawkins, M. J., and Borden, E. C.,** Modulation of human peripheral blood monocyte HLA-DR expression by recombinant IFN-γ, *Proc. Am. Assoc. Cancer Res.,* 26, 77, 1985.

54. **Matsui, M., Nakaniski, T., Noguchi, T., and Ferrone, S.,** Synergistic *in vitro* and *in vivo* anti-tumor effect of daunomycin-anti-96-kDa melanoma-associated antigen monoclonal antibody CL207 conjugate and recombinant IFN-γ, *J. Immunol.,* 141, 1410, 1988.

55. **Greiner, J. W., Guadagni, F., Noguchi, P., Pestka, S., Colcher, D., Fisher, P. B., and Schlom, J.,** Recombinant interferon enhances monoclonal antibody targeting of carcinoma lesions, *in vivo, Science,* 235, 895, 1987.

56. **Guadagni, F., Schlom, J., Pothen, S., Pestka, S., and Greiner, J.,** Parameters involved in the enhancement of monoclonal antibody targeting *in vivo* with recombinant interferon, *Cancer Immunol. Immunother.,* 26, 222, 1988.

57. **Greiner, J. W., Horan-Hand, P., Noguchi, P., Fisher, P. B., Pestka, S., and Schlom, J.,** Enhanced expression of surface tumor-associated antigens on human breast and colon tumor cells after recombinant human leukocyte α-interferon treatment, *Cancer Res.,* 44, 3208, 1984.

58. **Colcher, D., Zalutsky, M., Kaplan, W., Kufe, D., Austin, F., and Schlom, J.,** Radiolocalization of human mammary tumors in athymic mice by a monoclonal antibody, *Cancer Res.,* 43, 736, 1983.

59. **Sahagan, B. G., Dorai, H., Saltigaber-Muller, J., Tonegazza, F., Guidon, C. A., Lilly, S. P., McDonald, K. W., Morrissey, D. V., Stone, B. A., Davis, G. L., McIntosh, P. K., and Moore, G. P.,** A genetically engineered murine/human chimeric antibody retains specificity for human tumor-associated antigen, *J. Immunol.,* 137, 1066, 1986.

60. **Colcher, D., Horan-Hand, P., Nuti, M., and Schlom, J.,** A spectrum of monoclonal antibodies reactive with human mammary tumor cells, *Proc. Natl. Acad. Sci. U.S.A.,* 78, 3199, 1981.

61. **Johnson, V. G., Schlom, J., Paterson, A. J., Bennett, J., Magnani, J. L., and Colcher, D.,** Analysis of a human tumor-associated glycoprotein (TAG-72) identified by monoclonal antibody B72.3, *Cancer Res.,* 46, 850, 1986.

62. **Thor, A., Ohuchi, N., Szpak, C. A., Johnston, W. W., and Schlom, J.,** Distribution of oncofetal antigen tumor-associated glycoprotein-72 defined by monoclonal antibody B72.3, *Cancer Res.,* 46, 3118, 1986.

63. **Thor, A., Viglione, M. J., Muraro, R., Ohuchi, N., Schlom, J., and Gorstein, F.,** Monoclonal antibody B72.3 reactivity with human endometrium: a study of normal and malignant tissues, *Int. J. Gynecol. Pathol.,* 6, 235, 1987.

64. **Sheer, D. G., Schlom, J., and Cooper, H. L.,** Purification and composition of the human tumor-associated glycoprotein (TAG-72) defined by monoclonal antibodies CC49 and B72.3, *Cancer Res.,* 48, 6811, 1988.

65. **Itzkowitz, S. H., Yuan, M., Montgomery, C. K., Kjeldsen, T., Takahashi, H. K., Bigbee, W. L., and Kim, Y. S.,** Expression of Tn, sialosyl-Tn and T-antigens in human colon cancer, *Cancer Res.,* 49, 197, 1989.

66. **Klug, T. L., Sattler, M. A., Colcher, D., and Schlom, J.,** Monoclonal antibody immunoradiometric assay for an antigenic determinant (CA 72) on a novel pancarcinoma antigen (TAG-72), *Int. J. Cancer,* 38, 661, 1986.

67. **Colcher, D., Esteban, J. E., Carrasquillo, J. A., Sugarbaker, P., Reynolds, J. C., Bryant, G., Larson, S. M., and Schlom, J.,** Quantitative analysis of selective radiolabeled monoclonal antibody localization in metastatic lesions of colorectal cancer patients, *Cancer Res.,* 47, 1185, 1987.

68. **Colcher, D., Esteban, J. E., Carrasquillo, J. A., Reynolds, J. C., Bryant, G., Larson, S. M., and Schlom, J.,** Complementation of intracavitary and intravenous administration of a monoclonal antibody (B72.3) in patients with carcinoma, *Cancer Res.,* 47, 4218, 1987.

69. **Johnston, W. W., Szpak, C. A., Lottich, S. C., Thor, A., and Schlom, J.,** Use of a monoclonal antibody (B72.3) as an immunocytochemical adjunct to diagnosis of adenocarcinoma in human effusions, *Cancer Res.,* 45, 1894, 1985.

70. **Szpak, C. A., Johnston, W. W., Roggli, U., Kolbeck, J., Lottich, S. C., Vollmer, R., Thor, A., and Schlom, J.,** The diagnostic distinction between malignant mesothelioma of the pleura and adenocarcinoma of the lung as defined by a monoclonal antibody (B72.3), *Am. J. Pathol.,* 122, 252, 1986.

71. **Guadagni, F., Schlom, J., Johnston, W. W., Szpak, C. A., Goldstein, D., Smalley, R., Simpson, J. F., Borden, E. C., Pestka, S., and Greiner, J. W.,** Selective interferon-induced enhancement of tumor-associated antigens on a spectrum of freshly isolated human adenocarcinoma cells, *J. Natl. Cancer Inst.,* 81, 502, 1989.

72. **Liao, S-K., Kwong, P. C., Khosravi, M., and Dent, P. B.**, Enhanced expression of melanoma-associated antigens and β_2-microglobulin on cultured human melanoma cells by interferon, *J. Natl. Cancer Inst.*, 68, 19, 1982.

73. **Herlyn, M., Guerry, D., and Koprowski, H.**, Recombinant γ-interferon induces changes in expression and shedding of antigens associated with normal human melanocytes, nevus cells and primary and metastatic melanoma cells, *J. Immunol.*, 134, 4226, 1984.

74. **Giacomini, P., Imberti, L., Aguzzi, A., Fisher, P. B., Trinchieri, G., and Ferrone, S.**, Immunochemical analysis of the modulation of human melanoma-associated antigens by DNA recombinant immune interferon, *J. Immunol.*, 135, 2887, 1985.

75. **Ziai, M. R., Imberti, L., Tongsen, A., and Ferrone, S.**, Differential modulation by recombinant immune interferon of the expression and shedding of HLA antigen and melanoma-associated antigens by a melanoma cell line resistant to the antiproliferative activity of immune interferon, *Cancer Res.*, 45, 5877, 1985.

76. **Gross, N., Beck, D., Favre, S., and Carrel, S.**, *In vitro* antigenic modulation of human neuroblastoma cells induced by IFN-γ, retinoic acid and dibutyryl cyclic AMP, *Int. J. Cancer*, 39, 521, 1987.

77. **Real, F. X., Carrato, A., Schuessler, M. H., Welt, S., and Oettgen, H. F.**, IFN-γ-regulated expression of a differentiation antigen of human cells, *J. Immunol.*, 140, 1571, 1988.

78. **Rosenblum, M. G., Lamki, L. M., Murray, J. L., Carlo, D. J., and Gutterman, J. U.**, Interferon-induced changes in pharmacokinetics and tumor uptake of ^{111}In-labeled antimelanoma antibody 96.5 in melanoma patients, *J. Natl. Cancer Inst.*, 80, 160, 1988.

79. **Tran, R., Horan-Hand, P., Greiner, J. W., Pestka, S., and Schlom, J.**, Enhancement of surface antigen expression on human breast carcinoma cells by recombinant interferons, *J. Interferon Res.*, 8, 75, 1988.

80. **Iacobelli, S., Scambin, G., Natali, C., Benedetti-Panici, P., Baiocchi, G., Perrone, L., and Mancuso, S.**, Recombinant human leukocyte interferon-α2b stimulates the synthesis and release of a 90K tumor-associated antigen in human breast cancer cells, *Int. J. Cancer*, 42, 182, 1988.

81. **Guadagni, F., Kantor, J., Schlom, J., and Greiner, J. W.**, unpublished observations.

Chapter 3

INDUCTION OF TERMINAL DIFFERENTIATION OF HUMAN LEUKEMIC CELLS BY CHEMOTHERAPEUTIC AND CYTOTOXIC DRUGS

Kenji Yamato and H. Phillip Koeffler

TABLE OF CONTENTS

I. INTRODUCTION

Leukemia is a malignant disease caused by transformation of hematopoietic progenitor cells at various stages of differentiation. Acute leukemia cells have lost the capacity to differentiate to normal functioning cells and they continue to proliferate in medullary and extramedullary hematopoietic tissues. This abnormal proliferation impairs the growth of normal functioning cells. The stage of differentiation at which leukemic cells are blocked determines the phenotype of the leukemic cells and the disease process. The mechanisms which uncouple proliferation and differentiation in these cells remain unclear. However, many physiological agents, including retinoic acid, 1,25 dihydroxyvitamin D_3, and a variety of cytokines, and nonphysiological agents such as drugs, have been shown to induce terminal differentiation of the acute leukemic cells of patients and cell lines. Most chemotherapeutic drugs have predominantly cytodestructive activity. However, some can induce differentiation of leukemic cells. Differentiation of these leukemic cells is associated with the loss of self-renewal capacity, leading to a reduction in leukemic cell mass. Research in identifying inducers of differentiation of leukemic cells may help us understand the fundamentals of differentiation, may provide clues to the mechanism of leukemogenesis, and may possibly be used as therapy.

II. MYELOID LEUKEMIA CELL LINES

Leukemic cell lines are essential tools for studying the cell differentiation of hematopoietic cells.[1] They are frozen at various stages of differentiation. KG-1 cells[2] are early myeloblasts that have the capacity to differentiate to macrophage-like cells in the presence of potent phorbol diesters such as *12-0-tetradecanoyl-phorbol 13-acetate* (TPA).[3] ML-1 and -3 are myelomonoblasts that can mature to macrophage-like cells in the presence of TPA, dimethylsulfoxide (DMSO), or cytosine arabinoside.[4] The best-studied cell line is HL-60, which is derived from patients with acute myeloblastic leukemia (M-2 subclass). The cells have many characteristics of promyelocytic cells.[5] In the presence of various agents, they can be induced to differentiate to granulocytes, monocytes/macrophages, and eosinophils.[6-8] U937 cells are monoblastic cells which also can differentiate toward monocytes/macrophages.[9] K562 cells were established from patients with chronic myelocytic leukemia in blastic crisis.[10] These cells have characteristics of several cell lineages, expressing both globins[11] and megakaryocytic antigens.[12] They can be induced to switch their expression of globin.[13]

Differentiation of leukemic cells can be determined by morphological examination of the cells after their fixation and staining with Giemsa. With macrophage-like differentiation, the cells develop strong nonspecific esterase (NE) positivity, which is a ubiquitous esterase that can be detected by using α-naphthol-acetate and -butyrate.[14,15] The production of superoxide is detected by the reduction of nitroblue tetrazolium dye (NBT). Differentiation of blast cells to granulocytes or monocytes induces their ability to produce superoxide, as measured by reduction of NBT.[16,17] On the other hand, blast cells treated with TPA show no reduction of NBT, probably because of exhaustion of the superoxide pathway in the presence of the phorbol ester.[18] Receptors for the Fc portion of IgG (FcR) and complement (C_3) are expressed on HL-60 cells and increase with differentiation.[19,20] Two subclasses of FcR are differentially expressed on HL-60 cells: high-avidity FcR (Type I) are detectable on wild-type HL-60 cells and their expression decreases with granulocytic differentiation; expression of low-avidity FcR (Type III) increases with maturation toward granulocytes.[21,22]

Production of hemoglobin can be detected by benzidine staining after induction of K562 cells.[23] Isoelectric focusing and/or RNA analysis determines the globin subtypes.

III. INDUCERS OF DIFFERENTIATION OF LEUKEMIC CELLS

A. ANTHRACYCLINES

Anthracyclines have a common structure of a planar anthraquinone nucleus attached to an amino sugar. Cytotoxic action of these drugs is probably through intercalation into the DNA molecule,[24,25] although some of the cytotoxic effect of the drug may be exerted through interaction with the cell-surface membrane.[26,27] Among them, marcellomycin, aclacinomycin, and musettamycin induce differentiation of HL-60 cells.[28,29] More than 70% of the HL-60 cells mature toward granulocytes after exposure to these anthracyclines (40 to 80 nM, 7 d). Other members of the anthracyclines, such as adriamycin and carminomycin, inhibit the growth of HL-60 cells more than the first three anthracyclines, but can only partially trigger HL-60 cells to mature. Adriamycin (70 nM) increases expression of Fc and C_3 receptors and lysozyme, but it is unable to stimulate superoxide production or morphological differentiation.[30] Paradoxically, exposure to adriamycin (40 nM, 3 d) triggers 40% of K562 cells to express easily detectable levels of hemoglobin, as measured by benzidine stain. Induction of differentiation of K562 cells by adriamycin differs from hemin, another inducer of K562.[31] Hemin causes partial switching of hemoglobin from Hb Grower 1 ($\zeta_2\epsilon_2$) and Hb Portland ($\gamma_2\zeta_2$) to Hb X ($\epsilon_2\gamma_2$) and Hb Grower 2 ($\alpha_2\epsilon_2$). On the other hand, adriamycin does not cause hemoglobin switching, but, rather, only increases the production of embryonic hemoglobins that are normally expressed by K562.

Anthracycline can be divided at least into two groups. The first, which includes marcellomycin, aclacinomycin, and mussetamycin, has disaccharide or trisaccharide chains and are exceedingly potent inhibitors of the synthesis of asparagine-linked glycoprotein.[29] The second group of anthracyclines, such as adriamycin and carminomycin, has monosaccharide chains and has no effect on glycoprotein synthesis.[29] The significance of inhibition of glycosylation in triggering cell differentiation is unknown. The following observations suggest that inhibition of glycoprotein synthesis may be an important trigger for differentiation: (1) exposure of HL-60 cells to tunicamycin, an inhibitor of N-glycosidically linked glycoprotein biosynthesis, results in cellular differentiation,[32] and (2) differentiation of leukemic cells[33-35] is associated with a change in the expression of surface glycoprotein. One of the interesting modifications during differentiation is a rapid and dramatic decrease in the expression of transferrin receptors which have asparagine-linked oligosaccharides.[36-38] Only the group of anthracyclines with di- and trisaccharide chains causes downregulation of these transferrin receptors. However, it is unclear whether this downregulation is necessary for differentiation or is the result of differentiation.

B. TUNICAMYCIN

Tunicamycin selectively inhibits the transfer of *N*-acetylglucosamine from UDP-*N*-acetylglucosamine to dolichol phosphate, and thereby inhibits the formation of *N*-acetylglucosamine pyrophosphoryl dolichol.[39-43] Since *N*-acetylglucosamine lipid participates in the assembly of the oligosaccharides of many glycoproteins, inhibition of this lipid complex results in blockade of protein glycosylation.[44] Tunicamycin (0.1 to 1 μ/ml) induces morphological and functional differentiation of HL-60 cells and murine M1 leukemic cells.[32] Tunicamycin markedly decreases synthesis of glycoprotein in these cells, and induction of differentiation is blocked by the presence of UDP-*N*-acetylglucosamine.[32] These observations again suggest a close relationship between cell differentiation and inhibition of protein glycosylation.

C. 6-THIOGUANINE

A purine antimetabolite, 6-thioguanine (6-TG) exerts its cytotoxic effect by conversion to 6-thioguanosine 5′-monophosphate (6-TGMP) using hypoxanthine-guanine phosphoribosyltransferase (HGPRT). 6-TGMP is incorporated into DNA and RNA. 6-TG is cytotoxic without inducing differentiation of Friend murine erythroleukemia cells and triggers slight to moderate differentiation of HL-60 cells.[45-47] Paradoxically, HGPRT-deficient subclones of HL-60 cells (HGPRT − /HL-60) and Friend leukemic cells differentiate toward granulocytes and red cells, respectively, after exposure to 6-TG.[45,46] 6-TG is not metabolized to 6-TGMP in these HGPRT-deficient cells, suggesting that this purine analogue, itself, may be the active agent that induces cellular differentiation. Furthermore, wild-type HL-60 cells can be induced to differentiate by 6-TG in the presence of compounds that inhibit formation of 6-TG-derived nucleotides, such as hypoxanthine, or its nucleotides, such as inosine and deoxyinosine.[48] These observations suggest that induction of differentiation and cytotoxicity are separable and caused by 6-TG and 6-TG-derived nucleotides, respectively.

How 6-TG induces cellular differentiation is unknown. 6-TG has been shown to deplete intracellular GTP by inhibiting *de novo* purine nucleotide biosynthesis and inosine monophosphate (IMP) dehydrogenase.[49] Other inhibitors of IMP dehydrogenase, such as tiazofurin[50,51] and 6-methylmercaptopurine ribonucleotide (6-MMCPR), also can reduce *de novo* purine nucleotide biosynthesis[52,53] and induce cell differentiation of Friend erythroleukemia and HL-60 cells. 6-MMCPR is not incorporated into cellular nucleotides, and induction of differentiation can be blocked by adenine, which repletes depressed intracellular purine nucleotide pools.[53] Taken together, these observations suggest that depletion of intracellular purine nucleotides may trigger cellular differentiation of leukemic cells.

D. 5-AZA-2′-DEOXYCYTIDINE AND 5-AZACYTIDINE

The compounds 5-aza-2′-deoxycytidine (5-aza-CdR) and 5-azacytidine (5-aza-CR), are analogues of 2′-deoxycytidine in which the 5th carbon of the pyrimidine ring is replaced by nitrogen. The compound, 5-aza-CdR, is incorporated into cellular DNA after phosphorylation.[54] Pretreatment of mice with 5-aza-CdR caused a marked inhibition of DNA synthesis, as measured by incorporation of radioactive thymidine.[55,56] Both compounds are potent inhibitors of DNA methyltransferases. *In vitro* differentiation can be induced by 5-aza-CdR in the Friend erythroleukemia cells,[57] acute monoblastic and myeloblastic leukemic cells, from some patients,[58] and HL-60 cells.[59] Differentiation of Friend cells, as well as HL-60[60] and also K562 cells[61] can be caused by 5-aza-CR.

In mammalian cells, approximately 5% of the cytosine (C) residues are modified by methylation at the 5th position of the cytidine ring.[62] The methylation occurs almost exclusively at the cytosine-guanosine (CpG) sequences to yield 5-methylcytosine-guanosine (mCpG). CpG sites in the vicinity of many genes are undermethylated in tissues where the gene is actively transcribed, compared to tissues where the gene is inactive.[63] Undermethylation may be either cause or consequence of the activation of the gene. The methylation is catalyzed by DNA methyltransferases. Exposure of Friend erythroleukemia cells to 5-aza-CdR inhibits their DNA methylation and the enzymatic activity of their DNA methyltransferase.[57] Induction of hemoglobin synthesis in K562 cells after exposure to 5-aza-CdR is associated with hypomethylation of CpG sequences surrounding the gamma globin gene.[61] Therefore, activation of some genes may be controlled in part through alteration of the methylation of CpG. Methylation of cytidine in the regulatory sequence of the gene could change the interaction between DNA and transcriptional factors, since the 5-methyl group protrudes into the major groove of the double helix where specific recognition of DNA and *trans*-acting protein take places. The addition of a 5-methyl group to cytidine also tends to shift the conformational equilibrium of DNA from the beta form to the zeta form; the change in the sugar backbone structure in the zeta form makes *trans*-acting protein unable to bind to specific bases.

E. CYTOSINE ARABINOSIDE

Cytosine arabinoside (Ara-C) is one of the most effective agents in the treatment of adult acute nonlymphocytic leukemia.[64] Ara-C is converted to ara-cytidine triphosphate, which is incorporated into DNA.[65,66] Incorporation of Ara-C stops DNA replication by behaving as a relative DNA chain terminator.[67] Termination of DNA synthesis causes fragmentation of DNA, leading to the loss of proliferating capacity of the leukemic cells.[68-71] HL-60 cells can mature to monocyte-like cells when cultured with Ara-C (100 nM to 1 μM).[72] The mechanisms of this induction of cellular differentiation remains unclear. Aphidocolin, another analog of deoxycytidine which inhibits DNA synthesis without incorporation into DNA (100 nM to 1 μM), can also trigger differentiation of HL-60 cells toward monocytic-like cells.[72] In addition, Ara-C (500 nM) as well as aphidocolin stimulates K562 cells to synthesize hemoglobin and lose self-renewing capacity.[73] These results suggest that inhibition of DNA synthesis may be closely associated with the induction of differentiation of these cells. The inhibition of DNA replication by Ara-C results in aberrant forms of DNA synthesis; in addition, certain segments of DNA may be replicated more than others, with a resultant alteration of gene expression.[74]

The level of Ara-C (50 nM) which allows incorporation into DNA and slows the replication of DNA of human leukemic cells *in vitro* can be achieved by administration of "low-dose" Ara-C to patients (10 to 20 mg/ml/m^2/d).[75] Low-dose Ara-C is used for the treatment of selected patients with acute nonlymphocytic leukemia.[76,77] Some patients treated with low-dose Ara-C develop hematologic remission, which could result from the ability of Ara-C to kill leukemic cells and possibly to induce their differentiation. Some leukemic patients who achieve and remain in a hematologic remission while receiving Ara-C have persistence of their leukemic clone.[78] Mature granulocytes have been detected in this leukemic clone, suggesting that some of the leukemic cells are differentiating to mature cells in the presence of Ara-C.

F. OTHER PURINE AND PYRIMIDINE ANALOGUES AND METHOTREXATE

5-Bromo-2'-deoxyuridine (5-BrdU) is a thymidine analogue which is incorporated into DNA. It triggers cellular differentiation, including HL-60 leukemic cells,[78-81] and attenuates differentiation of Friend erythroleukemia cells.[82,83] About 30 to 40% of the HL-60 cells mature when cultured with 5-BrdU (3.0 μg/ml) for 7 days.

Maturation of HL-60 cells induced by 5-BrdU correlates with the incorporation of 5-BrdU into the DNA of HL-60 cells.[81] Thymidine competitively inhibits both incorporation of 5-BrdU into DNA and maturation of the cells. Thymidine kinase-deficient HL-60 cells, which cannot phosphorylate thymidine, are unable to incorporate either thymidine or 5-BrdU and cannot be triggered by 5-BrdU to mature, even though they still retain the ability to differentiate in response to other inducers.[81] These results suggest that 5-BrdU may trigger the maturation of HL-60 cells through its incorporation into the DNA. In contrast, other studies suggest that the cell membrane may be the target of 5-BrdU which triggers differentiation.[84,85]

3-Deazauridine is an antipyrimidine which inhibits CTP synthetase and thus depletes cells of CTP and dCTP. It is a potent inducer of granulocytic maturation of HL-60 cells. A 6-d exposure to 3-deazauridine (25 μM) causes nearly 100% of the HL-60 cells to become more mature.[60] Pyrazofurin (1 μM), virazole (40 μM), puromycin aminonucleotide (17 μM) and tricyclic nucleotide 3-amino-1,5-dihydro-5-methyl-1-b-D-ribofuranosyl-1,4,5,6,8-pentaazaacenaphthylene (10 μM) are all nucleotide analogues that are moderately potent inducers of granulocytic differentiation.[60] 5-Iodo-2'-deoxyuridine (50 μM), thymidine (1 mM), and methotrexate (14 nM) are less potent inducers; they induce 30% or less of HL-60 cells to differentiate.[60]

G. ACTINOMYCIN D

Actinomycin D (Act D) is an antibiotic isolated from *Streptomyces,* which is a potent inducer of the differentiation of some myeloid leukemic cells.[30,86] HL-60 cells cultured in the presence of 8 nM Act D increase their expression of Fc and C_3' receptors and lysozyme, and 90% of the cells morphologically mature.[30] K562 erythroleukemia cells can be induced to produce more hemoglobin after culture with Act D (0.5 nM).[86]

Actinomycin D has an interesting structure of a phenoxazone ring and two cyclic polypeptides. The polypeptide rings of this antibiotic intercalate into deoxyguanosine of double-strand DNA.[88] This intercalation impairs DNA as a template for both DNA and RNA synthesis.[88] Act D can also cause single-strand DNA breaks.[89] At low drug concentrations, inhibition of RNA synthesis predominates, whereas at higher concentrations both RNA and DNA synthesis are affected.[88] The mechanism by which Act D effects differentiation of leukemic cells is not understood.

H. MITOMYCIN C AND BLEOMYCIN

Mitomycin C is an antitumor antibiotic causing alkylation of DNA with intra- and interstrand cross-linking, resulting in inhibition of DNA synthesis.[90] It partially induces differentiation of HL-60, with expression of Fc and C_3' receptors and synthesis of lysozyme, but the cells do not undergo morphological maturation.[30] In contrast, mitomycin C potently induces hemoglobin synthesis of K562, increasing benzidine positivity from 5 to 70% after culture with the drug.[91] Bleomycin is a small molecular-weight peptide that produces single- and double-strand breaks in DNA.[92] It also can induce K562 cells to increase the synthesis of hemoglobin.[91] The relationship between their effect on DNA and on cellular differentiation is unclear.

I. HARRINGTONINE

Harringtonine (HT) is a tree alkaloid isolated from *Cephalotaxis harringtonia.* HT and the related *Cephalotaxus* alkaloid, homoharringtonine, have been used in cancer treatment in China with favorable results in several neoplastic diseases, including leukemias.[93-95]

Harringtonine (30 ng/ml) arrests cellular division and triggers differentiation of HL-60.[96] The cells differentiate toward monocytes, becoming positive for nonspecific esterase (85% of cells), phagocytotic (60%), and adherent (25%). Occasionally, leukemic cells from patients with acute myelomonoblastic leukemia are also triggered to differentiate by HT. HT blocks initiation of protein synthesis and partially inhibits DNA synthesis. How HT induces differentiation is not understood.

J. DEFEROXAMINE

Deferoxamine (DFO) is a chelator of ferric iron and is widely used for decreasing iron overload in patients.[97] This drug also inhibits DNA synthesis in human T- and B-lymphocytes[98] and HL-60 cells.[99] DFO decreases intracellular iron, which is essential for the action of ribonucleotide reductase in mammalian cells.[98] Low levels of deoxyribonucleotide triphosphate, resulting from decreased activity of ribonucleotide reductase, prevents cells from undergoing DNA replication.[98,99]

DFO inhibits clonal growth of HL-60 cells and induces them to mature to monocyte-like cells.[99] The mechanism of induction of differentiation is not clear. A similar effect is observed after HL-60 cells are cultured with hydroxyurea, another inhibitor of ribonucleotide reductase.[99] Hydroxyurea is a partial inducer of the differentiation of HL-60 cells. It induces expression of receptors for Fc and C_3[30] and antigens specific for monocytic cells,[99] but has no effect on the morphology of HL-60 cells. These results suggest that the inhibition of ribonucleotide reductase may be associated in part with the partial induction of maturation of leukemic cells.

TABLE 1
Chemotherapeutic and Cytotoxic Compounds that Induce Leukemic Cell Differentiation

Compound	Treatment[a]	Cell line	Induced cell type[b]	Mature cells(%)	Ref.
Anthracyclines					
Aclacinomycin	80 nM, 7d	HL-60	G	70	28
Marcellomycin	40 nM, 7d	HL-60	G	95	28
Adriamycin	70 nM, 7d	HL-60	—	0[c]	30
	40 nM, 4d	K562	E	40	31
Purines, pyrimidines, and their analogues					
Cytosine arabinoside	500 nM, 3d	HL-60	M	80	72
	500 nM, 3d	K562	E	80	73
Aphidocolin	1 μM, 3d	HL-60	M	85	72
	50 μM, 5d	K562	E	50	73
6-Thioguanine	0.5 μM, 7d	HL-60	G	60	47
Tiazofurin	10 μM, 7d	HL-60	G	70	51
6-Mercaptopurine	2 mM, 5d	K562	E	30	93
6-Methylmercaptopurine ribonucleotide	3 μM, 7d	HL-60	G	50	53
5-Azacytidine	30 μM, 6d	HL-60	G	50	60
5-Aza-2′-deoxycytidine	200 nM, 2d	HL-60	G	30	59
	25 μM, 7d	K562	E	70	61
5-Iodo-2′-deoxyuridine	50 μM, 6d	HL-60	G	40	60
5-Bromo-2′-deoxyuridine	25 μM, 6d	HL-60	G	30	60, 81
Fluorouracil	10 μM, 5d	K562	E	60	93
Hypoxanthine	5 mM, 6d	HL-60	G	85	6
Other anticancer antibiotics					
Actinomycin D	8 nM, 8d	HL-60	G	85	30
	0.5 nM, 4d	K562	E	35	87
Mitomycin C	60 nM, 5d	HL-60	—	0[c]	30
	370 nM, 5d	K562	E	70	93
Bleomycin	2.4 μM, 5d	K562	E	65	93
Miscellaneous agents					
Tunicamycin	0.5 μg/ml, 6d	HL-60	G	50	32
Hydroxyurea	65 μM, 5d	HL-60	—	—[d]	30, 99
	40 μM, 5d	K562	E	20	73
Methotrexate	14 nM, 6d	HL-60	G	30	60
Harringtonine	100 ng/ml, 4d	HL-60	M	60	97
Deferoxamine	100 μM, 3d	HL-60	M	90[d]	99

[a] Leukemic cells were cultured for various days (d) in liquid culture, harvested, and analyzed for percent of mature cells.
[b] G = granulocyte-like cells; M = macrophage-like cells; E = erythroid cells.
[c] Differentiation was detected as an increase in expression of Fc and C_3 receptors and synthesis of lysozyme.
[d] Monoclonal antibody (MO1), which is reactive with differentiated monocytes/macrophages, was used for a marker of differentiation.[100]

IV. SUMMARY

This chapter provided a moderately comprehensive list of chemotherapeutic and cytotoxic drugs that can induce differentiation of human leukemic cells. An understanding of the molecular means by which these diverse groups of agents trigger differentiation must await a further understanding of normal hematopoietic differentiation, including the identification of positive and negative *trans*-regulatory proteins which can initiate expression of a large array of tissue-specific genes.

A word of caution is necessary in the clinical interpretation of data presented in this chapter. HL-60 is blocked at a fairly mature stage of development; many agents can trigger further maturation of these cells. Most other myeloid cell lines are arrested at an early stage of development and are refractory to many of the inducing agents that are active with HL-60. Furthermore, leukemia in patients is often arrested at a more immature stage than HL-60 and these may also be refractory to putative inducing agents.

REFERENCES

1. **Lubbert, M. and Koeffler, H. P.,.** Myeloid cell lines: tools for studying differentiation of normal and abnormal hematopoietic cells, *Blood Rev.*, 2, 121, 1988.
2. **Koeffler, H. P. and Golde, D. W.,** Acute myelogenous leukemia: a human cell line response to colony stimulating activity, *Science*, 200, 1153, 1978.
3. **Koeffler, H. P., Bar-Eli, M., and Territo, M. C.,** Phorbol ester effect on differentiation of human myeloid leukemic cell lines blocked at different stages of maturation, *Cancer Res.*, 40, 563, 1980.
4. **Terada, K., Minowada, J., and Bloch, A.,** Kinetics of appearance of differentiation associated characteristics in ML-1, a line of human myeloblastic cells, after treatment with 12-O-tetradecanoyl phorbol-13-acetate, dimethyl sulfoxide, or 1-b-D-arabinofuranosyl cytosine, *Cancer Res.*, 42, 5152, 1982.
5. **Collins, S. J., Gallo, R. C., and Gallagher, R. E.,** Continuous growth and differentiation of human myeloid leukemic cells in culture, *Nature*, 270, 347, 1977.
6. **Collins, S. J., Bodner, A., Ting, R., and Gallo, R. C.,** Induction of morphological and functional differentiation of human promyelocytic leukemic cells (HL-60) by compounds which induce differentiation of murine leukemic cells, *Int. J. Cancer*, 25, 213, 1980.
7. **Rovera, G., O'Briene, T. A., and Diamond, L.,** Induction of differentiation in human promyelocytic leukemia cells by tumor promoting agents, *Science*, 204, 1979.
8. **Fischkof, S. A., Pollak, A., Gleich, G. J., Testa, J. R., Misawa, S., and Reber, T. J.,** Eosinophilic differentiation of the human promyelocytic leukemia cell line HL-60, *J. Exp. Med.*, 160, 179, 1984.
9. **Sundstorm, C. and Nilsson, K.,** Establishment and characterization of a human histiocytic lymphoma cell line (U937), *Int. J. Cancer*, 17, 565, 1976.
10. **Lozzio, C. B. and Lozzio, B. B.,** Human chronic myelogenous leukemia cell line with positive Philadelphia chromosome, *Blood*, 45, 321, 1975.
11. **Andersson, L. L., Zoknell, M., and Gahnberg, C. G.,** Induction of erythroid differentiation in the human leukemia cell line K562, *Nature*, 278, 364, 1979.
12. **Tetterro, P. A. T., Massaro, E., Muller, A., Jan Gelder, R. S., and Jon Dem Borne, A. E. G. K.,** Megakaryocytic differentiation of proerythroblastic K562 cell line cells, *Leuk. Res.*, 8, 197, 1984.
13. **Cioe, L., McNab, A., Hubbell, M. R., Meo, P., Curtis, P., and Rovera, G.,** Differential expression of the globin genes in human leukemia K562 (S) cells induced to differentiate by hemin or butyric acid, *Cancer Res.*, 41, 237, 1981.
14. **Elias, L., Wogenrich, F., Wallace, J., and Longmire, J.,** Altered patterns of proliferation and differentiation of HL-60 human promyelocytic leukemia cells in the presence of leukocyte-conditioned medium, *Leuk. Res.*, 4, 301, 1980.
15. **Rovera, G., Santoli, D., and Damsky, C.,** Human promyelocytic leukemia cells in culture differentiate into macrophage-like cells when treated with phorbol diester, *Proc. Natl. Acad. Sci. U.S.A.*, 76, 2779, 1979.
16. **Collins, S., Ruscetti, F., Gallangher, R., and Gallo, R.,** Normal functional characteristic of cultured human promyelocytic leukemia cells after induction of differentiation by dimethylsulfoxide, *J. Exp. Med.*, 149, 669, 1979.

17. **Ollson, I., Olofsson, T., and Mauritzon, N.,** Characterization of mononuclear blood cells derived differentiation inducing factor for human promyelocytic leukemia cell line HL-60, *J. Natl. Cancer Inst.,* 67, 1225, 1981.
18. **Newburger, P., Baker, R., Hansen, S., Duncan, R., and Greenberger, J.,** Functionally deficient differentiation of HL-60 promyelocytic leukemia cells induced by phorbol myristate acetate, *Cancer Res.,* 41, 1861, 1981.
19. **Miyaura, C., Abe, E., Kuribayashi, T., Tanaka, H., Konno, K., Nishi, Y., and Suda, T.,** 1 alpha,25 dihydroxyvitamin D_3 induces differentiation of human myeloid leukemia cells, *Biochem. Biophys. Res. Commun.,* 102, 937, 1981.
20. **Harris, P., Ralph, P., Gabrilove, J., Welte, K., Karmali, R., and Moore, M.,** Distinct differentiation-inducing activities of gamma interferon and cytokine factors acting on the human promyelocytic leukemia cell in HL-60, *Cancer Res.,* 45, 3090, 1985.
21. **Fleit, H. B., Wright, S. D., Durie, C. J., Valinsky, J. E., and Unkeless, J. C.,** Ontogeny of Fc receptors and complement receptors (C_3) during human myeloid differentiation, *J. Clin. Invest.,* 73, 516, 1984.
22. **Unkeles, J. C.,** Function and heterogeneity of human Fc receptors for immunoglobulin G, *J. Clin. Invest.,* 83, 355, 1989.
23. **Orkin, S. H., Harosi, F. L., and Leder, P.,** Differentiation in erythroleukemia cells and their somatic hybrids, *Proc. Natl. Acad. Sci. U.S.A.,* 72, 98, 1975.
24. **DiMarco, A. and Arcamone, F.,** DNA complexing antibiotics: daunomycin and their derivatives, *Arzneim. Forsch.,* 25, 368, 1975.
25. **Manfait, M., Alix, A. J. P., Jeannesson, P., Jardillier, J. C., and Theophanides, T.,** Interaction of adriamycin with DNA as studied by resonance Raman spectroscopy, *Nucleic Acids Res.,* 10, 3803, 1982.
26. **Tokes, Z. A., Roger, K. E., and Rembaum, A.,** Synthesis of adriamycin-coupled polyglutalaldehyde microspheres and evaluation of their cytostatic activity, *Proc. Natl. Acad. Sci. U.S.A.,* 79, 2026, 1982.
27. **Tirtton, T. R. and Yee, G.,** The anticancer agent adriamyin can be actively cytotoxic without entering cells, *Science,* 217, 248, 1982.
28. **Schwartz, E. L., Ishiguro, K., and Sartorelli, A. C.,** Induction of leukemia cell differentiation by chemotherapeutic agents, *Adv. enzyme Regul.,* 21, 3, 1983.
29. **Morin, M. J. and Sartorelli, A. C.,** Induction of glycoprotein biosynthesis by inducers of HL-60 cell differentiation, aclacinomycin A and marcellomycin, *Cancer Res.,* 44, 2807, 1984.
30. **Loten, J. and Sachs, L.,** Potential pre-screening for therapeutic agents that induce differentiation in human myeloid leukemia cells, *Int. J. Cancer,* 25, 561, 1980.
31. **Jeanesson, P., Ginot, L., Manfait, M., and Jardillier, J. C.,** Induction of hemoglobin synthesis in human leukemic K562 cells by adriamycin, *Anticancer Res.,* 4, 47, 1984.
32. **Nakayasu, M., Terada, M., Tamura, G., and Sugimura, T.,** Induction of differentiation of human and murine myeloid leukemia cells in culture by tunicamycin, *Proc. Natl. Acad. Sci. U.S.A.,* 77, 409, 1980.
33. **Cossu, G., Kuo, L., Pessano, S., Warren, L., and Cooper, R. A.,** Decrease synthesis of high molecular weight glycoprotein in human promyelocytic leukemia cells (HL-60) during phorbol ester-induced macrophage differentiation, *Cancer Res.,* 43, 2754, 1983.
34. **Felsted, R. L., Gupta, S. K., Glover, C. J., Fischkoff, S. A., and Gallagher, R. E.,** Cell surface membrane protein changes during the differentiation of cultured human promyelocytic leukemia HL-60 cells, *Cancer Res.,* 43, 2754, 1983.
35. **Skubitz, K. M. and August, J. T.,** Analysis of cell-surface protein changes accompanying differentiation of HL-60 cells, *Arch. Biochem. Biophys.,* 266, 1, 1983.
36. **Tei, D., Makino, Y., Sakagami, H., Kanamara, I., and Kanno, K.,** Decrease of transferrin receptor during mouse myeloid leukemia (M1) cell differentiation, *Biochem. Biophys. Res. Commun.,* 107, 1419, 1982.
37. **Yeh, C. J., Papamichiel, M., and Foulk, W. P.,** Loss of transferrin receptors following induced differentiation of HL-60 promyelocytic cells, *Exp. Cell Res.,* 138, 429, 1982.
38. **Omary, M. B. and Trowbridge, I. S.,** Biosynthesis of the human transferrin receptor in culture cells, *J. Biol. Chem.,* 256, 12888, 1981.
39. **Tkacz, J. S. and Lampen, L. O.,** Tunicamycin inhibition of polyisoprenyl N-acetylglucosamyl phosphate formation in calf liver microsome, *Biochem. Biophys. Res. Commun.,* 65, 248, 1975.
40. **Ericson, M. C., Gafford, J. T., and Elbein, A. D.,** Tunicamycin inhibits GlcNAc-lipid formation in plants, *J. Biol. Chem.,* 252, 7431, 1977.
41. **Struck, D. K., Siuta, P. B., Lane, M. D., and Kornfield, S.,** Effect of tunicamycin on the secretion of serum protein by primary culture of rat and chick hepatocytes, *J. Biol. Chem.,* 253, 5332, 1978.
42. **Hickman, S., Kulczycki, A., Jr., Lynch, R. G., and Kornfield, S.,** Studies of the mechanism of tunicamycin inhibition of IgA and IgE secretion by plasma cells, *J. Biol. Chem.,* 252, 4402, 1977.
43. **Hart, G. W. and Lennarz, W. J.,** Effects of tunicamycin on the biosynthesis of glucosaminoglycans by embryonic chick cornea, *J. Biol. Chem.,* 253, 5795, 1978.

44. **Takatsuki, A. and Tamura, G.,** Effect of tunicamycin on the synthesis of macromolecules in cultures of chick embryo fibroblasts infected with newcastle disease virus, *J. Antibiot.,* 24, 785, 1971.

45. **Gusella, J. F. and Housman, D.,** Induction of erythroid differentiation *in vitro* by purine and purine analogs, *Cell,* 8, 263, 1976.

46. **Ishiguro, K., Schwartz, E. L., and Sortorelli, A. C.,** Characterization of the metabolic forms of 6-thioguanine responsible for cytotoxicity and induction of differentiation of HL-60 acute promyelocytic leukemia cells, *J. Cell. Physiol.,* 121, 383, 1984.

47. **Papac, R. J., Brown, A. E., Schwartz, E. L., and Sartorelli, A. C.,** Differentiation of human promyelocytic leukemia cells *in vitro* by 6-thioguanine, *Cancer Lett.,* 10, 33, 1980.

48. **Ishiguro, K. and Sartorelli, A. C.,** Enhancement of the differentiation-inducing property of 6-thioguanine by hypoxanthine and its nucleotides in HL-60 promyelocytic leukemia cells, *Cancer Res.,* 45, 91, 1985.

49. **Nelson, J. A., Carpenter, J. W., Rose, L. M., and Adamson, D. J.,** Mechanism of action of 6-thioguanine, 6-mercaptopurine, and 8-azaguanine, *Cancer Res.,* 35, 2872, 1975.

50. **Lucas, D. L., Webster, H. K., and Wright, D. G.,** Purine metabolism in myeloid precursor cells during maturation, study with the HL-60 cell line, *J. Clin. Invest.,* 72, 1889, 1983.

51. **Sokoloski, J. A., Blair, O. C., and Sartorelli, A. C.,** Alterations in glycoprotein synthesis and guanosine triphosphate levels associated with the differentiation of HL-60 leukemia cells produced by inhibitors of inosine 5'-phosphate dehydrogenase, *Cancer Res.,* 46, 2314, 1986.

52. **Lin, T. S., Cheng, J. C., Ishiguro, K., and Sartorelli, J. A.,** Purine and 8-substituted purine arabinofuranosyl and ribofuranosyl nucleotide derivatives as potential inducers of the differentiation of Friend erythroleukemia, *J. Med. Chem.,* 28, 1481, 1985.

53. **Socoloski, J. A. and Sartorelli, A. C.,** Inhibition of the synthesis of glycoprotein and induction of the differentiation of HL-60 promyelocytic leukemia cells by 6-methylmercaptopurine ribonucleotide, *Cancer Res.,* 47, 6283, 1987.

54. **Momparler, R. L., Vesely, J., Momparler, L. F., and Rivards, G. E.,** Synergistic action of 5-aza-2'-deoxycytidine and 3-deazauridine on L1210 leukemic cells and EMT 6 tumor cells, *Cancer Res.,* 39, 3822, 1979.

55. **Cihak, A. and Vesely, J.,** Depression of DNA synthesis in mouse spleen after treatment with 5-aza-2'-deoxycytidine, *J. Natl. Cancer Inst.,* 63, 1035, 1979.

56. **Cihak, A., Vesely, J., and Hynie, S.,** Transformation and metabolic effect of 5-aza-2'-deoxycytidine in mice, *Biochem. Pharmacol.,* 29, 2929, 1980.

57. **Creusot, F., Acs, G., and Christman, J. K.,** Inhibition of DNA methyltransferase and induction of Friend erythroleukemia cell differentiation by 5-azacytidine and 5-aza-2'-deoxycytidine, *J. Biol. Chem.,* 257, 2041, 1982.

58. **Pinto, A., Attadia, V., Fiusco, A., Firrara, F., Spada, O. A., and DiFiere, P.,** 5-Aza-2'-deoxycytidine induces terminal differentiation of leukemic blasts from patients with acute myeloid leukemia, *Blood,* 64, 922, 1984.

59. **Momparler, R. L., Bouchard, J., and Samson, J.,** Induction of differentiation and inhibition of methylation of HL-60 myeloid leukemic cells by 5-aza-2'-deoxycytidine, *Leuk. Res.,* 9, 1361, 1985.

60. **Bodner, A. J., Ting, R. C., and Gallo, R. C.,** Induction of differentiation of human promyelocytic leukemia cells (HL-60) by nucleotides and methotrexate, *J. Natl. Cancer Inst.,* 67, 1025, 1981.

61. **Gambari, R., Senno, L., Barbieri, R., Viola, L., Tripodi, M., Raschella, G., and Antoni, A.,** Human leukemia K562 cells: induction of erythroid differentiation by 5-azacytidine, *Cell Differ.,* 14, 87, 1984.

62. **Razin, A. and Riggs, A. D.,** DNA methylation and gene function, *Science,* 210, 604, 1980.

63. **Doerfler, W.,** DNA methylation and gene activity, *Annu. Rev. Biochem.,* 52, 93, 1983.

64. **Frei, E., Bichers, J. N., Hewlett, J. S., Lane, M., Leary, W. V., and Talley, R. W.,** Dose schedule and antitumor studies of arabinosyl cytidine, *Cancer Res.,* 29, 1325, 1969.

65. **Major, P., Egan, E. M., Beardsley, G., Miden, M. D., and Hufe, D. W.,** Lethality of human myeloblasts correlates with the incorporation of ara-C into DNA, *Proc. Natl. Acad. Sci. U.S.A.,* 78, 3235, 1981.

66. **Kufe, D., Spriggs, D., and Egan, E. M.,** Relationships between Ara-CTP pools, formation of (Ara-C) DNA and cytotoxicity of human leukemia cells, *Blood,* 64, 54, 1984.

67. **Major, P., Egan, E. M., Herrick, D., and Kufe, D. W.,** The effect of ara-C incorporation on DNA synthesis, *Biochem. Pharmacol.,* 31, 2937, 1982.

68. **Buick, R. N., Till, J. E., and McCulloch, E. A.,** Colony assay for proliferative blast cells circulating in myeloblastic leukemia, *Lancet,* 1, 862, 1977.

69. **Killman, S. A., Cronkite, E. P., Robertson, J. S., Fliedner, T. M., and Bond, V. P.,** Estimation of phases of the life cycle of leukemic cells from labeling human beings *in vivo* with tritiated thymidine, *Lab. Invest.,* 12, 671, 1962.

70. **Aye, M. T., Till, J. E., and McCulloch, E. A.,** Interacting populations affecting proliferation of leukemic cells in culture, *Blood,* 45, 485, 1975.

71. **Minden, M. D., Till, J. E., and McCulloch, E. A.,** Proliferative stage of blast cell progenitors in acute myeloblastic leukemia (AML), *Blood,* 52, 592, 1978.

72. **Griffin, J. M., Munroe, D., Majkor, P., and Kufe, D. W.,** Induction of differentiation of human myeloid leukemia cell by inhibitors of DNA synthesis, *Exp. Hematol.,* 10, 774, 1982.

73. **Luida-Deluca, C., Mitchell, T., Springs, D., and Kufe, D. W.,** Induction of terminal differentiation in human K562 erythroleukemia cells by arabinofuranosyl cytidine, *J. Clin. Invest.,* 74, 821, 1984.

74. **Woodcock, D., Fox, R., and Cooper, I.,** Evidence for new mechanism of cytotoxicity of 1-b-D arabinofuranosylcytidine, *Cancer Res.,* 39, 1416, 1979.

75. **Kufe, D. W., Griffin, J. D., and Spriggs, D. R.,** Cellular and clinical pharmacology of low-dose ara-C, *Semin. Oncol.,* 12 (Suppl.), 200, 1985.

76. **Griffin, J. D., Spring, D., Wisch, J. S., and Kufe, D. W.,** Treatment of preleukemic syndrome with continuous intravenous infusion of low-dose cytidine arabinoside, *J. Clin. Oncol.,* 3, 982, 1985.

77. **Wisch, J. S., Griffin, J. D., and Kufe, D. W.,** Response of pre-leukemic syndrome to continuous infusion of low-dose cytarabine, *N. Engl. J. Med.,* 309, 1599, 1983.

78. **Hossfeld, D. K., Weh, J. J., and Kleeberg, U. R.,** Low-dose cytarabine: chromosome finding suggesting its cytostatic as well as differentiating effect, *Leuk. Res.,* 9, 329, 1985.

79. **Davidsonm, R. L. and Kaufman, E. R.,** Deoxycytidine reverses the suppression of pigmentation caused by 5-Brd-Urd without changing the amount of 5-BrdUrd in DNA, *Cell,* 1, 923, 1977.

80. **Biswas, D., Abdullah, K., and Brennessel, B.,** On the mechanism of 6-bromodeoxyuridine induction of prolactin synthesis in rate pituitary tumor cells, *J. Cell Biol.,* 81, 1, 1979.

81. **Koeffler, H. P., Yen, J., and Charlson, J.,** The study of human myeloid differentiation using bromodeoxyuridine (BrdU), *J. Cell. Physiol.,* 116, 111, 1983.

82. **Bick, M. D.,** Bromodeoxyuridine: inhibitor of Friend leukemia cell induction, *Biochem. Biophys. Acta,* 475, 279, 1977.

83. **Ashmon, C. and Davidson, R.,.** Inhibition of Friend erythroleukemia cell differentiation by bromodeoxyuridine in DNA, *J. Cell. Physiol.,* 102, 45, 1980.

84. **Schurbert, D. and Jacob, F.,** 5-Bromodeoxyuridine-induced differentiation of neuroblastoma, *Proc. Natl. Acad. Sci. U.S.A.,* 67, 247, 1970.

85. **Brown, J. C.,** Surface glycoproteins characteristic of differentiated state of neuroblastoma C-1300 cells, *Exp. Cell Res.,* 69, 440, 1971.

86. **Okabe-Kado, J., Hayashi, M., Honma, Y., Hozumi, M., and Tsuruo, T.,** Effects of inducers of erythroid differentiation of human leukemia K562 cells on vincristine resistant K562/VCR cells, *Leuk. Res.,* 7, 1983.

87. **Sobell, H. M., Jain, S. C., and Sakere, T. D.,** Stereochemistry of actinomycin-DNA binding, *Nature New Biol.,* 231, 200, 1971.

88. **Reich, E., Franklin, R. E., Shatkin, A. J., and Tatum, E. L.,** Action of actinomycin-D on animal cells and viruses, *Proc. Natl. Acad. Sci. U.S.A.,* 48, 1238, 1962.

89. **Ross, W. E., Glaubiger, D. L., and Kohn, K. W.,** Quantitative, qualitative aspects of intercalator-induced DNA damage, *Biochem. Biophys. Acta,* 502, 41, 1979.

90. **Iyer, V. and Szybalski, W.,** Mitomycin and porfiromycin: chemical mechanism of activation and cross-linking of DNA, *Science,* 145, 55, 1964.

91. **Rowley, P. T., Ohlsson-Wilhelm, B. M., Farley, B. A., and La Bella, S.,** Inducers of erythroid differentiation in K562 human leukemia cells, *Exp. Hematol.,* 9, 32, 1981.

92. **Takeshita, M., Grollman, A. P., Ohtsubo, E., and Ohtsubo, H.,** Interaction of bleomycin with DNA, *Proc. Natl. Acad. Sci. U.S.A.,* 75, 5983, 1978.

93. Cephalotaxus Research Coordinating Group, Cephalotoxinesters in the treatment of acute leukemia, preliminary clinical assessment, *Clin. Med. J.,* 2, 263, 1976.

94. Chinese People's Liberation Army 187th Hospital, Homoharringtonine in acute leukemia: clinical analysis of 31 cases, *Clin. Med. J.,* 3, 319, 1976.

95. **Warrell, R. P., Jr., Coonley, C. J., and Gee, T. S.,** Homoharringtonine: an effective new drug for remission induction in refractory nonlymphoblastic leukemia, *J. Clin. Oncol.,* 3, 617, 1985.

96. **Boyd, A. W. and Sullivan, J. R.,** Leukemic cell differentiation in vivo and in vitro: arrest of proliferation parallels the differentiation induced by antileukemic drug harringtonine, *Blood,* 63, 384, 1984.

97. **Modell, B., Letsky, E. A., Flynn, D. M., Peto, R., and Weatherall, D. J.,** Survival and desferrioxamine in thalassemia major, *Br. Med. J.,* 284, 1081, 1982.

98. **Lederman, H. M., Cohen, A., Lee, J. W. W., Freeman, M. H., and Gelfand, E. W.,** Deferoxamine: a reversible S-phase inhibitor of human lymphocytic proliferation, *Blood,* 64, 784, 1984.

99. **Kaplinsky, C., Estrov, Z., Freedman, M. J., Gelfand, E. W., and Cohen, A.,** Effect of deferoxamine in DNA synthesis, DNA repair, cell proliferation and differentiation of HL-60 cells, *Leukemia,* 1, 437, 1987.

100. **Told, R., Griffin, J., Ritz, J., Nadler, L., Abrams, T., and Schlossman, S.,** Expression of normal monocyte-macrophage differentiation induced by leukocyte conditioned medium and phorbol diester, *Leuk. Res.,* 5, 491, 1981.

Chapter 4

INDUCED DIFFERENTIATION OF TRANSFORMED CELLS WITH POLAR/APOLAR COMPOUNDS AND THE REVERSIBILITY OF THE TRANSFORMED PHENOTYPE*

Richard A. Rifkind, Ronald Breslow, and Paul A. Marks

TABLE OF CONTENTS

* Original studies performed in the DeWitt Wallace Research Laboratories, Memorial Sloan-Kettering Cancer Center, Cornell University, NY, were supported in part by a grant from The National Cancer Institute (CA-31768 and CA-08748), and in the Department of Chemistry, Columbia University, NY, by Merck and Company.

I. INTRODUCTION

Considerable evidence supports the hypothesis that neoplastic transformation of cells in a developmental lineage does not necessarily abrogate the potential of cancer cells to display their normal developmental potential.[1-3] The most compelling clinical evidence to date derives from the demonstration[4,5] that clonal markers of transformed hematopoietic precursors (leukemic blast cells), including both chromosomal markers and restriction length polymorphisms, can be found in functionally differentiated blood cells during clinical remissions. A variety of environmental conditions are also recognized which can induce or permit transformed cells to express their developmental program. A broad spectrum of chemical agents with apparently diverse properties, including a family of relatively simple hybrid polar/apolar compounds, vitamin D derivatives, retinoids, proteases, inhibitors of nucleic acid synthesis, steroid hormones, tumor promoters, and growth factors, can induce a variety of transformed cell lines and even some primary human tumor explants to express their differentiated phenotypes, in part or in whole.[1-3,6-13] Murine teratocarcinoma cells, introduced into normal blastocysts and reimplanted in a pregnant mouse, can participate in normal, noncancerous histogenesis, generating chimeric progeny.[14] Study of these models of cancer cell differentiation can provide insight into the nature of the transformed phenotype, the mechanisms subverting normal differentiation in cancer cells, and the controls modulating the coordinated expression of genes implicated in the developmental process. Recent evidence, as well, suggests that insight may be obtained toward the design of novel therapeutic interventions in clinical cancer, based upon the induction of phenotypic changes and differentiation.

Although no single cell system satisfies all requirements for an experimental model of these phenomena, the murine erythroleukemia cell (MELC),[15] a rapidly proliferating, virus-transformed erythroid precursor which can be induced to display essentially normal erythroid differentiation, has provided one extremely useful experimental system. MELC are developmentally unipotential and do not provide a model for examining the factors governing developmental decisions during multilineage differentiation or early embryonic development. But they do permit studies directed at elucidating the mechanisms controlling coordinated gene expression leading to terminal stages of differentiation, the onset of terminal cell division, and cessation of proliferation. This review will emphasize our own studies and the current understanding of induced differentiation of MELC by polar/apolar compounds, but it will also examine studies with other cells which provide the basis for ongoing clinical trials and the prospects for designing compounds with improved clinical potential.

II. POLAR/APOLAR DIFFERENTIATION-INDUCING COMPOUNDS

Following the initial observations of Friend et al.[15] on the differentiation-inducing properties of dimethyl sulfoxide, our early studies identified a series of polar compounds which were more effective inducers of the differentiation of MELC and a number of other transformed cell lines.[16,17] Although the mechanism of action of these compounds was not, and is not yet, known, it was considered that if, at the unknown cellular target site of action, more than one polar, solvent-like molecular group could be brought to interact in the proper steric relationships, then more effective inducing compounds might be developed. From this concept has come some remarkably effective cytodifferentiating agents, the best studied of which to date is hexamethylene *bis*-acetamide (HMBA, Table 1, compound 1), consisting of two acetamide molecules linked at nitrogen by a six-carbon polymethylene chain.[16,18] Although the optimal apolar chain contains six methylenes, shorter chains with branching methyl sidechains are also effective — Table 1, compounds (cmpds) 3 through 5 — indicating the importance, as well, of the number of hydrophobic hydrocarbons.

TABLE 1
Some Polar Compounds Effective as Cytodifferentiating Agents

CPD #	Structure	Optimal Concentration (mM)	Benzidine Reactive Cells (%)	Commitment (%)
1.	$CH_3\text{-}C\text{-}N\text{-}(CH_2)_6\text{-}N\text{-}C\text{-}CH_3$ (O,H; H,O)	5	>95	>95
2.	$CH_3\text{-}N\text{-}C\text{-}CH_3$ (H, O)	50	~70	ND
3.	$CH_3\text{-}C\text{-}N\text{-}(CH_2)_2\text{-}C\text{-}(CH_2)_2\text{-}N\text{-}C\text{-}CH_3$ (O,H; CH_3,H; H,O)	5	>95	>90
4.	$CH_3\text{-}C\text{-}N\text{-}(CH_2)_3\text{-}C\text{-}CH_2\text{-}N\text{-}C\text{-}CH_3$ (O,H; CH_3,H; H,O)	5	>95	>90
5.	$CH_3\text{-}C\text{-}N\text{-}CH_2\text{-}C\text{-}C\text{-}CH_2\text{-}N\text{-}C\text{-}CH_3$ (O,H; H_3C,H; H,O; CH_3)	5	>95	ND
6.	$CH_3\text{-}N\text{-}C\text{-}(CH_2)_6\text{-}C\text{-}N\text{-}CH_3$ (H,O; O,H)	5	>95	>90
7.	$(CH_3)_2\text{-}N\text{-}C\text{-}(CH_2)_6\text{-}C\text{-}N\text{-}(CH_3)_2$ (H,O; O,H)	5	>90	>90
8.	$(CH_3)_2\text{-}N\text{-}C\text{-}(CH_2)_7\text{-}C\text{-}N\text{-}(CH_3)_2$ (H,O; O,H)	2	>90	>90
9.	$HN\text{-}(CH_2)_6\text{-}N\text{-}(CH_2)_6\text{-}NH$ with $C\text{-}CH_3$ (=O) on each N	1	>90	>90

Dimerization of acetamide by linkage through methyl groups also generates active compounds, such as suberic acid *bis-N*-methylamide (SBDA, Table 1, compound 6), which can be considered *N*-methylacetamide linked at the acetyl group by four methylene units. The inducing activity of SBDA is equivalent to that of HMBA, yet its catabolism is completely different, supporting the idea that the compounds themselves, rather than their catabolic products, are the principal active agents. Additional hybrid compounds have also been synthesized and tested for differentiation-inducing capacity. The most active of these to date is a dimer of HMBA. *Bis*-hexamethylene triacetamide (BHTA, Table 1, compound 9) is severalfold more active than HMBA, on a molar basis, and points the way toward developing other more effective agents.

Our present understanding suggests that, for optimal activity, two, or better yet, three uncharged polar groups of limited bulk must be connected by apolar chains of about six carbons. Ongoing studies are directed at evaluating whether compounds of this type will demonstrate improved properties in clinical tests (see below) of differentiation inducers for the control of cancer.

III. CHARACTERISTICS OF HMBA-MEDIATED MELC DIFFERENTIATION

MELC are virus-transformed hematopoietic precursors blocked, it would appear, at a developmental stage in the erythroid lineage approximating the colony-forming cell for erythropoiesis (CFU-E).[19,20] They display a number of characteristics which make them particularly useful for studying the coordinated pattern of modulated cellular functions and gene expression observed during terminal differentiation. They can be passaged and cloned, essentially indefinitely, retaining their potential to respond to inducers of differentiation. Variants blocked at apparently independent steps in the developmental process have been isolated.[21,22] Upon exposure to inducers, MELC initiate a developmental program similar in many essentials to that observed during normal erythropoiesis.[19]

When HMBA is added to MELC (line 745A-DS19, the standard inducible cell line in our laboratories, derived from Friend's original isolate), there ensues a latent period of 10 to 12 h (approximately one cell cycle for MELC) before commitment to terminal erythroid differentiation can be detected. Commitment is defined, in this system, as acquisition of the irreversible capacity to express the differentiated phenotype, including morphogenesis, biochemical changes (such as hemoglobin synthesis), and the cessation of cell division, despite removal of the inducer.[23,24] With continued exposure to HMBA, the latent period is followed by a period during which there is progressive recruitment of MELC to the committed state, which is complete by about 48 to 60 h, by which time over 95% of the cells are committed and expressing the phenotypic program of differentiation.

HMBA-mediated commitment to terminal differentiation is a multistep process.[25,26] It can be interrupted, at apparently independent steps, by exposure to the glucocorticoid, dexamethasone,[25,27] or the tumor promoter, 12-O-tetradecanoylphorbol-13-acetate (TPA).[28] During the latent period, HMBA initiates a number of changes, including altered membrane fluidity, changes in sodium, potassium, and calcium flux, and a transient increase in cyclic AMP concentration, as well as modulation in the expression of a number of genes, including c-*myb*, c-*myc*, c-*fos*, and the gene for the p53 protein.[1,29-35] A series of studies on the possible role of nuclear protooncogene expression, examining both mRNA and protein levels, suggest that persistent inducer-mediated suppression of c-*myb* gene expression, but not c-*myc*, is required for recruitment of MELC to terminal differentiation,[36-38] as, likewise, appears to be the case for normal erythropoiesis.[39]

Among the earliest events during the latent period initiated by exposure of MELC to HMBA is a rapid translocation of protein kinase C (PKC) activity to the plasma membrane, and the generation by proteolysis of a soluble protein kinase activity that is catalytically active in the absence of Ca^{2+} and phospholipid (PL).[39] Agents such as the tumor promoter TPA, which downregulates PKC activity, or protease inhibitors such as leupeptin, which blocks formation of the Ca^{2+}/PL-independent kinase activity, also block inducer-mediated suppression of c-*myb* and inhibit induction of erythroid differentiation in MELC.[39]

Following the latent period, during the period of progressive recruitment of MELC to differentiate, the most striking biochemical change is the accumulation of globin mRNA, its translation into globin, activation of the heme-synthetic enzyme pathway, and the accumulation of hemoglobin.[1,40] The primary regulatory mechanism engaged in this transition is at the transcriptional level.[41,42] Nevertheless, in uninduced MELC certain structural features of chromatin and DNA, in the globin gene domains, which are characteristic of active transcriptional regions, are already detected, and there is evidence of a low level of globin mRNA accumulation.[42-44] These include a pattern of hypomethylation of DNA about both the α1 and βmaj globin gene loci, a two- to threefold increase in the sensitivity of the globin gene domains to DNaseI digestion, compared with lineage-inactive genes, and some disruption of the nucleosome pattern.[42,43] HMBA initiates further changes in chromatin structure, including the appearance of specific DNaseI hypersensitivity sites, which appear to precede and to be prerequisite for the dramatic increase in globin gene transcription which characterizes differentiation.[41,42] Although further hypomethylation at the globin genes does not appear to be required for accelerated transcription,[43] a transient hypomethylation at other unspecified genetic loci does appear to be critical for the differentiation process.[45] This system has also proved uniquely valuable in defining and isolating several of the chromatin-binding proteins specifically interacting with the globin gene promoter regions.[46]

IV. MELC WITH INCREASED SENSITIVITY TO POLAR/ APOLAR AGENTS

It has already been pointed out that in the induction of the standard line of MELC there is a 10 to 12 h latent period before the onset of progressive commitment to terminal differentiation. The duration of this latent period is subject to modulation by a number of conditions. In the presence of HMBA, simultaneous exposure of MELC to dexamethasone blocks completion of the latent period, effectively prolonging it for as long as the cells are exposed to the steroid and HMBA (Figure 1).[25] Upon removal of the inhibiting steroid, in the continuous presence of HMBA, MELC rapidly (<1 h) display the fully committed phenotype. Recently, it has been observed that MELC which have developed a low level of resistance to vincristine (2 to 5 ng/ml) show a markedly increased sensitivity to the cytodifferentiating action of polar/apolar compounds, including HMBA.[47] In particular, these vincristine-resistant lines are induced to commitment with little or no latent period (Figure 1), are recruited more rapidly to differentiation and terminal cell division, and are sensitive to significantly lower concentrations of inducer.

The molecular mechanisms underlying this accelerated response to HMBA and the abrogation of the latent period is as yet unknown. The vincristine-resistant cells do share some, but not all, features of cells exhibiting the multidrug resistance phenotype (MDR).[48] Vincristine-resistant MELC display (1) decreased accumulation of ³H-vinblastine, compared to sensitive cells, (2) cross-resistance to colchicine (but not to adriamycin), and (3) resensitization to the cytotoxic effects of vincristine by exposure to verapamil.[49] Nevertheless, at levels of vincristine resistance sufficient to sensitize them to the effects of HMBA, these cells show no increase in either mdr mRNA or p-glycoprotein, characteristic of most tumor cell lines made resistant to these chemotherapeutic agents. We speculate that sensitivity to inducer reflects the constitutive expression in vincristine-resistant MELC of a factor (or factors) critical for completing the multistep process leading to the commitment to differentiate.

V. CLINICAL EXPERIENCE WITH HMBA IN THE TREATMENT OF CANCER

Evidence that HMBA (and related compounds) can overcome the developmental block characteristic of transformation in a number of different cell lines from experimental animals and humans, as well as a number of primary tumor cell explants *in vitro* (Table 2), has provided the basis for further exploration of the potential clinical role of polar/apolar differentiation-inducing agents. Toxicity of the agent has been examined in several phase I clinical trials.[50-57] These studies have revealed that dose escalation is limited by a number of reversible toxicities, the most significant of which is thrombocytopenia. In the most recent study, a 10-d continuous infusion, repeated every 3 to 4 weeks, achieved a maximum tolerated dose of about 20 g/m²/d and mean steady-state HMBA plasma concentrations ranging from 0.75 to 96 mM. Despite the limited duration of administration and the clearly suboptimal

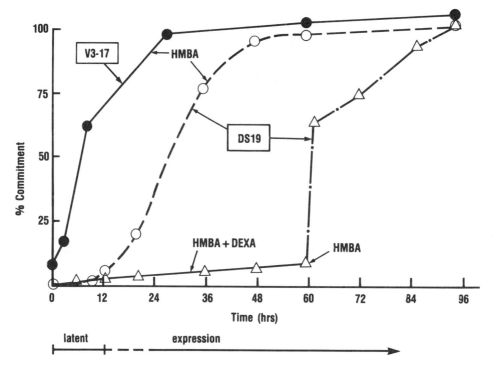

FIGURE 1.   Kinetics of HMBA-induced commitment of MELC (DS19 strain) to terminal differentiation (○——○). There is a latent period of about 12 h, followed by progressive recruitment of cells to commitment, approaching 100% by 48 to 60 h. In the presence of dexamethasone (DEXA), commitment is suppressed (△——△); when DEXA is removed (but HMBA remains), there is rapid expression of commitment in a large proportion of cells (△-·-·-△). Operationally, this can be viewed as a DEXA-mediated prolongation of the latent period. Contrariwise, vincristine-resistant MELC (V3.17 cells) are induced to commitment without a detectable latent period (●——●).

TABLE 2
Cell Lines Which Are Induced to Differentiate by HMBA or Related
Polar/Apolar Compounds[1]

Cell lines
  Mouse
    Erythroleukemia
    Teratocarcinoma
    Hepatic carcinoma
    Neuroblastoma
  Rat
    Mammary carcinoma
    LB myoblast
  Canine
    Kidney epithelial carcinoma
  Human
    HL-60 promyelocytic leukemia
    Melanoma
    Colon carcinoma
    Bladder carcinoma
    Glioblastoma multiforme
    Lymphoma
    Primary explants from colon and bladder carcinoma, melanoma, and acute myeloid leukemia

plasma concentration (5 mM is the consistently optimal level in *in vitro* experiments), partial objective responses, in terms of measurable tumor regression, were noted in 5 of 33 patients. The responding patients had metastatic cancers of breast, colon, and lung. Additional positive, but transient, hematologic responses have been observed in some patients with the preleukemic myelodysplastic syndrome.[57] Although intriguing, these results fall short of providing evidence for a predictable and sustained therapeutic benefit. Current strategies include the design and synthesis of agents with better therapeutic indices, and approaches to the augmentation of tumor cell sensitivity to differentiation inducers, following the lead provided by the observations on the HMBA sensitivity of vincristine-resistant MELC.

VI. SUMMARY

The development of agents which effectively induce transformed cells to initiate expression of a coordinated program of differentiation, including the termination of proliferation, has important implications both for an understanding of the molecular controls active in terminal differentiation and for the treatment of cancer. Among the cytodifferentiation-inducing agents, the hybrid polar/apolar compound, HMBA, is one of the best characterized with respect to its activity on a number of transformed cell lines, as well as in the clinical setting. Clearly, HMBA is not yet the agent which will satisfy the requirements for a truly effective clinical test of cytodifferentiation as an anticancer strategy. Newer agents, related structurally to HMBA, have been identified which are as active as or more so than HMBA, and whose structures suggest that they will have quite different pharmacokinetics. The fact that vincristine-resistant cells are strikingly more sensitive to this class of compounds provides yet another approach to improving the potential for a clinical effect. Only further study will determine whether this class of relatively simple multifunctional compounds can be made therapeutically useful, and whether they can help elucidate the nature of the complex interaction between malignant transformation and differentiation.

REFERENCES

1. **Marks, P. A., Sheffery, M., and Rifkind, R. A.,** Induction of transformed cells to terminal differentiation and the modulation of gene expression, *Cancer Res.,* 47, 659, 1987.
2. **Sporn, M. B., Roberts, A. B., and Driscoll, J. S.,** Principles of cancer biology: growth factor and differentiation, in *Cancer: Principles and Practice of Oncology,* Hellman, S., Rosenberg, S. A., and DeVita, V. T., Jr., Eds., Lippincott, Philadelphia, 1985, 49.
3. **Sachs, L.,** Control of normal cell differentiation and the phenotypic reversion of malignancy in myeloid leukaemia, *Nature,* 274, 535, 1978.
4. **Jacobson, R. J., Temple, M. J., Singer, J. W., Raskind, W., Powell, J., and Fialkow, P. J.,** A clonal complete remission in a patient with acute nonlymphocytic leukemia originating in a multipotent stem cell, *N. Engl. J. Med.,* 310, 1513, 1984.
5. **Fearon, E. R., Burke, P. J., Schiffer, C. A., Zehnbauer, B. A., and Vogelstein, B.,** Differentiation of leukemia cells to polymorphonuclear leukocytes in patients with acute nonlymphocytic leukemia, *N. Engl. J. Med.,* 315, 15, 1986.
6. **Marks, P. A.,** Genetics, cell differentiation and cancer, in *Genetics, Cell Differentiation and Cancer,* Bristol-Myers Cancer Symposium, Academic Press, New York, 1985, 209.
7. **Sartorelli, A. C., Moran, M. J., and Ishiguro, K.,** Cancer chemotherapeutic agents as inducers of leukemia cell differentiation, in *Bristol-Myers Cancer Symposia,* Academic Press, New York, 1987, 205.
8. **Francis, G. E., Guimaraes, J. E., Berney, J. J., and Wing, M. A.,** Differentiation in myelodysplastic, myeloid leukaemic and normal hemopoietic cells: a new approach exploiting the synergistic interaction between differentiation inducers and DNA synthesis inhibitors, in *Modern Trends in Human Leukemia, Vol. 6, Haematology and Blood Transfusion 29,* Gallo, R., Greaves, and Janks, Eds., Springer-Verlag, Berlin, 1985, 402.

9. **Nilsson, K., Ivhed, I., and Forsbeck, K.,** Induced differentiation in human malignant hematopoietic cell lines, in *Gene Expression During Normal and Malignant Differentiation,* Andersson, L. C., Gahmberg, C. C., and Eteblom, P., Eds., Academic Press, New York, 1985, 57.

10. **Hozumi, M.,** Established leukemia cell lines: their role in the understanding and control of leukemia proliferation, *CRC Crit. Rev. Oncol. Hematol.,* 3, 235, 1985.

11. **Block, A.,** Induced cell differentiation in cancer therapy, *Cancer Treat. Rep.,* 68, 199, 1984.

12. **Koeffler, H. P.,** Induction of differentiation of human acute myelogenoces leukemia cells: therapeutic implications, *Blood,* 4, 709, 1983.

13. **Metcalf, D.,** The granulocyte-macrophage colony stimulating factors, *Science,* 229, 16, 1985.

14. **Mintz, B. and Fleischman, R. A.,** Teratocarcinomas and neoplasms as developmental defects in gene expression, *Cancer Res.,* 34, 211, 1981.

15. **Friend, C., Scher, W., Holland, J., and Sato, T.,** Hemoglobin synthesis in murine erythroleukemia cells *in vitro:* stimulation of erythroid differentiation by dimethylsulfoxide, *Proc. Natl. Acad. Sci. U.S.A.,* 68, 378, 1971.

16. **Reuben, R. C., Wife, R. L., Breslow, R., Rifkind, R. A., and Marks, P. A.,** A new group of potent inducers of differentiation in murine erythroleukemia cells, *Proc. Natl. Acad. Sci. U.S.A.,* 73, 862, 1976.

17. **Tanaka, M., Levy, J., Terada, M., Breslow, R., Rifkind, R. A., and Marks, P. A.,** Induction of erythroid differentiation in murine virus infected erythroleukemia cells by highly polar compounds, *Proc. Natl. Acad. Sci. U.S.A.,* 72, 1003, 1975.

18. **Reuben, R., Khanna, P. L., Gazitt, Y., Breslow, R., Rifkind, R. A., and Marks, P. A.,** Inducers of erythroleukemic differentiation: relationship of structure to activity among planar-polar compounds, *J. Biol. Chem.,* 253, 4214, 1978.

19. **Marks, P. A. and Rifkind, R. A.,** Erythroleukemic differentiation, *Annu. Rev. Biochem.,* 47, 419, 1978.

20. **Tsiftsoglou, A. S. and Robinson, S. H.,** Differentiation of leukemic cell lines: a review focusing on murine erythroleukemia and human HL-60 cells, *Int. J. Cell Cloning,* 3, 349, 1985.

21. **Ohta, Y., Tanaka, M., Terada, M., Miller, O. J., Bank, A., Marks, P. A., and Rifkind, R. A.,** Erythroid cell differentiation: murine erythroleukemia cell variant with unique pattern of induction by polar compounds, *Proc. Natl. Acad. Sci. U.S.A.,* 73, 1232, 1976.

22. **Marks, P. A., Chen, Z. X., Banks, J., and Rifkind, R. A.,** Erythroleukemia cells: variants inducible for hemoglobin synthesis without commitment to terminal cell division, *Proc. Natl. Acad. Sci. U.S.A.,* 80, 2281, 1983.

23. **Fibach, E., Reuben, R. C., RIfkind, R. A., and Marks, P. A.,** Effect of hexamethylene bisacetamide on the commitment to differentiation of murine erythroleukemia cells, *Cancer Res.,* 37, 440, 1977.

24. **Gusella, J. F., Geller, R., Clarke, B., Weeks, V., and Housman, D.,** Commitment to erythroid differentiation by Friend erythroleukemia cells: a stochastic analysis, *Cell,* 9, 221, 1976.

25. **Chen, Z. X., Banks, J., Rifkind, R. A., and Marks, P. A.,** Inducer-mediated commitment of murine erythroleukemia cells to differentiation: a multistep process, *Proc. Natl. Acad. Sci. U.S.A.,* 79, 471, 1982.

26. **Murate, T., Kaneda, T., Rifkind, R. A., and Marks, P. A.,** Inducer-mediated commitment of murine erythroleukemia cells to terminal division: the expression of commitment, *Proc. Natl. Acad. Sci. U.S.A.,* 81, 3394, 1984.

27. **Kaneda, T., Murate, T., Sheffery, M., Brown, K., Rifkind, R. A., and Marks, P. A.,** Gene expression during terminal differentiation: dexamethasone suppression of inducer-mediated α_1- and β^{maj}-globin gene expression, *Proc. Natl. Acad. Sci. U.S.A.,* 82, 5020, 1985.

28. **Yamasaki, H., Fibach, E., Nudel, U., Weinstein, I. B., Rifkind, R. A., and Marks, P. A.,** Tumor promoters inhibit spontaneous and induced differentiation of murine erythroleukemia cells in culture, *Proc. Natl. Acad. Sci. U.S.A.,* 74, 3451, 1977.

29. **Mager, D. and Bernstein, A.,** The program of Friend cell erythroid differentiation: early changes in Na^+/K^+ATPase function, *J. Supermol. Struct.,* 8, 431, 1978.

30. **Cantley, L., Bernstein, A., Hunt, D. M., Crichley, V., and Mak, T. W.,** Induction by ouabain of hemoglobin synthesis in cultured Friend erythroleukemic cells, *Cell,* 9, 375, 1976.

31. **Bridges, K., Levenson, R., Housman, D., and Cantley, L.,** Calcium regulates commitment of murine erythroleukemia cells to terminal erythroid differentiation, *J. Cell Biol.,* 90, 542, 1981.

32. **Lyman, G., Papahajopoulos, D., and Preisler, H.,** Phospholipid membrane stabilization by dimethylsulfoxide and other inducers of Friend leukemic cell differentiation, *Biochem. Biophys. Acta,* 448, 460, 1976.

33. **Muller, C. P., Volloch, Z., and Shinitzky, M.,** Correlation between cell density, membrane fluidity and the availability of transferring receptors in Friend erythroleukemia cells, *Cell Biophys.,* 2, 233, 1980.

34. **Gazitt, Y., Deitch, A. D., Marks, P. A., and Rifkind, R. A.,** Cell volume changes in relation to the cell cycle of differentiating erythroleukemic cells, *Exp. Cell Res.,* 117, 413, 1978.

35. **Gazitt, Y., Reuben, R. C., Deitch, A. D., Marks, P. A., and Rifkind, R. A.,** Changes on cyclic adenosine 3^1:5^1 — monophosphate levels during induction of differentiation in murine erythroleukemic cells, *Cancer Res.,* 38, 3779, 1978.

36. **Ramsay, R. G., Ikeda, K., Rifkind, R. A., and Marks, P. A.,** Changes in gene expression associated with induced differentiation of erythroleukemia: proto-oncogenes, globin genes and cell division, *Proc. Natl. Acad. Sci. U.S.A.,* 83, 6849, 1986.

37. **Richon, V., Ramsay, R. G., Rifkind, R. A., and Marks, P. A.,** Modulation of the c-myb, c-myc, and p53 mRNA and protein levels during induced murine erythroleukemia cell differentiation, *Oncogenes,* 4, 165, 1989.

38. **Clarke, M. F., Kukowska-Latallo, J. F., Westin, E., Smith, M., and Prochownik, E. V.,** Constitutive expression of a c-myb cDNA blocks Friend murine erythroleukemia cell differentiation, *Mol. Cell. Biol.,* 8, 884, 1988.

39. **Todokoro, K., Watson, R. J., Higo, H., Amanuma, H., Kuramochi, S., Yanagisawa, H., and Ikawa, Y.,** Down-regulation of c-myb gene expression is a prerequisite for erythropoietin-induced erythroid differentiation, *Proc. Natl. Acad. Sci. U.S.A.,* 85, 8900, 1988.

40. **Sassa, S.,** Sequential induction of heme pathway enzymes during erythroid differentiation of mouse Friend erythroleukemia cells in culture, *J. Exp. Med.,* 143, 305, 1976.

41. **Salditt-Georgieff, M., Sheffery, M., Krauter, K., Darnell, J. E., Rifkind, R. A., and Marks, P. A.,** Induced transcription of the mouse β-globin transcription unit in erythroleukemia cells: time course of induction and changes in chromatin structure, *J. Mol. Biol.,* 172, 437, 1984.

42. **Cohen, R. B. and Sheffery, M.,** Nucleosome disruption precedes transcription and is largely limited to the transcribed domain of globin genes in murine erythroleukemia cells, *J. Mol. Biol.,* 182, 109, 1985.

43. **Sheffery, M., Rifkind, R. A., and Marks, P. A.,** Murine erythroleukemia cell differentiation: DNase I hypersensitivity and DNA methylation near the globin genes, *Proc. Natl. Acad. Sci. U.S.A.,* 79, 1180, 1982.

44. **Weich, N., Marks, P. A., and Rifkind, R. A.,** Regulation of murine α-, β^{maj}-, and β^{min}-globin gene expression, *Biochem. Biophys. Res. Commun.,* 150, 204, 1988.

45. **Razin, A., Levine, A., Kafri, T., Agostini, S., Gomi, T., and Cantoni, G. L.,** Relationship between transient DNA hypomethylation and erythroid differentiation of murine erythroleukemia cells, *Proc. Natl. Acad. Sci. U.S.A.,* 85, 9003, 1988.

46. **Barnhart, K. M., Kim, C. G., Banerji, S. S., and Sheffery, M.,** Identification and characterization of multiple erythroid cell proteins that interact with the promoter of the murine α-globin gene, *Mol. Cell. Biol.,* 8, 3215, 1988.

47. **Melloni, E., Pontremoli, S., Damiani, G., Viotti, P., Weich, N., Rifkind, R. A., and Marks, P. A.,** Vincristine-resistant erythroleukemia cells have marked increased sensitivity to hexamethylene bisacetamide induced differentiation, *Proc. Natl. Acad. Sci. U.S.A.,* 85, 3835, 1988.

48. **Ling, V. and Riordan, J. R.,** Genetic and biochemical characterization of multidrug resistance, *Pharmacol. Ther.,* 28, 51, 1985.

49. **Michaeli, J., Chen, Z. X., Marks, P. A., and Rifkind, R. A.,** unpublished data, 1989.

50. **Egorin, M. J., Sigman, L. M., VanEcho, D. A., Forrest, A., Whitacre, M. Y., and Aisner, J.,** Phase I clinical and pharmacokinetic study of hexamethylene bisacetamide (NSC 95580) administered as a five-day continuous infusion, *Cancer Res.,* 47, 617, 1987.

51. **Rowinsky, E. W., Ettinger, D. S., Grochow, L. B., Brundrett, R. B., Cates, A. E., and Donehower, R. C.,** Phase I and pharmacologic study of hexamethylene bisacetamide (HMBA) in patients with advanced cancer, *J. Clin. Oncol.,* 4, 1835, 1986.

52. **Rowinsky, E. K., McGuire, W. P., Anhalt, G. J., Ettinger, D. S., and Donehower, R. C.,** Hexamethylene bisacetamide-induced cutaneous vasculitis, *Cancer Treat. Rep.,* 71, 471, 1987.

53. **Rowingsky, E. L., Ettinger, D. S., McGuire, W. P., Noe, D. A., Grochow, L. B., and Donehower, R. C.,** Prolonged infusion of hexamethylene bisacetamide: a phase I and pharmacological study, *Cancer Res.,* 47, 5788, 1987.

54. **Egorin, M. J., Snyder, S. W., Cohen, A. S., Zuhowski, E. G., Subramanyam, B., and Callery, P. S.,** Metabolism of hexamethylene bisacetamide and its metabolites in leukemic cells, *Cancer Res.,* 48, 1712, 1988.

55. **Callery, P. S., Egorin, M. J., Geelhaar, L. A., and Nayar, M. S. B.,** Identification of metabolites of the cell-differentiating agent hexamethylene bisacetamide in humans, *Cancer Res.,* 46, 4900, 1986.

56. **Young, C. W., Fanucchi, M. P., Walsh, T. B., Blatzer, L., Yaldaei, S., Stevens, Y. W., Gordon, C., Tong, W., Rifkind, R. A., and Marks, P. A.,** Phase I trial and clinical pharmacologic evaluation of hexamethylene bisacetamide by 10-day continuous infusion at 28 day intervals, *Cancer Res.,* 48, 7304, 1988.

57. **Andreeff, M., Young, C., Clarkson, B., Fetten, J., Rifkind, R. A., and Marks, P. A.,** Treatment of myelodysplastic syndromes (MDS) with hexamethylene bisacetamide, *Blood,* 72 (Abstr.), 186a, 1988.

Chapter 5

5-AZACYTIDINE

Peter A. Jones

TABLE OF CONTENTS

I. INTRODUCTION

The nucleoside analogues, 5-azacytidine (5-Aza-CR) and 5-aza-2'-deoxycytidine (5-Aza-CdR), were first synthesized in Czechoslovakia in 1964.[1,2] These agents were initially demonstrated to have activity as antileukemic compounds in mice,[3] and early studies also demonstrated that 5-aza-CR might be useful as a chemotherapeutic agent in patients with acute myelogenous leukemia.[4] However, much of the current interest in the analogs has focused on their abilities to function as powerful inhibitors of DNA methylation in prokaryotic and eukaryotic cells. They also have the remarkable property of inducing dramatic heritable changes in cellular phenotype, and this has stimulated their use in systems to understand the molecular control of cellular differentiation. The drugs have, therefore, been useful in understanding the relationship between DNA methylation and gene expression.

II. METABOLISM OF AZANUCLEOSIDES

The metabolism of 5-aza-CR and 5-aza-CdR have been extensively reviewed in the past.[5,6] The initial metabolism of the two compounds is quite different. 5-Aza-CR is phosphorylated by the uridine-cytidine kinase enzyme to the monophosphate and can subsequently enter both the DNA and RNA of treated cells. In contrast, 5-aza-CdR is metabolized by deoxycytidine kinase and enters DNA, but not RNA. The s-triazine rings of 5-aza-CR and 5-aza-CdR are relatively unstable in aqueous solutions and decompose, with half-lives of the order of 12 h, with the loss of a formyl group to form guanylurea derivatives.[7,8] The drugs can also be inactivated intracellularly by cytidine deaminase and deoxycytidylate deaminase, respectively.[9,10] Cells differ in their sensitivities to these drugs, based primarily upon the levels of enzymes responsible for their primary activation[11] and by the activity of deaminating enzymes which form nontoxic azauridine derivatives.

Several studies have demonstrated that both analogs are incorporated into DNA following metabolism to 5-aza-2'-deoxycytidine triphosphate (5-aza-dCTP).[5,6] 5-Aza-CdR does not inhibit DNA replication, in contrast to analogs such as cytosine arabinoside.[12] Bouchard and Momparler[13] showed that purified DNA polymerase alpha extracted from calf thymus efficiently catalyzed the incorporation of 5-aza-dCTP into DNA. The Km for 5-aza-dCTP was 3 μM, which compared well with the value of 2 μM established for dCTP, and detailed kinetic analysis showed that 5-aza-dCTP was almost as good as dCTP as a substrate for the polymerase. It has also been reported that incorporation of 5-aza-2-deoxycytidine into DNA does not inhibit the rate of chain elongation.[14] Cultured cells are quite effective in the incorporation of these analogs into DNA, and as many as 5% of newly synthesized cytosine residues in actively growing cells may be replaced by the 5-azacytosine moiety.[15] There is, therefore, little immediate inhibition of growth during the first cell cycle of exposure to 5-aza-CdR, but rather, a delayed inhibitory effect presumably due to effects of the drug on other cellular parameters.[16]

5-Azacytosine appears to be relatively stable once incorporated into DNA.[15,17] We have been unable to detect any evidence for breakdown of the 5-azacytosine ring in DNA even after protracted time periods. This suggests that the cytotoxicity induced by azanucleosides is not due to the inherent lability of the pyrimidine ring, but, rather, due to other biological effects subsequent to analog incorporation.

The incorporation of the s-triazine ring into DNA leads to the appearance of alkali labile sites.[14,18] While the drugs may be capable of inducing alkali labile sites in DNA, it is unclear whether these exist in the cell or are created as a result of the analytical method used for their detection. This is because the incorporated azacytosine ring would be expected to be especially sensitive to alkaline hydrolysis.[7]

Overall, the data show that the azanucleosides can be incorporated effectively into DNA where they are relatively stable. The triazine ring appears to be protected from the rapid hydrolysis seen in the nucleoside and nucleotide forms. There is no evidence that 5-aza-CdR inhibits DNA replication; however, DNA extracted from treated cells shows the presence of alkali-labile sites which are presumably reflections of sites in which the drug has been incorporated *in vivo,* but it is unclear from the alkaline elution experiments whether the ring has opened within DNA *in vivo.*

III. INHIBITION OF NUCLEIC ACID METHYLATION BY 5-AZACYTIDINE

One of the most striking biochemical effects of 5-azanucleosides is a strong inhibition of nucleic acid methylation. Lu et al.[19] and Lu and Randerath[20] were the first to observe that 5-aza-CR inhibited the methylation of transfer RNA *in vivo.* In a later study, Lu and Randerath[21] showed that the administration of 5-aza-CR to mice led to a rapid time- and dose-dependent decrease in the level of the tRNA cytosine 5-methyltransferase activity in liver. RNA synthesis was a prerequisite for the inhibition of the enzyme by 5-aza-CR, and this resulted in the synthesis of tRNA specifically lacking 5-methylcytidine. The authors also made the important observation that it was the nucleic acid containing the incorporated base which acted as an inhibitor of the RNA methyltransferase enzyme.

The inhibition of DNA methyltransferase by azacytosine-containing DNA was first demonstrated by Friedman,[22] who found that the fraudulent base could inhibit cytosine methylation in *Escherichia coli.* Friedman also showed that the DNA extracted from bacteria grown in the presence of 5-aza-CR inhibited the DNA cytosine 5-methyltransferase activity extracted from *E. coli* K12 cells without affecting the adenine methyltransferase enzyme present in the same cells.[23] *E. coli* DNA containing 5-azacytosine irreversibly inhibited the DNA methyltransferase activity, and it was shown in later work that this was due to the formation of a stable complex between the azacytosine-containing DNA and the methyltransferase enzymes. The complex formed between bacterial methyltransferases and aza-cytosine-containing DNA appears to be covalent, and this results in a depletion of active enzyme from the cells.[24,25]

Jones and Taylor[26] were the first to discover that 5-aza-CR and 5-aza-CdR were highly effective inhibitors of DNA methylation in mammalian cells, and this observation was confirmed subsequently in several other systems.[27-29] Comparative studies between the azan-ucleosides and other inhibitors of DNA methylation, such as cycloleucine, ethionine, and 5′-deoxy-5′-methylthioadenosine, confirmed that 5-aza-CR and 5-aza-CdR were the most potent inhibitors of DNA modification known.[15,29] These experiments also demonstrated that DNA extracted from 5-aza-CR-treated cells was a very effective acceptor of methyl groups *in vitro* in reactions catalyzed by crude preparations of methyltransferase.[15,28] The drug-containing DNA was therefore not inherently incapable of accepting methyl groups so that its demethylated state was presumably due to a depletion of active enzyme within the cells prior to extraction. Therefore, many hemimethylated sites remained in the DNA and these could subsequently be modified in an *in vitro* reaction mix. It was also found that incor-poration of 5-azacytosine into DNA was necessary for it to be an effective inhibitor of the reaction,[30] and the presence of the azanucleoside base led to a marked decrease in the amount of active methyltransferase which could be extracted from treated cells.[27,30,31]

These experiments led to the development of a model by Santi et al.[32] that the presence of the modified nucleoside resulted in the formation of a covalent bond between the fraudulent base and the six position of the triazine ring. 5-Aza-CR therefore acts as a mechanism-based inhibitor of the methyltransferase, which leads to the formation of a strong, possibly covalent bond with the enzyme, resulting in a substantial demethylation of the genome.

IV. MECHANISMS OF CYTOTOXICITY

5-Aza-CR and 5-aza-CdR were originally developed as cancer chemotherapeutic agents and there has been considerable effort devoted toward defining the mechanisms of cytotoxicity induced by the chemicals. The biological effects of the analogs require their incorporation into DNA after activation of the nucleosides to the nucleotide level. Cells resistant to 5-aza-CR sometimes show reduced levels of uridine-cytidine kinase, which is responsible for the initial activation of the drug.[33] Cells resistant to 5-aza-CdR can show decreased levels of deoxycytidine kinase and thus do not phosphorylate 5-aza-CdR to the monophosphate.[34]

Initially, it was thought that the cytotoxicity elicited by 5-azanucleosides was due to the instability of the ring in the DNA. However, as discussed earlier, there is little evidence for this supposition. It seems more likely that some of the cytotoxic effects of the drugs are due to the formation of stable protein complexes with DNA containing azanucleosides, as originally described by Christman et al.[31] Studies in our laboratory have suggested that the formation of these stable protein complexes requires the incorporation of azacytosine residues opposite methylated CG doublets in order for the appropriate protein-DNA complexes to be formed.[35] We found that repeated treatments of mouse 10T^1/$_2$ cells with 5-aza-CdR resulted in the formation of cells with drastically decreased levels of DNA methylation.[16] These cells became resistant to further drug treatment and formed decreased amounts of stable complexes between the DNA methyltransferase and other proteins which interact with DNA containing 5-azacytosine. It seems quite probable that the strong complexes would prove inhibitory to cellular functions (e.g., DNA synthesis and transcription), and these would be highly deleterious to treated cells.

Studies with bacterial systems have shown that the presence of DNA methyltransferases, which could presumably form covalent complexes with DNA containing azacytosine, substantially increases the toxicity of the drug to bacteria.[36,37] These experiments have also suggested that the cytotoxic activity of azacytosine in DNA is most likely due to the formation of strong protein-DNA complexes.

V. EFFECT OF 5-AZACYTIDINE ON CHROMOSOME STRUCTURE

Banerjee and Benedict[38] first showed that 5-aza-CR induced a two- to threefold increase in the number of sister chromatid exchanges (SCEs) in hamster fibrosarcoma cell lines at biologically effective doses, and these studies were later confirmed.[39,40] Bianchi et al.[41] have recently examined the mechanism for 5-aza-CR-induced induction of SCEs. Treatment of CHO cells with 5-aza-CR induced an approximately threefold increase in the number of SCEs, but had little effect on the number of SCEs occurring in mosquito cells. On the other hand, both cell types showed the same increases in chromatid exchanges when treated with other compounds, such as mitomycin C. The authors suggested that it was the production of asymmetric methylation of CG palindromic dinucleotides following 5-aza-CR treatment which was responsible for the increase in SCEs formed, since vertebrate DNA is heavily methylated, whereas mosquito DNA is not. They proposed that the increase in exchanges was due to the presence of hemimethylated DNA in the second cell cycle following drug treatment, and that hemimethylation resulted in the formation of the increased rate of exchanges. This appears to be a reasonable explanation of the results; however, it should also be remembered that greater binding of proteins to the DNA of treated CHO cells would be expected, based upon their 5-methycytosine content. The formation of DNA protein complexes may, therefore, also be responsible for the increased rate of chromatid exchanges observed.

Viegas-Pequignot and Dutrillaux[42,43] demonstrated that 5-aza-CR causes considerable delay in the condensation of chromosome segments corresponding to the G bands in treated cells. The drug, therefore, elicits immediate observable effects on chromatin structure. These observations have been confirmed by Schmid et al.[44] and Haaf et al.,[45] who showed that the incorporation of the cytosine analogs causes very distinct undercondensation of heterochromatin on specific chromosomes in treated cultures. Ferraro and Lavia[46] demonstrated that differential activity of ribosomal gene clusters in human lymphocytes could be visualized by silver staining of the chromosomes soon after 5-aza-CR treatment. The data suggested that the activation of the clusters was related directly to the hypomethylating activity of the analog.

The incorporation of 5-azacytosine also changes the replication order of DNA segments within treated cells. Schafer and Priest[47] showed that the drug could induce changes in the timing of replication, including that of the heterocyclic inactive X chromosome, whose replication time was shortened by 5-aza-CR. These observations have recently been extended by Selig et al.,[48] who found that the drug could alter dramatically the replication time of satellite DNA sequences located near the centromeric regions of mouse chromosomes in both fibroblasts and lymphocytes.

VI. MUTAGENIC ACTIVITY OF 5-AZACYTIDINE

Many assays for mutagenic activity have established that 5-aza-CR is mutagenic in systems ranging from viruses to *Drosophila* (Table 1). Since many of these target organisms do not contain 5-methylcytosine within their DNA, it must always be considered a possibility that some of the effects of 5-aza-CR on treated cells are due to this mutagenic activity.

In contrast to the reports with nonvertebrate cells, there have been no clear demonstrations of mutagenic activity of 5-aza-CR in mammalian cells. We were unable to detect mutagenic activity in 10T^1/$_2$ or hamster V79 cells using two different markers,[56] and Kerbel et al.[57] also observed no mutagenic activity in mouse tumor cells. Therefore, while 5-aza-CR might be mutagenic in prokaryotic nonvertebrate cell systems, it is not strongly mutagenic in cultured vertebrate cells. It is also important that the mutagenic and differentiation-inducing effects of the drug be compared at equivalent dose levels. Changes in gene expression induced by 5-aza-CR are often many orders of magnitude higher than that which would be anticipated for agents which act by purely mutagenic mechanisms. Nevertheless, it is necessary to exclude the possibility that the drug acts through mutagenic mechanisms, particularly in systems which appear to respond at low frequency to drug-induced gene expression.

VII. 5-AZACYTIDINE AND X-CHROMOSOME REACTIVATION

A potential role for DNA methylation in X inactivation was first suggested in a pioneering paper by Riggs.[58] Considerable evidence has now accumulated to suggest that the promoter regions of genes located on inactive X chromosomes are methylated relative to the alleles located on the active chromosome. For example, Hpall and Hha1 sites in the promoter of the human HPRT gene on the active X chromosome are hypomethylated relative to the inactive X chromosome,[59] and similar results have been reported for the mouse HPRT gene.[60] Keith et al.[61] examined the 5' control region and first exon of human phosphoglycerate kinase and determined the methylation status of all the Hpall sites within the CG island in the 5' region of this gene. They found that the Hpall sites were entirely methylated on an inactive X chromosome, whereas they were completely unmethylated when situated on an active chromosome. These, and other experiments, have confirmed that the methylation of the CG islands of X-linked genes is strongly linked to transcriptional inactivity.

Further evidence for a role of DNA methylation in controlling the activity of X-linked genes came from the experiments using mouse-human somatic cell hybrids containing a

TABLE 1
Mutagenic Activity of 5-Azacytidine

System	Result	Ref.
Arbovirus	Formation of large plaques	49
S. typhymurium	Reverse mutations	50
E. coli	Various mutations	51, 37
S. cerevisiae	Mitotic recombination and point mutation	52
A. nidulans	Conversion to "fluffy" phenotype	53, 54
D. melanogaster	Recombinogenic activity; also somatic mutations	55
Mouse 10T$^1/_2$ cells	No induction of ouabain resistance	56
Hamster V79 cells	No induction of ouabain resistance or 8-azaguanine resistance	56
Mouse tumor cells	No induction of ouabain resistance	57

structurally normal inactive human X chromosome.[62] It was found that 5-aza-CR induced the expression of X-linked genes at a high frequency, suggesting that they were kept inactive by a methylation-linked mechanism. Further studies by Venolia et al.[63] demonstrated that DNA extracted from these cells could be used to transfer the activated gene to recipient cells, thus providing that the change in expression induced by 5-aza-CR was elicited at the level of the DNA molecule.

Since the initial report by Mohandas et al.,[62] there have been several studies which have shown that inactive X-linked genes can be reactivated by azanucleoside treatment.[64-67] Jones et al.[65] demonstrated that mouse cells containing a human inactive X chromosome were most sensitive to 5-aza-CdR-induced reexpression if the cells were treated in the latter part of the S phase when the chromosome was replicating. They also demonstrated that at least two divisions were necessary before the gene was expressed in the treated cells, suggesting that symmetrical demethylation of both DNA strands was necessary for expression. The requirement for two replication cycles after drug treatment has been confirmed by Homman et al;[67] however, the degree of expression of the induced gene was increased considerably if the cells were allowed to divide further before being placed in selective medium. Additional chromatin reorganization may therefore be needed after drug-induced hypermethylation for maximal changes in expression to become apparent.

In contrast to the studies with mouse cells, or mouse cells containing human X chromosomes, it has proven to be more difficult to induce the expression of X-linked genes in normal human cells. Wolf and Migeon[68] failed to induce expression of the HPRT gene from normal diploid fibroblasts by 5-aza-CR treatment. There may therefore be more plasticity for gene expression in the mouse cells, and demethylation alone does not seem to be sufficient to induce the expression of X-linked genes in the normal human diploid cells.

VIII. ACTIVATION OF SPECIFIC GENE PRODUCTS

Azanucleosides have a remarkable ability to induce the expression of suppressed genetic information in diverse mammalian cells. The number of genes which can be activated by drug treatment is extraordinary. Most of these were last reviewed in 1985,[69,70] and a more recent, but not comprehensive, list is shown in Table 2.

Richardson and colleagues[71,72] have shown that cloned T-cells treated with 5-aza-CR lose the requirement for antigen and can be activated by autologous HLA-D molecules alone, thus becoming autoreactive. Their more recent studies have shown that hydrolazine and procainamide, two drugs associated with a lupus-like autoimmune disease, also inhibit DNA methylation and induce autoreactivity in cloned T-cell lines. Thus, drug-induced autoimmune disease may be due to activation of unidentified genes by a DNA demethylation mechanism.

Hsiao et al.[73] found that 5-aza-CR induced the expression of endogenous retrovirus-related sequences in 10T$^1/_2$ cells in a mechanism associated with hypomethylation of the

TABLE 2
Recent Reports on Induction of Gene
Expression by 5-Azacytidine

System	Ref.
Autoreactivation of T-cells	71, 72
Endogenous retrovirus expression	73
Human metallothionein	74
DNA repair genes	76
Hepatitis B core protein	77
Harvey-*ras* gene	79
Herpes simplex type 2	81
Adenovirus late promoter	82
Suppression of papillomavirus transeription	83

corresponding cellular DNA. Heguy et al.[74] showed that human metallothionein 1B is expressed in a tissue-specific manner which involves *cis*-acting methylation and that treatment of nonexpressing cells with 5-aza-CR resulted in the expression of the gene. This observation extended the original studies of Compere and Palmiter[75] that the drug could induce metal-lothionein inducibility in treated cells. 5-Aza-CR can induce the expression of DNA repair genes in CHO cells,[76] and this observation extends the number of genes which can be activated in CHO cells by the nucleoside.[69]

5-Aza-CR is very effective at activating suppressed viral information. Korba et al.[77] showed that the expression of the hepatitis B virus core gene could be prevented in human cells by methylation of a single HpaII site 280 base pairs upstream from the structural gene. The methylation-induced suppression could be reversed easily by azanucleoside treatment. Similar results have been reported for the *ras* gene, which is inactive when transfected in a methylated state into recipient cells[78] and which can be reactivated by 5-aza-CR.[79] Altanerova[80] was the first to show that 5-aza-CR induced the expression of latent viral information, and recent experiments by Stephanopolous et al.[81] have confirmed these observations. Guinea pig dorsal root ganglia and spinal cords infected with latent herpes simplex virus II could be induced to express virus after drug treatment, suggesting a role for methylation in herpes simplex virus latency and reactivation. 5-Aza-CR can also activate the major late promoter in adenovirus-transformed cells in which the promoter is not normally active.[82] Rosl et al.[83] have recently obtained an opposite result for the transcription of human papilloma virus type 18 in gilal-fibroblast or gilal-keratinocyte hybrids. The downregulation of HPV18 mRNA within the cells correlates directly with the cessation of cellular growth and supports a model of a postulated intracellular control mechanism directed against papilloma viruses which can be induced by 5-aza-CR. The activation of this putative controlling region by 5-aza-CR might, therefore, result in the downregulation of the expression of the viral sequences.

Table 2 and earlier studies[69,70] illustrate the large number of systems which respond to 5-aza-CR by the expression of suppressed genetic information. Presumably, the activity of these sequences has been influenced by *de novo* methylation events within cells, which can subsequently be reversed by hypomethylation. Many of the studies listed in Table 2, and other observations in the literature, have shown directly that 5-aza-CR induces the under-methylation of the gene which is subsequently expressed. Experiments with transfected genes have shown that methylation before transfection can silence gene activity and that the transcriptional block can be reversed by 5-aza-CR, strongly suggesting that the mechanism of activation is due to a demethylation event. The ability of a gene to respond to 5-aza-CR induction is, therefore, evidence that it has become inactive due to a hypermethylation event.

IX. EFFECTS OF 5-AZACYTIDINE ON GENE EXPRESSION *IN VIVO*

Most of our knowledge concerning the mechanism of action of 5-aza-CR comes from studies with cultured cells. However, there have been several investigations which have established that the drug can modify gene expression in intact organisms. Cook and Jen-Fu[84] found that intraperitoneal administration of the drug to neonatal rats decreased the serum alpha fetoprotein level on day 6 after birth, when compared to saline injected controls. Thus, the drug appeared to be hastening the normal process of development within these animals. Rothrock et al.[85] treated 20-d-old rat fetuses *in utero* with 5-aza-CR, and this caused the precocious expression of tyrosine aminotransferase activity. Similar results were not found with cytosine arabinoside and 6-azacytidine, suggesting that the effects on gene expression were due to a DNA methylation-linked mechanism. In more recent studies, Rothrock et al.[86] found significant changes in fetal livers, including a dramatic maturation of hepatocyte morphology. The changes in morphology were reflected in the strong expression of genes normally associated with later development, including those coding for tyrosine aminotransferase and phosphenolpyruvate carboxykinase. These results confirmed the earlier studies of Benvenisty et al.,[87] who had observed the precocious expression of phosphenolpyruvate carboxykinase within rat fetuses injected *in utero*. Overall, these data using rat fetuses and neonatal animals suggest that undermethylation of specific sequences may hasten precocious gene expression.

Initial reports that 5-aza-CR could stimulate fetal hemoglobin synthesis *in vivo* in baboons[88] or patients with thalassemia[89] and sickle cell anemia[90] attributed the changes in globin expression to a hypomethylation of the globin genes. However, subsequent studies demonstrated that similar effects could be obtained with drugs, such as cytosine arabinoside, which do not act by inducing DNA hypermethylation.[91] Hence, it has not been easy to determine the mechanism by which 5-aza-CR induces the changes in hemoglobin expression within these systems.[92]

Although it seems likely that 5-aza-CR can induce changes in gene expression in whole animals, the results have to be interpreted with a great deal of caution because of the inherent cytotoxicity of the agent. It has not always been simple to separate cytotoxicity and its complex effects on cellular homeostasis in complete organisms, from the effects of the analog of DNA methylation.

X. INDUCTION OF CELLULAR DIFFERENTIATION

5-Aza-CR has marked effects on the phenotypic stability of cultured cells. Cells of the $10T^1/_2$ mouse line of embryo fibroblasts yield fully differentiated muscle, fat, and cartilage cells several days or weeks after exposure to the analog.[93-95] These new phenotypes, which are seen at much lower frequencies or not at all in untreated control cultures, are fully normal by all biochemical and morphological criteria applied to them. Cells with the appearance of striated muscle develop myosin ATPase activity, express the appropriate form of creatine phosphokinase,[96] develop functional acetylcholine receptors on their surfaces, and twitch either spontaneously or in response to added acetylcholine. Cells with the morphological appearance of cultured adipocytes stain positively with Oil Red O and express the key enzymes required for fatty acid biosynthesis.[95] Cartilage cells express the Type II collagen gene[95] and elaborate a matrix containing the appropriate proteoglycan molecules. Additionally, 5-aza-CR is capable of inducing the oncogenic transformation of $10T^1/_2$ cells.[97,98]

The new phenotypes which arise in treated cultures are seen several divisions after exposure to the analogs. Furthermore, it is possible to isolate colonies from treated dishes which have become determined to the muscle or adipocyte lineages. For example, cells with the properties of myoblasts may be obtained as clonal populations,[96,99,100] and these will grow indefinitely in culture. Cells with the biochemical properties of preadipocytes may also be obtained.[101] These two cell types have properties distinct from the original 10T$\frac{1}{2}$ cultures from which they were derived. Since the changes in phenotype are seen in cells which no longer contain 5-azacytosine in their DNA, the effects of the drug on cellular differentiation do not require the continued presence of the fraudulent nucleotide in DNA.

Several lines of evidence support the hypothesis that the changes in differentiation induced by 5-aza-CR are due to its ability to inhibit DNA methylation. For example, biologically effective doses of the drug cause strong inhibition of the methylation of newly incorporated cytosine residues in DNA. Other analogs of cytosine which have modifications in the 5 position of the pyrimidine ring and which inhibit DNA methylation also elicit the same changes in cellular differentiation. On the other hand, analogs which do not inhibit methylation do not change the differentiated state of 10T$\frac{1}{2}$ cells.[26]

The changes elicited by 5-aza-CR are not restricted to 10T$\frac{1}{2}$ cells. Early experiments established that the drug could also induce the formation of muscle cells in 3T3 cells, transformed 10T$\frac{1}{2}$, and in cells derived from an adult mouse prostate.[95,102] The drug can induce myogenesis in differentiation-defective myoblasts[103] or the formation of preadipocytes in Chinese hamster embryo fibroblasts.[104] Darmon et al.[105] found that it could induce the formation of epithelial cells from mesenchymal cells derived from teratocarcinomas. The ability to change cellular differentiation therefore appears to be quite general and is not restricted to the 10T$\frac{1}{2}$ line. However, this cell line exhibits the greatest and most consistent alterations in phenotype and is the one which has found most usage for the molecular dissection of the determination process.

The high frequency of induction of the muscle phenotype by 5-aza-CR and the fact that the drug could elicit changes in differentiation with as little as a 5-min treatment in the sensitive part of the S phase[102] prompted Konieczny and Emerson[99] to suggest that the drug induced the demethylation of one or a small number of genes, whose subsequent expression defined the determined myogenic state. Evidence for the existence of such genes first came from the work of Lassar et al.,[100] who showed that the muscle phenotype could be transfected from myogenic derivatives of 10T$\frac{1}{2}$ cells by DNA-mediated gene transfer. This demonstration was followed by the cloning of a gene expressed in myogenic derivatives of 10T$\frac{1}{2}$ cells, but not in the parent cells, which was called MyoD1.[106] Importantly, Davis et al.[106] demonstrated that the expression of this gene in untreated 10T$\frac{1}{2}$ cells resulted in the acquisition of the myogenic phenotype. In more recent studies, Pinney et al.[107] isolated another determination gene called *myd*, which is distinct from MyoD1 and may act as part of a hierarchy of genes whose expression defines myogenesis.[108] The isolation and characterization of these determination genes illustrates the value of the 10T$\frac{1}{2}$ system primed to differentiate with 5-aza-CR.

The MyoD1 gene has several interesting properties which suggest that it functions as a DNA binding protein to control the expression of genes necessary for myogenesis. Tapscott et al.[109] have shown that the gene product is a phosphoprotein which has a region of homology to the *myc* gene, and that it is localized in the nucleus. The gene is located on human chromosome 11 and mouse chromosome 7, and is the first example of a determination gene whose function appears to be the control of expression of the subset of genes necessary for myogenesis.

Current hypotheses suggest that the changes in differentiation and determination elicited by 5-aza-CR proceed directly to the formation of muscle, fat, and cartilage cells. However, there is some evidence for the existence of intermediate states following drug treatment. Early experiments showed that the progeny of a single 10T$\frac{1}{2}$ cell treated with 5-aza-CR could form all three phenotypes, suggesting the existence of a multipotential intermediate cell type.[102] Some of the cell lines isolated after drug treatment appear to be bipotential and are capable of expressing more than one phenotype. For example, the adipocyte line isolated by Chapman et al.[101] has been shown to form muscle cells spontaneously.[106] We have isolated an unstable, multipotential derivative of 10T$^1/_2$ cells after 5-aza-CR treatment which can form muscle, fat, and cartilage at a low-passage number.[110] The cells lose the ability to differentiate with increased passaging and may then be reinduced to express new developmental pathways following additional drug treatments. The existence of these cells suggests that 5-aza-CR may initially act to induce the expression of an unstable multipotential cell which can subsequently differentiate into either of the three pathways mentioned earlier, or revert back to a 10T$^1/_2$-like phenotype.

The use of cells primed to differentiate with 5-aza-CR has therefore opened many new avenues of research into the molecular control of cell determination. Dissection of the hierarchies of genes controlling developmental pathways and the potential role of DNA methylation in this process promise to be particularly exciting research areas.

ACKNOWLEDGMENTS

The research reported in this chapter was supported by grant R35CA49758 from the National Cancer Institute.

REFERENCES

1. **Piskala, A. and Sorm, F.,** Nucleic acids components and their analogues. L1. Synthesis of l-glycosyl derivatives of 5-azauracil and 5-azacytosine, *Collect. Czech. Chem. Commun.,* 29, 2060, 1964.
2. **Pliml, J. and Sorm, F.,** Synthesis of 2-deoxy-D-ribofuranosyl-5-azacytosine, *Collect. Czech. Chem. Commun.,* 29, 2576, 1964.
3. **Sorm, F. and Vesely, J.,** Effect of 5-aza-2'-deoxycytidine against leukemic and hemopoietic tissues in AKR mice, *Neoplasma,* 15, 339, 1968.
4. **Von Hoff, D. D., Slavik, M., and Muggia, F. M.,** 5-Azacytidine, a new anticancer drug with effectiveness in acute myelogenous leukemia, *Ann. Int. Med.,* 85, 237, 1976.
5. **Vesely, J. and Cihak, A.,** 5-Azacytidine: mechanism of action and biological effects in mammalian cells, *Pharmacol. Ther.,* 2, 813, 1978.
6. **Momparler, R. L.,** Molecular, cellular and animal pharmacology of 5-aza-2'-deoxycytidine, *Pharmacol. Therm.,* 30, 287, 1986.
7. **Notari, R. E. and deYoung, J. L.,** Kinetics and mechanism of degradation of the antileukemic agent 5-azacytidine in aqueous solution, *J. Pharm. Sci.,* 64, 1148, 1975.
8. **Lin, K. T., Momparler, R. L., and Rivard, G. E.,** High performance liquid chromatographic analysis of chemical stability of 5-aza-2'-deoxycytidine, *J. Pharm. Sci.,* 70, 1228, 1981.
9. **Chabner, B. A., Drake, J. C., and Johns, D. G.,** Deamination of 5-azacytidine by a human leukemia cell cytidine deaminase, *Biochem. Pharmacol.,* 22, 2763, 1973.
10. **Momparler, R. L., Rossi, M., Bouchard, J., Vaccaro, C., Momparler, L. F., and Bartolucci, S.,** Kinetic interaction of 5-aza-2'-deoxycytidine-5'-monophosphate with deoxycytidylate deaminase, *Mol. Pharmacol.,* 25, 436, 1984.
11. **Vesely, J., Cihak, A., and Sorm, F.,** Association of decreased uridine and deoxycytidine kinase with enhanced RNA and DNA polymerase in mouse leukemic cells resistant to 5-azacytidine and 5-aza-2'-deoxycytidine, *Cancer Res.,* 30, 2180, 1970.
12. **Momparler, R. L. and Goodman, J.,** *In vitro* cytotoxic and biochemical effects of 5-aza-2'-deoxycytidine, *Cancer Res.,* 37, 1636, 1977.
13. **Bouchard, J. and Momparler, R. L.,** Incorporation of 5-aza-2'-deoxycytidine-5'-triphosphate into DNA: interactions with mammalian DNA polymerase α and DNA methylase, *Mol. Pharmacol.,* 24, 109, 1983.
14. **Covey, J. M., D'Incalci, M., Tilchen, E. J., Zaharko, D. S., and Kohn, K. W.,** Differences in DNA damage produced by incorporation of 5-aza-2'-deoxycytidine or 5,6-dihydro-5-azacytidine into DNA of mammalian cells, *Cancer Res.,* 46, 5511, 1986.
15. **Jones, P. A. and Taylor, S. M.,** Hemimethylated duplex DNAs prepared from 5-azacytidine treated cells, *Nucleic Acids Res.,* 9, 2933, 1981.

16. **Flatau, E., Gonzales, F. A., Michalowsky, L. A., and Jones, P. A.,** DNA methylation in 5-aza-2'-deoxycytidine-resistant variants of C3H10T$^{1}/_{2}$ Cl 8 cells, *Mol. Cell. Biol.,* 4, 2098, 1984.
17. **Townsend, A. J. and Cheng, Y.-C.,** Sequence-specific effects of ara-5-aza-CTP and ara-CTP on DNA synthesis by purified human DNA polymerases *in vitro:* visualization of chain elongation on a defined template, *Mol. Pharmacol.,* 32, 330, 1987.
18. **D'Incalci, M., Covey, J. M., Zaharko, D. S., and Kohn, K. W.,** DNA alkali labile sites induced by incorporation of 5-aza-2'-deoxycytidine into DNA of mouse leukemia L1210 cells, *Cancer Res.,* 45, 3197, 1985.
19. **Lu, L. W., Chiang, G. H., Medina, D., and Randerath, K.,** Drug effects on nucleic acid modification. I. Specific effect of 5-azacytidine on mammalian transfer RNA modification *in vivo, Biochem. Biophys. Res. Commun.,* 68, 1094, 1976.
20. **Lu, L.-J. W. and Randerath, K.,** Effects of 5-azacytidine on transfer RNA methyltransferases, *Cancer Res.,* 39, 940, 1979.
21. **Lu, L.-J. W. and Randerath, K.,** Mechanism of 5-azacytidine-induced transfer RNA cytosine-5-methyltransferase deficiency, *Cancer Res.,* 40, 2701, 1980.
22. **Friedman, S.,** The effect of azacytidine on *E. coli* DNA methylase, *Biochem. Biophys. Res. Commun.,* 89, 1324, 1979.
23. **Friedman, S.,** The inhibition of DNA (cytosine-5) methylases by 5-azacytidine — the effect of azacytosine-containing DNA, *Mol. Pharmacol.,* 19, 314, 1981.
24. **Santi, D. V., Norment, A., and Garrett, C. E.,** Covalent bond formation between a DNA-cytosine methyltransferase and DNA containing 5-azacytosine, *Proc. Natl. Acad. Sci. U.S.A.,* 81, 6993, 1984.
25. **Friedman, S.,** The irreversible binding of azacytosine-containing DNA fragments to bacterial DNA (cytosine-5) methyltransferases, *J. Biol. Chem.,* 260, 5698, 1985.
26. **Jones, P. A. and Taylor, S. M.,** Cellular differentiation, cytidine analogs and DNA methylation, *Cell,* 20, 85, 1980.
27. **Tanaka, M., Hibasami, H., Nagai, J., and Ikeda, T.,** Effects of 5-azacytidine on DNA methylation in Ehrlich's ascites tumor cells, *Aust. J. Exp. Biol. Med. Sci.,* 58, 391, 1980.
28. **Creusot, F., Acs, G., and Christman, J. K.,** Inhibition of DNA methyltransferase and induction of Friend erythroleukemia cell differentiation by 5-azacytidine and 5-aza-2'-deoxycytidine, *J. Biol. Chem.,* 257, 2041, 1982.
29. **Woodcock, D. M., Adams, J. K., Allan, R. G., and Cooper, I. A.,** Effect of several inhibitors of enzymatic DNA methylation on the *in vivo* methylation of different classes of DNA sequences in a cultured human cell line, *Nucleic Acids Res.,* 11, 489, 1983.
30. **Taylor, S. M. and Jones, P. A.,** Mechanism of action of eukaryotic DNA methyltransferase: use of 5-azacytosine containing DNA, *J. Mol. Biol.,* 162, 679, 1982.
31. **Christman, J. K., Schneiderman, N., and Acs, G.,** Formation of highly stable complexes between 5-azacytosine-substituted DNA and specific non-histone nuclear proteins, *J. Biol. Chem.,* 260, 4059, 1985.
32. **Santi, D. V., Garrett, C. E., and Barr, P. J.,** On the mechanism of inhibition of DNA-cytosine methyltransferases by cytosine analogs, *Cell,* 39, 9, 1983.
33. **Vesely, J., Cihak, A., and Sorm, F.,** Biochemical mechanisms of drug resistance, development of resistance to 5-azacytidine and simultaneous depression of pyrimidine metabolism in leukemic mice, *Int. J. Cancer,* 2, 639, 1967.
34. **Vesely, J., Cihak, A., and Sorm, F.,** Characteristics of mouse leukemia cells resistant to 5-azacytidine and 5-aza-2'-deoxycytidine, *Cancer Res.,* 28, 1995, 1968.
35. **Michalowsky, L. A. and Jones, P. A.,** Differential nuclear protein binding to 5-azacytosine-containing DNA as a potential mechanism for 5-aza-2'-deoxycytidine resistance, *Mol. Cell. Biol.,* 7, 3076, 1987.
36. **Bhagwat, A. S. and Roberts, R. J.,** Genetic analysis of the 5-azacytidine sensitivity of Escherichia coli K-12, *J. Bacteriol.,* 169, 1537, 1987.
37. **Lal, D., Som, S., and Friedman, S.,** Survival and mutagenic effects of 5-azacytidine in *E. coli, Mutat. Res.,* 193, 229, 1988.
38. **Banerjee, A. and Benedict, W. F.,** Production of sister chromatid exchanges by various cancer chemotherapeutic agents, *Cancer Res.,* 39, 797, 1979.
39. **Hori, T.-A.,** Induction of chromosome decondensation, sister chromatid exchanges and endoreduplications by 5-azacytidine, an inhibitor of DNA methylation, *Mutat. Res.,* 121, 47, 1983.
40. **Shipley, J., Sakai, K., Tantravahi, U., Fendrock, B., and Latt, S. A.,** Correspondence between effects of 5-azacytidine on SCE formation, cell cycling and DNA methylation in Chinese hamster cells, *Mutat. Res.,* 150, 333, 1985.
41. **Bianchi, N. O., Larramendy, M., and Bianchi, M. S.,** The asymmetric methylation of CG palindrome dinucleotides increases sister chromatid exchanges, *Mutat. Res.,* 197, 151, 1988.
42. **Viegas-Pequignot, E. and Dutrillaux, B.,** Segmentation of human chromosomes induced by 5-ACR (5-azacytidine), *Hum. Genet.,* 34, 247, 1976.

43. **Viegas-Pequignot, E. and Dutrillaux, B.,** Detection of G-C rich heterochromatin by 5-azacytidine in mammals, *Hum. Genet.,* 57, 134, 1981.

44. **Schmid, M., Grunert, D., Haaf, T., and Engel, W.,** A direct demonstration of somatically paired heterochromatin of human chromosomes, *Cytogenet. Cell Genet.,* 36, 554, 1983.

45. **Haaf, T., Ott, G., and Schmid, M.,** Differential inhibition of sister chromatid condensation induced by 5-azadeoxycytidine in human chromosomes, *Chromosoma,* 94, 389, 1986.

46. **Ferraro, M. and Lavia, P.,** Differential gene activity visualized on sister chromatids after replication in the presence of 5-azacytidine, *Chromosoma,* 91, 307, 1985.

47. **Shafer, D. A. and Priest, J. H.,** Reversal of DNA methylation with 5-azacytidine alters chromosome replication patterns in human lymphocyte and fibroblast cultures, *Am. J. Hum. Genet.,* 36, 534, 1984.

48. **Selig, S., Ariel, M., Goitein, R., Marcus, M., and Cedar, H.,** Regulation of mouse satellite DNA replication time, *EMBO J.,* 7, 419, 1988.

49. **Halle, S.,** 5-Azacytidine as a mutagen for arboviruses, *J. Virol.,* 2, 1228, 1968.

50. **Podger, D. M.,** Mutagenicity of 5-azacytidine in *Salmonella typhimurium, Mutat. Res.,* 121, 1, 1983.

51. **Fucik, V., Zandrad, S., Sormova, Z., and Sorm, F.,** Mutagenic effect of 5-azacytidine in bacteria, *Collect. Czech. Chem. Commun.,* 30, 2883, 1965.

52. **Zimmerman, F. K. and Scheel, I.,** Genetic effects of 5-azacytidine in *Saccharomyces cerevisiae, Mutat. Res.,* 139, 21, 1984.

53. **Tamame, M., Antequera, F., Villanueva, J. R., and Santos, T.,** High-frequency conversion to a "fluffy" developmental phenotype in *Aspergillus* supp. by 5-azacytidine treatment: evidence for involvement of a single nuclear gene, *Mol. Cell. Biol.,* 3, 2287, 1983.

54. **Tamame, M., Antequera, F., and Santos, E.,** Developmental characterization and chromosomal mapping of the 5-azacytidine-sensitive HuF locus of *aspergillus nidulans, Mol. Cell. Biol.,* 8, 3043, 1988.

55. **Katz, A. J.,** Genotoxicity of 5-azacytidine in somatic cells of Drosophila, *Mutat. Res.,* 143, 195, 1985.

56. **Landolph, J. R. and Jones, P. A.,** Mutagenicity of 5-azacytidine and related nucleosides in C3H/10T^1/$_2$ Cl8 and V79 cells, *Cancer Res.,* 42, 817, 1982.

57. **Kerbel, R. S., Frost, P., Liteplo, R., Carlow, D., and Elliott, B. E.,** Possible epigenetic mechanisms of tumor progression: induction of high frequency heritable but phenotypically unstable changes in the tumorigenic and metastatic properties of tumor cell populations by 5-azacytidine treatment, *J. Cell. Physiol.,* Suppl. 3, 87, 1984.

58. **Riggs, A. D.,** X-inactivation, differentiation and DNA methylation, *Cytogenet. Cell Genet.,* 14, 9, 1975.

59. **Wolf, S. F., Jolly, D. J., Lunnen, K. D., Friedman, T., and Migeon, B. R.,** Methylation of the hypoxanthine phosphoribosyltransferase locus on the human X chromosome: implications for X chromosome inactivation, *Proc. Natl. Acad. Sci. U.S.A.,* 81, 2806, 1984.

60. **Lock, L. F., Melton, D. W., Caskey, C. T., and Martin, G. R.,** Methylation of the mouse HPRT gene differs on the active and inactive X chromosomes, *Mol. Cell. Biol.,* 6, 914, 1986.

61. **Keith, D. H., Singer-Sam, J., and Riggs, A. D.,** Active X chromosome DNA is unmethylated at eight CCGG sites clustered in a guanine-plus-cytosine-rich island at the 5′ end of the gene for phosphoglycerate kinase, *Mol. Cell. Biol.,* 6, 4122, 1986.

62. **Mohandas, T., Sparkes, R. S., and Shapiro, L. J.,** Reactivation of an inactive human X chromosome: evidence for X inactivation by DNA methylation, *Science,* 211, 393, 1981.

63. **Venolia, L., Gartler, S. M., Wassman, E. R., Yen, P., Mohandas, T., and Shapiro, L. J.,** Transformation with DNA from 5-azacytidine reactivated X chromosomes, *Proc. Natl. Acad. Sci. U.S.A.,* 79, 2352, 1982.

64. **Graves, J. A. M.,** 5-Azacytidine-induced reexpression of alleles on the inactive X-chromosome in a *Mus musculus* × *M. caroli* cell line, *Exp. Cell Res.,* 141, 99, 1982.

65. **Jones, P. A., Taylor, S. M., Mohandas, T., and Shapiro, L. J.,** Cell cycle specific reactivation of an active X-chromosome locus by 5-azadeoxycytidine, *Proc. Natl. Acad. Sci. U.S.A.,* 79, 1215, 1982.

66. **Paterno, G. D., Adra, C. N., and McBurney, M. W.,** X chromosome reactivation in mouse embryonal carcinoma cells, *Mol. Cell. Biol.,* 5, 2705, 1985.

67. **Homman, N., Heuertz, S., and Hors-Cayla, M. C.,** Time dependence of X gene reactivation induced by 5-azacytidine: possible progressive restructuring of chromatin, *Exp. Cell Res.,* 172, 481, 1987.

68. **Wolf, S. F. and Migeon, B. R.,** Studies of X chromosome DNA methylation in normal human cells, *Nature,* 295, 667, 1982.

69. **Jones, P. A.,** Effects of 5-azacytidine and its 2′-deoxyderivative on cell differentiation and DNA methylation, *Pharmacol. Ther.,* 28, 17, 1985.

70. **Jones, P. A.,** Altering gene expression with 5-azacytidine, *Cell,* 40, 485, 1985.

71. **Richardson, B.,** Effect of an inhibitor of DNA methylation on T cells. II. 5-azacytidine induces self-reactivity in antigen-specific T4$^+$ cells, *Hum. Immunol.,* 17, 456, 1986.

72. **Cornacchia, E., Golbus, J., Maybaum, J., Strahler, J., Hanash, S., and Richardson, B.,** Hydralazine and procainamide inhibit T cell DNA methylation and induce autoreactivity, *J. Immunol.,* 140, 2197, 1988.

73. **Hsiao, W.-L., W., Gattoni-Celli, S., and Weinstein, I. B.,** Effects of 5-azacytidine on expression of endogenous retrovirus-related sequences in C3H/10T1/2 cells, *J. Virol.,* 57, 1119, 1986.

74. **Heguy, A., West, A., Richards, R. I., and Karin, M.,** Structure and tissue-specific expression of the human metallothionein 1_B gene, *Mol. Cell. Biol.,* 6, 2149, 1986.

75. **Compere, S. J. and Palmiter, R. D.,** DNA methylation controls the inducibility of the mouse metallothionein-1 gene in lymphoid cells, *Cell,* 25, 233, 1981.

76. **Jeggo, P. A. and Holliday, R.,** Azacytidine-induced reactivation of a DNA repair gene in Chinese hamster ovary cells, *Mol. Cell. Biol.,* 6, 2944, 1986.

77. **Korba, B. E., Wilson, V. L., and Yoakum, G.,** Induction of hepatitis B virus core gene in human cells by cytosine demethylation in the promoter, *Science,* 228, 1103, 1985.

78. **Borrello, M. G., Pierotti, M. A., Bongazone, I., Donghi, R., Mondellini, P., and Porta, G. D.,** DNA methylation affecting the transforming activity of the human Ha-ras oncogene, *Cancer Res.,* 47, 75, 1987.

79. **Borrello, M. G., Pierotti, M. A., Donghi, R., Bongarzone, I., Cattadori, M. R., Traversari, C., Mondellini, P., and Della Porta, G.,** Modulation of the human Harvey-*ras* oncogene expression by DNA methylation, *Oncogene Res.,* 2, 197, 1988.

80. **Altanerova, V.,** Virus production induced by various chemical carcinogens in a virogenic hamster cell line transformed by Rous sarcoma virus, *J. Natl. Cancer Inst.,* 49, 1375, 1972.

81. **Stephanopoulos, D. E., Kappes, J. C., and Bernstein, D. I.,** Enhanced *in vitro* reactivation of herpes simplex virus type 2 from latently infected guinea-pig neural tissues by 5-azacytidine, *J. Gen. Virol.,* 69, 1079, 1988.

82. **Robert, N., Toth, M., Tellier, R., Horvath, J., Dery, C. V., and Weber, J. M.,** Activation of the major late promoter in adenovirus transformed cells by 5-azacytidine, *Virology,* 165, 296, 1988.

83. **Rosl, F., Durst, M., and zur Hausen, H.,** Selective suppression of human papillomavirus transcription in non-tumorigenic cells by 5-azacytidine, *EMBO J.,* 7, 1321, 1988.

84. **Cook, J. R. and Jen-Fu, C.,** Effect of 5-azacytidine on rat liver alpha-fetoprotein gene expression, *Biochem. Biophys. Res. Commun.,* 116, 939, 1983.

85. **Rothrock, R., Perry, S. T., Isham, K. R., Lee, K.-L., and Kenney, F. T.,** Activation of tyrosine aminotransferase expression in fetal liver by 5-azacytidine, *Biochem. Biophys. Res. Commun.,* 113, 645, 1983.

86. **Rothrock, R., Lee, K.-L., Isham, K. R., and Kenney, F. T.,** Changes in hepatic differentiation following treatment of rat fetuses with 5-azacytidine, *Arch. Biochem. Biophys.,* 263, 237, 1988.

87. **Benvenisty, N., Szyf, M., Mencher, D., Razin, R., and Reshet, L.,** Tissue-specific hypomethylation and expression of rat phosphoenolpyruvate carboxykinase gene induced by *in vivo* treatment of fetuses and neonates with 5-azacytidine, *Biochemistry,* 24, 5015, 1985.

88. **DeSimone, J., Heller, P., Hall, L., and Zwiers, D.,** 5-Azacytidine stimulates fetal hemoglobulin synthesis in anemic baboons, *Proc. Natl. Acad. Sci. U.S.A.,* 79, 4428, 1982.

89. **Ley, T. J., DeSimone, J., Anagnou, N. P., Keller, G. H., Humphries, R. K., Turner, P. H., Young, N. S., Heller, P., and Nienhuis, A. W.,** 5-Azacytidine selectively increases γ-globin synthesis in a patient with B$^+$ thalassemia, *N. Engl. J. Med.,* 307, 1469, 1982.

90. **Ley, T. J., DeSimone, J., Noguchi, C. T., Turner, P. H., Schechter, A. N., Heller, P., and Nienhuis, A. W.,** 5-Azacytidine increases γ-globin synthesis and reduces the proportion of dense cells in patients with sickle cell anemia, *Blood,* 62, 370, 1983.

91. **Papayannopoulou, T., de Ron, A. T., Veith, R., Knitter, G., and Stamatoyannopoulos, G.,** Arabinosylcytosine induces fetal hemoglobin in baboons by perturbing erythroid cell differentiation kinetics, *Science,* 224, 617, 1984.

92. **Galanello, R., Stamatoyannopoulos, G., and Papayannopoulou, T.,** Mechanism of HbF stimulation by S-stage compounds. *In vitro* studies with bone marrow cells exposed to 5-azacytidine, ara-C or hydroxyurea, *J. Clin. Invest.,* 81, 1209, 1988.

93. **Constantinides, P. G., Jones, P. A., and Gevers, W.,** Functional striated muscle cells from non-myoblast precursors following 5-azacytidine treatment, *Nature,* 267, 364, 1977.

94. **Constantinides, P. G., Taylor, S. M., and Jones, P. A.,** Phenotypic conversion of cultured mouse embryo cells by aza pyrimidine nucleosides, *Dev. Biol.,* 66, 57, 1978.

95. **Taylor, S. M. and Jones, P. A.,** Multiple new phenotypes induced in 10T1/2 and 3T3 cells treated with 5-azacytidine, *Cell,* 17, 771, 1979.

96. **Liu, L., Harrington, M., and Jones, P. A.,** Characterization of myogenic cell lines derived by 5-azacytidine treatment, *Dev. Biol.,* 117, 331, 1986.

97. **Benedict, W. F., Banerjee, A., Gardner, A., and Jones, P. A.,** Induction of morphological transformation in mouse C3H/10T1/2 clone 8 cells and chromosomal damage in hamster A(T$_1$) Cl-3 cells by cancer chemotherapeutic agents, *Cancer Res.,* 37, 2202, 1977.

98. **Rainier, S. and Feinberg, A. P.,** Capture and characterization of 5-aza-2'-deoxycytidine treated C3H10T1/2 cells prior to transformation, *Proc. Natl. Acad. Sci. U.S.A.,* 85, 6384, 1988.

99. **Konieczny, S. F. and Emerson, C. P.,** 5-Azacytidine induction of stable mesodermal stem cell lineages from 10T$^1/_2$ cells: evidence for regulatory genes controlling determination, *Cell,* 38, 791, 1984.

100. **Lassar, A. B., Paterson, B. M., and Weintraub, H.,** Transfection of DNA locus that mediates the conversion of 10T$^1/_2$ fibroblasts to myoblasts, *Cell,* 47, 649, 1986.
101. **Chapman, A. B., Knight, D. M., Dieckmann, B. S., and Ringold, G. M.,** Analysis of gene expression during differentiation of adipogenic cells in culture and hormonal control of the developmental program, *J. Biol. Chem.,* 259, 15548, 1984.
102. **Taylor, S. M. and Jones, P. A.,** Changes in phenotypic expression in embryonic and adult cells treated with 5-azacytidine, *J. Cell. Physiol.,* 111, 187, 1982a.
103. **Walker, C. and Shay, J. W.,** 5-Azacytidine induced myogenesis in a differentiation defective cell line, *Differentiation,* 25, 259, 1984.
104. **Sager, R. and Kovac, P.,** Pre-adipocyte determination either by insulin or by 5-azacytidine, *Proc. Natl. Acad. Sci. U.S.A.,* 79, 480, 1982.
105. **Darmon, M., Nicolas, J. F., and Lamblin, D.,** 5-Azacytidine is able to induce the conversion of kera-tocarcinoma-derived mesenchymal cells into epithelial cells, *EMBO J.,* 3, 961, 1984.
106. **Davis, R. L., Weintraub, H., and Lassar, A. B.,** Expression of a single transfected cDNA converts fibroblasts to myoblasts, *Cell,* 51, 987, 1987.
107. **Pinney, D. F., Pearson-White, S. H., Konieczny, S. F., Latham, K. E., and Emerson, C. P., Jr.,** Myogenic lineage determination and differentiation: evidence for a regulatory gene pathway, *Cell,* 53, 781, 1988.
108. **Blau, H. M.,** Hierarchies of regulatory genes may specify mammalian development, *Cell,* 53, 673, 1988.
109. **Tapscott, S. J., Davis, R. L., Thayer, M. J., Cheng, P. F., Weintraub, H., and Lassar, A. B.,** MyoD1: a nuclear phosphoprotein requiring a *myc* homology region to convert fibroblasts to myoblasts, *Science,* 242, 405, 1988.
110. **Harrington, M. A., and Jones, P. A.,** Mesodermal determination genes: evidence from DNA methylation studies, *Bioessays,* 8, 100, 1988.

Chapter 6

INDUCTION OF DIFFERENTIATION IN FIBROBLAST CELLS BY IRRADIATION AND CHEMICAL CARCINOGENS*

Duane L. Guernsey

TABLE OF CONTENTS

*   Supported by the Dalhousie University Faculty of Medicine, the National Cancer Institute of Canada, and National Institutes of Health — grant CA36483.

I. MESODERMAL/MESENCHYMAL LINEAGE DIFFERENTIATION

A. FIBROBLASTS

The mesoderm is the middle layer of the three germ layers of the mammalian embryo. This layer is responsible for forming most of the muscle, skeletal, and connective tissue structures. Mesenchymal cells in prenatal life occupy the sites where loose connective tissue develops. It is believed that mesenchymal cells differentiate into all the cell types of loose connective tissue, including fat, muscle, and fibroblasts.[1] Fibroblasts synthesize molecules destined for secretion into the extracellular space, including numerous collagen types, elastic fibers, ground substance, and proteoglycans.[2]

Embryonic cells develop into the specialized cell types of the body by the two sequential processes of determination and differentiation. Determination is the conversion of embryonic cells into specific lineages of stem cells that subsequently differentiate into functionally specific cell types. Mesodermal lineage differentiation would reflect a fibroblastic-like embryonic cell committed to the determination of a myoblast, preadipocyte, or prechondrocyte. These stem cell types would still retain the fibroblastic morphology until environmental factors would invoke differentiation to the specific cell types; muscle cells forming myotubes, adipocytes accumulating lipid, or the mature chondrocyte, respectively.

B. *IN VITRO* MODEL SYSTEMS TO INVESTIGATE MESODERMAL LINEAGE DIFFERENTIATION

The advent of several cell culture systems has enabled investigators to directly analyze cellular and molecular events of mesodermal lineage differentiation. These cell systems have generally been a fibroblastic cell type that can be induced to differentiate into one or multiple phenotypes of muscle, adipocytes, and chondrocytes. Additionally, there has been a report of the differentiation of fibroblasts into a macrophage phenotype.[3] The C3H/10T1/2 (clone 8) mouse embryo fibroblastic cell line originally established by Reznikoff et al.[4] has been found to be sensitive to induced differentiation into myoblasts, adipocytes, and chondrocytes.[5-7] Originally, mouse 3T3 cells were found to spontaneously differentiate into adipocytes;[8] more recently, they were induced to form muscle, adipocytes, and chondrocytes.[4] Sager and Kovac[9] demonstrated that the CHEF/18 diploid Chinese hamster embryo cell line, which is fibroblastic in morphology, behaves like a mesenchymal stem cell in its ability to differentiate into adipocytes, myoblasts, and chondrocytes. Kopelovich and colleagues have established the neodifferentiation of human skin fibroblasts to the adipocyte morphology by induction with various viruses and glucocorticoids.[10-12]

C. MOLECULAR BASIS OF THE COMMITMENT TO MESODERMAL LINEAGE DIFFERENTIATION

The involvement of DNA hypomethylating agents, such as 5-azacytidine, in the commitment of fibroblasts to mesodermal lineages is covered by Peter Jones in Chapter 5 of this text and therefore will not be discussed in detail here. Suffice it to say that in 1979 Taylor and Jones[5] demonstrated that 5-azacytidine induced cultures of C3H/10T1/2 cells into the three differentiated phenotypes of muscle, adipocytes, and chondrocytes. We will be concerned with the evidence that the commitment to the differentiated mesodermal phenotypes is mediated by genetic loci, and not what mechanism regulates their expression developmentally.

Konieczny and Emerson[7] demonstrated that 5-azacytidine converts C3H/10T1/2 cells into three stably determined, yet undifferentiated, stem cell lineages which can differentiate into muscle, chondrocytes, and adipocytes. The conversion from the undifferentiated stem cell to the new phenotype can only occur under an appropriate environmental stimulus, such as confluency-induced growth arrest. They showed that the conversion of C3H/10T$^1/_2$ cells is accompanied by specific changes in protein synthetic patterns unique for each lineage, but this does not involve a direct activation of differentiation-specific gene expression which, in the case of the muscle cell lineage, is activated only when differentiation is induced by appropriate culture conditions. They suggest that conversion of C3H/10T$^1/_2$ cells is by activation of "determination" regulatory loci which establish lineages of stem cells that have the potential to differentiate into only one of the three subsequent phenotypes. Their results further suggest that the three lineages are delineated by separate genetic loci, and that this may involve as few as one to three genetic events. Konieczny and Emerson subsequently provided evidence that myoblast lineage determination acts to establish a regulatory control system that mediates expression of differentiation-specific transcription signals.[13]

Evidence that a genetic loci is directly involved in the conversion of C3H/10T$^1/_2$ fibroblasts to myoblasts was reported by Lassar et al.[14] They demonstrated that transfection of C3H/10T$^1/_2$ cells with high-molecular-weight DNA from azacytidine-induced C3H/10T$^1/_2$ myoblasts resulted in the myogenic conversion of a significant number of transfected colonies. This was specifically induced by the DNA from myoblasts, since DNA from parental C3H/10T$^1/_2$ cells was not able to induce myogenic conversion subsequent to transfection. This was direct evidence that myoblast conversion is the result of a structural modification in a single (or two closely linked) DNA locus. This altered DNA locus is capable of inducing myoblast lineage determination in C3H/10T$^1/_2$ cells. These studies were extended by Davis et al.[15] using cDNAs prepared from mRNAs that were expressed exclusively in myoblasts, and not in parental C3H/10T$^1/_2$ cells. They showed that the expression of one myoblast-specific cDNA transfected into C3H/10T$^1/_2$ or 3T3 fibroblasts was capable of inducing myogenic conversion. They further demonstrated that this gene, called MyoD1, was expressed only in mouse skeletal muscle *in vivo,* and not in any other adult or newborn mouse tissue. This would suggest that MyoD1 expression alone may be sufficient for myogenic determination in these mouse cells.

II. RELATIONSHIP BETWEEN DIFFERENTIATION AND ONCOGENESIS

A. CELLULAR

The intimate relation between carcinogenesis and developmental differentiation has been of interest to cancer biologists for many years.[16-22] Much of this research has focused on the reversal of malignancy by induction of differentiation. Scott and colleagues have demonstrated this relationship in mesenchymal differentiation in a series of reports investigating the differentiation of preadipocyte stem cells to the adipocyte phenotype.[23-27] These studies have established that resistance to neoplastic transformation can be regulated in mammalian stem cells by the process of nonterminal differentiation.

B. ONCOGENES AND DIFFERENTIATION

At the molecular level, the relationships between oncogenesis and differentiation have emerged from numerous studies of protooncogenes and oncogenes. Investigations have reported changes in protooncogene expression during development,[28-31] tissue-specific expression of protooncogenes,[29,31-33] changes in protooncogene expression during differentiation,[34-39] and the influence of oncogenes on the induction or prevention of differentiation.[40-45]

C. ONCOGENES AND MESODERMAL/MESENCHYMAL DIFFERENTIATION

Falcone et al.[46] investigated the effects of the viral oncogenes *myc, erb, fps,* and *src* on the differentiation program of quail myogenic cells. Transformed cultures of myogenic cells failed to fuse into multinucleate myotubes and to express muscle-specific genes. However, under altered culture conditions, *src-*, *fps-*, and *erb-*transformed clones could partially differentiate into atypical myotubes. In contrast, the ability for terminal differentiation was completely blocked on transformation with *myc-*containing viruses. They suggest that the v-*myc-*induced block of differentiation is a consequence of relaxed growth control.

Transfection of a mouse myoblast cell line with the oncogenic forms of H-*ras* or N-*ras* completely prevented myoblast fusion and the induction of muscle-specific gene products.[47] In contrast, cells transfected with proto-oncogenic forms of these oncogenes had no effect on the ability of the cells to fully differentiate. Also of importance is their finding that myoblasts transfected with the activated *ras* genes demonstrated normal growth properties and ceased proliferating in the absence of mitogens, thus indicating that oncogenic *ras* inhibition of differentiation is mediated via a mechanism independent of cell proliferation. Schneider et al.[48] also reported that autonomous expression of c-*myc* in a transfected mouse myoblast cell line partially inhibited, but did not prevent, myogenic differentiation.

It is also important to note that the MyoD1 gene locus (discussed above as the gene inducing fibroblasts to myogenic commitment after transfection) has a region homologous to the c-*myc* protooncogene.[49] This *myc* homology region was found to be required to activate myogenesis in stably transfected fibroblasts.[49]

Proto-oncogene expression has been investigated during *in vitro* differentiation of L6 rat myoblasts.[50] Of the nine proto-oncogenes studied, only *sis, ras, myc,* and *abl* demonstrated changes during differentiation. The expression of *sis, ras, myc,* and *abl* ceased or was greatly reduced in the myotubes, compared to proliferating myoblasts.

Distel et al.[51] have reported the involvement of the c-*fos* proto-oncogene in adipocyte differentiation. A putative lipid-binding protein (termed adipocyte P2, aP2) is transcriptionally activated in association with adipocyte differentiation of 3T3 preadipocytes. The aP2 gene contains a regulatory region that binds *trans-*acting factors that inhibit the expression in preadipocytes. They found that c-*fos* is a component of the complex that is directly required for sequence-specific binding to DNA.

D. PROTO-ONCOGENE EXPRESSION AND DETERMINATION TO MESODERMAL PHENOTYPES

Our laboratory has been interested in whether there are any detectable changes in oncogene expression related to the determination step in mesodermal lineage differentiation.[52] We established a rat embryo cell line (X-REF-23) characterized as fibroblastic, immortal, nontumorigenic, and retaining a diploid karyotype. We found that during a defined period in the age of these cells in culture that they will spontaneously differentiate along mesodermal cell lineages to form myotubes, adipocytes, and a chondrocyte-like cell. We isolated and characterized lineage-determined (nondifferentiated) preadipocyte (G clone) and myoblast (AMB) lines. Using these cell lines and a nondetermined daughter cell, we investigated the expression of 15 protooncogenes concomitant with adipocyte and myogenic lineage-specific determination. Table 1 shows the results of dot-blot analysis of total cellular RNA, and Table 2 shows the results using poly (A)$^+$ RNA and a second myogenic clone. There were no significant differences seen in the expression of c-*fms*, c-N-*myc*, c-*fes*, c-*raf*, c-*mos*, c-H-*ras*, c-*fos*, or c-*myc* in either lineage-determined cell line, compared to the fibroblastic daughter cell or to each other. There were slight differences in the levels of c-K-*ras* and c-*abl*. The preadipocyte cell line showed a 20 to 30% elevation in c-K-*ras* RNA, compared to the fibroblast, while both myoblast clones demonstrated a 20 to 30% decrease in levels. This is consistent with a 60% change between the adipocyte lineage and the myogenic lineage. This was similar to the results obtained with c-*abl*. Since c-*abl* and c-K-*ras* are known to have two transcripts each, we analyzed their poly (A)$^+$ RNA by Northern analysis to determine whether lineage determination was associated with alterations in the multiple

TABLE 1
Proto-oncogene RNA Levels in Preadipocyte (G-clone) and Myoblast (AMB) Cell Lines Relative To A Nondetermined/Nondifferentiating Daughter Fibroblastic Cell (X-REF-23)[52]

Cell	Protooncogene					
	fos	*myc*	*mos*	*myb*	K-*ras*	H-*ras*
AMB	+10	+20	−30	−10	−20	0
G-cl	−20	0	+10	+10	+40	+10

Note: Data are expressed as the percent change from X-REF-23. RNA was total cellular RNA.

TABLE 2
Proto-oncogene RNA Levels in Preadipocyte (G-clone) and Myoblast (AMC) Cell Lines Relative To A Nondetermined/ Nondifferentiating Daughter Fibroblast Cell (X-REF-23)[52]

Cell	Proto-oncogene						
	erb-B	*abl*	*fms*	N-*myc*	*fes*	*raf*	*erb*-A
AMC	ns[a]	−40	0	−10	+10	−10	+10
G-cl	ns[a]	+30	0	+10	+20	0	+80

	K-*ras*	*mos*	*sis*	*src*	H-*ras*	*fos*	*myc*
AMC	−30	−10	ns[a]	ns[a]	0	+10	−10
G-cl	+30	+10	ns[a]	ns[a]	0	0	−10

Note: Data are expressed as the percent change from X-REF-23. RNA was poly (A)[+] RNA.

[a] ns = no detectable signal.

transcripts. The sizes and ratios of the transcripts were not altered in either lineage, and the amount of RNA was similar, as seen in the dot-blot analysis. The c-*erb*-A RNA levels were elevated 80% in the adipocyte lineage, compared to the fibroblasts, and 90% compared to the myoblast lineage. It was of interest that we observed higher *erb*-A RNA levels in the preadipocyte than in either the fibroblast or the myoblast. This may be part of the lineage program, since it is recognized that lipid metabolism is regulated by thyroid hormones,[53] and the c-*erb*-A protein product is the nuclear receptor for thyroid hormone.[54,55] However, we do not believe any of the changes in proto-oncogene expression to be significant. There was no dramatic appearance or disappearance of expression that would be more consistent with the genetic commitment for lineage determination. Our results do not rule out the possibility that other proto-oncogenes not tested here may play a role. Our results and those discussed above are more consistent with the role of oncogenes in assembling the multiple new pathways of transcriptional regulation to convert the determined cell to the new phenotype.

III. RADIATION AND CHEMICAL CARCINOGEN INDUCTION OF LINEAGE DETERMINATION

A. PERSPECTIVES
From the discussions above it is clear that there is a relationship between differentiation

TABLE 3
X-Irradiation Induction of Adipogenesis in C3H/10T$^1/_2$
Cells[60]

Treatment	6 Gy	0 Gy
Adipocyte colonies/total no. plates	103/122	0/88
Plates with adipocyte colonies/total no. plates	66/122	0/88
Surviving colonies	36,590	32,280
Adipogenic frequency	2.8 × 10^{-3}	—

Note: Adipogenic frequencies were calculated as the total number of adipocyte colonies/total number of surviving colonies. The total number of surviving colonies were calculated as the plating efficiency × the number of cells seeded × the total number of plates. Data are a composite of four separate experiments.

and oncogenesis that can be evidenced at the cellular and molecular level of investigation. The question arises as to whether carcinogenic agents can affect the differentiation process. There exists a lot of data demonstrating that carcinogens, as well as tumor promoters and tumor suppressors, can modulate the differentiation of lineage-determined cells. However, there is little data concerning whether carcinogens can induce, or suppress, lineage-specific determination. Our laboratory has investigated the effects of physical and chemical carcinogens on mesodermal lineage determination using both the C3H/10T$^1/_2$ mouse embryo cells and the X-REF-23 rat embryo cells.

B. MESODERMAL LINEAGE DETERMINATION OF C3H/10T$^1/_2$ CELLS INDUCED BY X-IRRADIATION AND BENZO[A]PYRENE

1. Induction of Determination

X-irradiation and benzo[a]pyrene are potent mutagens and their ability for oncogenic transformation of C3H/10T$^1/_2$ cells has been well established in our laboratory and others.[56-59] However, the effects of these carcinogenic agents on differentiation have not been investigated. Therefore, we undertook the following series of experiments.[60] C3H/10T$^1/_2$ cells were seeded 24 h prior to X-irradiation (6 Gy) or exposure to benzo[a]pyrene (1.2 μg/ml) at a low cell density. Two to three weeks after the cells reached confluency, the cultures were stained with toluidine blue and Oil Red-O. A focus of adipocytes was defined as 20 or more cells containing large lipid droplets that stained red with Oil Red-O. A focus of myotubes would be observed as fused multinucleated cells, as previously described.[5-7,52] The benzo[a]pyrene-treated cultures were also scored for oncogenic-transformed foci, as previously described.[61] Table 3 shows the results of four separate experiments investigating whether X-irradiation could induce differentiation in the C3H/10T$^1/_2$ cells. In all four experiments, adipocyte colony formation was observed at a consistent frequency, ranging from 2.3 to 3.2 × 10^{-3}, with a composite frequency of 2.8 × 10^{-3}. In contrast, X-irradiation was not able to induce the myogenic phenotype in any of the cultures. We further found that benzo[a]pyrene was also able to consistently induce the adipogenic phenotype, although at a decreased efficiency. The mean frequency of benzo[a]pyrene-induced adipogenesis was 1.1 × 10^{-3}. No myogenic colonies were observed at the end of the experiments or during the course of the experiments.

It was of interest to find that the frequency of benzo[a]pyrene-induced adipogenesis was identical to the frequency of induced neoplastic transformation (Table 4). It has generally been found that the frequency of *in vitro* oncogenesis is significantly higher than the frequency of mutagen-induced single somatic mutation.[16] We have found that benzo[a]pyrene induces similar frequencies of oncogenesis and differentiation. Additionally, we show that 6 Gy of

TABLE 4
Benzo[a]pyrene Induction of Adipogenesis in C3H/
10T$^1/_2$ Cells[60]

Treatment	B[a]P	No B[a]P
Adipocyte colonies/total no. plates	11/48	1/30
Transformed colonies/total no. colonies	11/48	0/30
Surviving colonies	10,250	15,400
Adipogenic frequency	1.1×10^{-3}	6.5×10^{-5}
Transformation frequency	1.1×10^{-3}	—

Note: Frequencies calculated as in Table 3. Data are a composite of two separate experiments.

TABLE 5
Effects of Corticosterone ($10^{-6} M$), RU 486 ($10^{-7} M$) and X-
Rays (4 Gy) on Adipogenic Determination of C3H/10T$^1/_2$ Cells[65]

Experimental conditions	No. adipocyte colonies/no. plates
Controls (FBS)[a]	4/50
X-rays	48/50
X-rays/corticosterone	1,296/50
X-rays/corticosterone/RU 486[b]	424/50
Corticosterone	977/50
Corticosterone/RU 486[b]	411/50
RU 486[b]	5/25

Note: Data are a composite of two separate experiments.

[a] FBS, 10% fetal bovine serum.
[b] RU 486, glucocorticoid receptor antagonist.

X-irradiation converts fibroblasts to adipocytes at a frequency of 2.8×10^{-3}. This is comparable to the 6 Gy X-ray-induced frequencies of the oncogenic transformation of C3H/10T$^1/_2$ cells reported by Miller and Hall[62] (2×10^{-3}) and Terzaghi and Little[56] (1.5×10^{-3}).

These results demonstrate that a physical carcinogen (X-rays) and a chemical carcinogen (benzo[a]pyrene) can concomitantly induce both the neoplastic and differentiated phenotypes in cell culture. The efficiency of the carcinogen to convert fibroblasts to adipocytes was similar to the efficiency for neoplastic transformation. This suggests a possible relationship between carcinogenesis and oncogenesis. Whether the target(s) for the carcinogen is the same or different remains unknown.

2. Corticosterone and X-Ray-Induced Adipogenesis

Since it had been found that glucocorticoids can increase,[63] decrease,[64] or have little effect[65] on the X-irradiation transformation of C3H/10T$^1/_2$ cells, we were interested in determining the possible glucocorticoid effects on the X-ray-induced adipogenic lineage determination of C3H/10T$^1/_2$ cells.[65] Table 5 demonstrates the results of investigating the effects of the endogenous rodent glucocorticoid, corticosterone, on the adipogenic conversion of C3H/10T$^1/_2$ fibroblasts. It is seen that X-rays alone are able to induce a low frequency of adipocyte determination. When the cells were incubated in 10^{-6} M corticosterone and X-irradiated, the number of adipocyte colonies was enhanced 27-fold. In the two experiments, the frequency of adipogenic conversion with X-rays and corticosterone was 9.6 and 8.3 $\times$

10^{-3}, respectively. However, the surprising results showed that corticosterone was not enhancing the adipogenic effects of X-rays, but, rather, was itself inducing a high frequency of adipocyte determination. It can also be seen that the antagonist to the glucocorticoid receptor, RU 486, significantly suppressed the corticosterone-induced adipocyte determination in both groups. This would suggest that at least a large part of the corticosterone effects on adipocytic determination is mediated through the glucocorticoid receptor complex. Perhaps if we had used a molar excess of RU 486, the suppression of differentiation would have been complete. It has previously been reported that glucocorticoids enhance the conversion of preadipocytes to adipocytes.[66,67] However, we have found that corticosterone may also induce the determination of C3H/10T$^1/_2$ fibroblasts to the adipocyte lineage.

The influence of glucocorticoid hormones on carcinogenesis and differentiation may, in part, reflect the actions of many hormones on their respective target tissues, such as the actions of thyroid hormone. Thyroid hormone is known to have profound effects on the growth and development of vertebrates,[68,69] while our laboratory also demonstrated dramatic effects of thyroid hormone on neoplastic transformation of cells in culture induced by X-irradiation,[57,58] chemicals,[59,70] viruses,[71] or DNA transfection.[72] In attempting to understand the molecular basis of altered erythroblast differentiation in viral *erb*-A infected cells,[73] it was determined that the protein product of the *erb*-A oncogene is the nuclear receptor for thyroid hormone.[54,55] It is easy to speculate that other hormone receptors are intimately involved in differentiation and, when inappropriately expressed, may mediate specific pathways to oncogenesis.

C. MESODERMAL LINEAGE DETERMINATION OF X-REF-23 RAT EMBRYO CELLS BY X-IRRADIATION

As previously mentioned, our laboratory has established a rat embryo fibroblastic cell line that is characterized as immortal, nontumorigenic, unable to grow in soft agar, and is a normal diploid karyotype.[74] This cell line is susceptible to carcinogen-induced neoplastic transformation,[74] and was found to spontaneously differentiate along mesodermal lineages to form muscle and adipocytes.[52]

We also performed a series of experiments to determine whether carcinogens could induce lineage determination in fibroblastic clones of X-REF-23 that were not able to spontaneously commit to lineage differentiation.[74] While ultraviolet light, 3-methylcholanthrene, and the tumor promoter TPA were not able to convert the fibroblasts to differentiated phenotypes, it was found that X-irradiation was a powerful inducer of lineage-determined differentiation. The x-irradiation-induced conversion of X-REF-23 fibroblasts to muscle cells and adipocytes exhibited a dose-dependent response. The muscle lineage was induced from 2 to 6 Gy of X-irradiation, reaching a plateau at 6 to 8 Gy. Higher doses of X-irradiation caused a high frequency of cell killing. At 6 to 8 Gy, approximately 35 to 40% of the surviving colonies were muscle phenotype. The frequency of adipogenic colonies increased from 2 to 8 Gy of X-irradiation, reaching 20 to 25% of the surviving colonies. The reasons for the specificity of X-rays, and not the other carcinogens tested, to induce lineage-determined differentiation is unknown.

IV. SUMMARY

A. OTHER CELL SYSTEMS THAT ARE INDUCED TO DIFFERENTIATE BY CARCINOGENS

There have been other reports of carcinogen-induced differentiation. Scher and Friend[75] reported that agents causing altered DNA structure were inducers of erythroid differentiation. They found that the inducers DMSO, X-rays, bleomycin, butyrate, and ultraviolet light caused DNA degradation. Their results suggested that single-stranded scission in DNA may be an early step in the control of differentiation. Prasad[76] demonstrated that X-rays induced morphological differentiation of mouse neuroblastoma cells *in vitro*. The carcinogen-induced differentiation of mouse epidermal cells has been developed as an assay for carcinogens.[77]

B. MESODERMAL LINEAGE DETERMINATION

It has become increasingly evident that understanding the molecular basis of cytodifferentiation will be crucial to understanding the etiology of neoplasia. The induction of neoplastic transformation and differentiation by carcinogens in the nontransformed mesodermal stem cells discussed above (3T3, CHEF/18, C3H/10T$^1/_2$, and X-REF-23) may provide valuable model systems.

The mechanism of action of carcinogens to induce conversion of fibroblasts to mesodermal lineages is not known at the present time. It is not clear whether induced differentiation is a genetic or epigenetic event, or whether the carcinogens have a direct or indirect action. Some of the agents inducing differentiation are mutagens and may create alterations in the genome, resulting in a lineage-determination pathway. It is also known that many of the inducing agents create oxygen-free radicals that induce cellular and DNA damage if not protected by endogenous radical scavenging enzymes, such as manganese-containing superoxide dismutase (SOD, EC.1.15.1.1). Oberley et al.[78] have provided substantial evidence for a role of SOD in the differentiation of the plasmodial slime mold. They found that differentiating cells underwent MnSOD induction, while nondifferentiation cells did not. We have recently observed a similar relationship in the conversion of the X-REF-23 rat preadipocytes and myoblasts to the mesodermal lineages of the adipocyte and muscle phenotypes.[78]

There is also increasing evidence that DNA methylation plays an important role in eukaryotic gene expression and differentiation.[5-7] It is clear that carcinogens can induce both neoplastic transformation and differentiation. It is important to note the report of Wilson and Jones,[79] who provide evidence that carcinogenic agents may cause heritable changes in DNA methylation patterns in some cells by a variety of mechanisms.

REFERENCES

1. **Ham, A. W.,** *Histology,* Lippincott, Philadelphia, 1974.
2. **Goldberg, B. and Rabinovitch, M.,** Connective tissue, in *Histology,* Weiss, L., Ed., Elsevier, New York, 1983.
3. **Krawisz, B. R., Florine, D. L., and Scott, R. E.,** Differentiation of fibroblast-like cells into macrophages, *Cancer Res.,* 41, 2891, 1981.
4. **Reznikoff, C. A., Brankow, D. W., and Heidelberger, C.,** Establishment and characterization of a cloned line of C3H mouse embryo cells sensitive to postconfluence inhibition of division, *Cancer Res.,* 33, 3231, 1973.
5. **Taylor, S. M. and Jones, P. A.,** Multiple new phenotypes induced in 10T$^1/_2$ and 3T3 cells treated with 5-azacytidine, *Cell,* 17, 771, 1979.
6. **Taylor, S. M. and Jones, P. A.,** Changes in phenotypic expression in embryonic and adult cells treated with 5-azacytidine, *J. Cell. Physiol.,* 111, 187, 1982.
7. **Konieczny, S. F. and Emerson, C. P.,** 5-Azacytidine induction of stable mesodermal stem cell lineages from 10$^1/_2$ cells: evidence for regulatory genes controlling determination, *Cell,* 38, 791, 1984.
8. **Green, H. and Kehinde, O.,** Sublines of mouse 3T3 cells that accumulate lipid, *Cell,* 1, 113, 1974.
9. **Sager, R. and Kovac, P.,** Pre-adipocyte determination either by insulin or by 5-azacytidine, *Proc. Natl. Acad. Sci. U.S.A.,* 79, 480, 1982.
10. **Kopelovich, L., Rich, R., and Wallace, A.,** Hydrocortisone promotes the neodifferentiation of Kirsten murine sarcoma virus transformed human skin fibroblasts to adipocytes: relevance to oncogenic mechanisms, *Exp. Cell Biol.,* 54, 25, 1986.

11. **Kopelovich, L., Wallace, A., and Rich, R. F.,** The kinetics of neodifferentiation of Kirsten murine sarcoma virus transformed human skin fibroblasts to adipocyte cells by hydrocortisone, *Cancer Invest.,* 5, 567, 1987.

12. **Kopelovich, L. and Fenyk, J.,** Transformation/neodifferentiation of human skin fibroblasts by acute oncornaviruses and dexamethasone, *Exp. Cell Biol.,* 56, 311, 1988.

13. **Konieczny, S. F. and Emerson, C. P.,** Differentiation, not determination, regulates muscle gene activation: transfection of troponin I genes into multipotential and muscle lineages of 10T1/2 cells, *Mol. Cell. Biol.,* 5, 2423, 1985.

14. **Lassar, A. B., Paterson, B. M., and Weintraub, H.,** Transfection of a DNA locus that mediates conversion of 10T1/2 fibroblasts to myoblasts, *Cell,* 47, 649, 1986.

15. **Davis, R. L., Weintraub, H., and Lassar, A. B.,** Expression of a single transfected cDNA converts fibroblasts to myoblasts, *Cell,* 51, 987, 1987.

16. **Straus, D. S.,** Somatic mutation, cellular differentiation, and cancer causation, *JNCI,* 67, 233, 1981.

17. **Pierce, G. B.,** Teratocarcinoma: model for a developmental concept of cancer, *Curr. Top. Dev. Biol.,* 2, 223, 1967.

18. **Markert, C. L.,** Neoplasia: a disease of cell differentiation, *Cancer Res.,* 28, 1908, 1968.

19. **Harris, H.,** Some thoughts about genetics, differentiation, and malignancy, *Somatic Cell. Genet.,* 5, 923, 1979.

20. **Pierce, G. B. and Wallace, C.,** Differentiation of malignant to benign cells, *Cancer Res.,* 31, 127, 1971.

21. **Rubin, H.,** Cancer as a dynamic developmental disorder, *Cancer Res.,* 45, 2935, 1985.

22. **Sachs, L.,** Constitutive uncoupling of pathways of gene expression that control growth and differentiation in myeloid leukemia: a model for the origin and progression of malignancy, *Proc. Natl. Acad. Sci. U.S.A.,* 77, 6152, 1980.

23. **Wille, J. J., Maercklein, P. B., and Scott, R. E.,** Neoplastic transformation and defective control of cell proliferation and differentiation, *Cancer Res.,* 42, 5139, 1982.

24. **Scott, R. E. and Maerclein, P. B.,** An initiator of carcinogenesis selectively and stably inhibits stem cell differentiation: support for a concept that the initiation of carcinogenesis involves multiple phases, *Proc. Natl. Acad. Sci. U.S.A.,* 82, 2995, 1985.

25. **Wier, M. L. and Scott, R. E.,** Defective control of terminal differentiation and its role in carcinogenesis in the 3T3 T proadipocyte stem cell line, *Cancer Res.,* 45, 3339, 1985.

26. **Sparks, R. L., Seibel-Ross, E. I., Weir, M. L., and Scott, R. E.,** differentiation, dedifferentiation, and transdifferentiating of BALB/c 3T3 T mesenchymal stem cells: potential significance in metaplasia and Neoplasia, *Cancer Res.,* 46, 5312, 1986.

27. **Estervig, D. N., Maercklein, P. B., and Scott, R. E.,** Resistance to neoplastic transformation induced by nonterminal differentiation, *Cancer Res.,* 49, 1008, 1989.

28. **Muller, R., Slamon, D. J., Adamson, E. D., Tremblay, J. M., Muller, D., Cline, M. J., and Verma, I. M.,** Transcription of c-*onc* genes c-*ras*^{Ki} and c-*fms* during mouse development, *Mol. Cell. Biol.,* 3, 1062, 1983.

29. **Propst, F., Rosenberg, M. P., Iyer, A., Kaul, K., and Vande Woude, G. F.,** c-*mos* proto-oncogene RNA transcripts in mouse tissues: structural features, developmental regulation, and localization in specific cell types, *Mol. Cell. Biol.,* 7, 1629, 1987.

30. **Fults, D. W., Towle, A. C., Lauder, J. M., and Maness, P. F.,** pp60^{c-src} in the developing cerebellum, *Mol. Cell. Biol.,* 5, 27, 1985.

31. **Muller, R., Slamon, D. J., Tremblay, J. M., Cline, M. J., and Verma, I. M.,** Differential expression of cellular oncogenes during pre- and postnatal development in the mouse, *Nature,* 299, 640, 1982.

32. **Cotton, P. C. and Brugge, J. S.,** Neural tissues express high levels of the cellular src gene product, *Mol. Cell. Biol.,* 3, 1157, 1983.

33. **Brugge, J. S., Cotton, P. C., Queral, A. E., Barrett, J. N., Nonner, D., and Keane, R. W.,** Neurons express high levels of a structurally modified, activated form of pp60^{c-src}, *Nature,* 316, 554, 1985.

34. **Gee, C. E., Griffin, J., Sastre, L., Miller, L. J., Springer, T. A., Piwnica-Worms, H., and Roberts, T. M.,** Differentiation of myeloid cells is accompanied by increased levels of pp60^{c-src} protein and kinase activity, *Proc. Natl. Acad. Sci. U.S.A.,* 83, 5131, 1986.

35. **Griep, A. E. and DeLuca, H. F.,** Decreased c-*myc* expression is an early event in retinoic acid-induced differentiation of F9 teratocarcinoma cells, *Proc. Natl. Acad. Sci. U.S.A.,* 83, 5539, 1986.

36. **Grosso, L. E. and Pitot, H. C.,** Transcriptional regulation of c-*myc* during chemically induced differentiation of HL-60 cultures, *Cancer Res.,* 45, 847, 1985.

37. **Yen, A. and Guernsey, D. L.,** Increased c-*myc* RNA levels associated with the precommitment state during HL-60 myeloid differentiation, *Cancer Res.,* 46, 4156, 1986.

38. **Thiele, C., Reynolds, C. P., and Isreal, M. A.,** Decreased expression of N-*myc* precedes retinoic acid-induced morphological differentiation of human neuroblastoma, *Nature (London),* 313, 404, 1985.

39. **Guernsey, D. L. and Yen, A.,** Retinoic acid induced modulation of c-*myc* not dependent on its continued presence: possible role in precommitment for HL-60 cells, *Int. J. Cancer,* 42, 50, 1989.

40. **Colletta, G., Pinto, A., Di Fiore, P. P., Fusco, A., Ferrentino, M., Avvedimento, V. E., Tsuchida, N., and Vecchio, G.,** Dissociation between transformed and differentiated phenotype in rat thyroid epithelial cells after transformation with a temperature-sensitive mutant of the Kirsten murine sarcoma viruses, *Mol. Cell. Biol.,* 3, 2099, 1983.

41. **Alema, S., Casalbore, P., Agostini, E., and Tato, F.,** Differentiation of PC12 phaeochromocytoma cells induced by v-*src* oncogene, *Nature (London),* 316, 557, 1985.

42. **Hankins, W. D. and Scolnick, E. M.,** Harvey and Kirsten sarcoma viruses promote the growth and differentiation of erythroid precursor cells *in vitro, Cell,* 26, 91, 1981.

43. **Bar-Sagi, D. and Feramisco, J. R.,** Microinjection of the *ras* oncogene protein into PC12 cells induces morphological differentiation, *Cell,* 42, 841, 1985.

44. **Pierce, J. H. and Aaronson, S. A.,** Myeloid cell transformation by *ras*-containing murine sarcoma viruses, *Mol. Cell. Biol.,* 5, 667, 1985.

45. **Noda, M., Ko, M., Ogura, A., Liu, D., Amano, T., Takano, T., and Ikawa, Y.,** Sarcoma viruses carrying *ras* oncogenes induce differentiation-associated properties in a neuronal cell line, *Nature (London),* 318, 1985.

46. **Falcone, G., Tato, F., and Alema, S.,** Distinctive effects of the viral oncogenes *myc, erb, fps,* and *src* on the differentiation program of quail myogenic cells, *Proc. Natl. Acad. Sci. U.S.A.,* 82, 426, 1985.

47. **Olson, E. N., Spizz, G., and Tainsky, M. A.,** The oncogenic forms of N-*ras* or H-*ras* prevent skeletal myoblast differentiation, *Mol. Cell. Biol.,* 7, 2104, 1987.

48. **Schneider, M. D., Perryman, M. B., Payne, P. A., Spizz, G., Roberts, R., and Olson, E. N.,** Autonomous expression of c-*myc* in BC₃H1 cells partially inhibits but does not prevent myogenic differentiation, *Mol. Cell. Biol.,* 7, 1973, 1987.

49. **Tapscott, S. J., Davis, R. L., Thayer, M. J., Cheng, P.-F., Weintraub, H., and Lassar, A. B.,** MyoD1: a nuclear phosphoprotein requiring a myc homology region to convert fibroblasts to myoblasts, *Science,* 242, 405, 1988.

50. **Sejersen, T., Wahrmann, J. P., Sumegi, J., and Ringertz, N. R.,** Change in expression of oncogenes (*sis, ras, myc,* and *abl*) during *in vitro* differentiation of L6 rat myoblasts, in *Molecular Biology of Tumor Cells,* Wahren, B., Ed., Raven Press, New York, 1985, 243.

51. **Distel, R. J., Ro, H.-S., Rosen, B. S., Groves, D. L., and Spiegelman, B. M.,** Nucleoprotein complexes that regulate gene expression in adipocyte differentiation: direct participation of c-*fos, Cell,* 49, 835, 1987.

52. **Ketelaar, D. A., Utesh, G. R., Sierra, E., and Guernsey, D. L.,** Mesodermal cell lineage determination and proto-oncogene expression, *Cell Differ. Dev.,* 25, 89, 1989.

53. **Goodridge, A. G.,** Regulation of malic enzyme in hepatocytes in culture: a model system for analyzing the mechanism of action of thyroid hormone, in *Molecular Basis of Thyroid Hormone Action,* Oppenheimer, J. H. and Samuels, H. H., Eds., Academic Press, New York, 1983, 246.

54. **Sap, J., Munoz, A., Damm, K., Goldberg, Y., Ghysdael, J., Leutz, A., Beug, H., and Vennstrom, B.,** The c-*erb*-A protein is a high-affinity receptor for thyroid hormone, *Nature (London),* 324, 635, 1986.

55. **Weinberger, C., Thompson, C. C., Ong, E. S., Lebo, R., Gruol, D. J., and Evans, R. M.,** The c-*erb*-A gene encodes a thyroid hormone receptor, *Nature (London),* 324, 641, 1986.

56. **Terzaghi, M. and Little, J. B.,** X-irradiation-induced transformation in a C3H mouse embryo-derived cell line, *Cancer Res.,* 36, 1367, 1976.

57. **Guernsey, D. L., Borek, C., and Edelman, I. S.,** Crucial role of thyroid hormone in x-ray-induced neoplastic transformation in cell culture, *Proc. Natl. Acad. Sci. U.S.A.,* 78, 5708, 1981.

58. **Guernsey, D. L., Ong, A., and Borek, C.,** Thyroid hormone modulation of x-ray-induced *in vitro* neoplastic transformation, *Nature (London),* 288, 591, 1980.

59. **Borek, C., Guernsey, D. L., Ong, A., and Edelman, I. S.,** Critical role played by thyroid hormone in induction of neoplastic transformation by chemical carcinogens in tissue culture, *Proc. Natl. Acad. Sci. U.S.A.,* 80, 5749, 1983.

60. **Guernsey, D. L., Leuthauser, S. W. C., and Koebbe, M. J.,** Induction of cytodifferentiation in C3H/10T1/2 mouse embryo cells by x-irradiation and benzo(a)pyrene, *Cell Differ.,* 16, 147, 1985.

61. **Reznikoff, C. A., Bertram, J. S., Brankow, D. W., and Heidelberger, C.,** Quantitative and qualitative studies of chemical transformation of cloned C3H mouse embryo cells sensitive to postconfluence inhibition of cell division, *Cancer Res.,* 33, 3239, 1973.

62. **Miller, R. and Hall, E. J.,** X-ray dose fractionation and oncogenic transformation in cultured mouse embryo cells, *Nature (London),* 272, 58, 1978.

63. **Kennedy, A. R. and Weichselbaum, R. R.,** Effects of dexamethasone and cortisone with x-irradiation on transformation of C3H/10T1/2 cells, *Nature (London),* 294, 97, 1981.

64. **Kennedy, A. R.,** Cortisol suppresses radiation transformation in vitro, *Cancer Lett.,* 29, 289, 1985.

65. **Guernsey, D. L. and Schmidt, T. J.,** Corticosterone effects on differentiation and x-ray-induced transformation of C3H/10T1/2 mouse cells, *Cell Differ.,* 24, 159, 1988.

66. **Chapman, A. B., Knight, D. M., Dieckman, B. S., and Ringold, G. M.,** Analysis of gene expression during differentiation of adipogenic cells in culture and hormonal control of the developmental program, *J. Biol. Chem.,* 259, 15548, 1984.

67. **Chapman, A. B., Knight, D. M., and Ringold, G. M.,** Glucocorticoid regulation of adipocyte differentiation: hormonal triggering of the developmental program and induction of a differentiation-dependent gene, *J. Cell Biol.,* 101, 1227, 1985.

68. **Schwartz, H. L.,** Effect of thyroid hormone on growth and development, in *Molecular Basis of Thyroid Hormone Action,* Oppenheimer, J. H. and Samuels, H. H., Eds., Academic Press, New York, 1983, chap. 14.

69. **Galton, V. A.,** Thyroid hormone action in amphibian metamorphosis, in *Molecular Basis of Thyroid Hormone Action,* Oppenheimer, J. H. and Samuels, H. H., Eds., Academic Press, New York, 1983, chap. 15.

70. **Guernsey, D. L. and Leuthauser, S. W. C.,** Correlation of thyroid hormone dose-dependent regulation of K-*ras* proto-oncogene expression with oncogene activation by methylcholanthrene: loss of thyroidal regulation in the transformed mouse cell, *Cancer Res.,* 47, 3052, 1987.

71. **Fisher, P. B., Guernsey, D. L., Weinstein, I. B., and Edelman, I. S.,** Modulation of adenovirus transformation by thyroid hormone, *Proc. Natl. Acad. Sci. U.S.A.,* 80, 196, 1983.

72. **Leuthauser, S. W. C. and Guernsey, D. L.,** Thyroid hormone affects the expression of neoplastic transformation induced by DNA-transfection, *Cancer Lett.,* 35, 321, 1987.

73. **Frykberg, L., Palmieri, S., Beug, H., Graf, T., Hayman, M. J., and Vennstrom, B.,** Transforming capacities of avian erythroblastosis virus mutants deleted in the *erb*A or *erb*B oncogenes, *Cell,* 32, 227, 1983.

74. **Sierra-Rivera, E.,** Multistep nature of neoplastic transformation in rat embryo fibroblasts, Ph.D. thesis, University of Iowa, Iowa City, 1987.

75. **Scher, W. and Friend, C.,** Breakage of DNA and alterations in folded genomes by inducers of differentiation in Friend erythroleukemic cells, *Cancer Res.,* 38, 841, 1978.

76. **Prasad, K. N.,** X-ray-induced morphological differentiation of mouse neuroblastoma cell *in vitro, Nature (London),* 234, 471, 1971.

77. **Kulesz-Martin, M. F., Koehler, B., Hennings, H., and Yuspa, S. H.,** Quantitative assay for carcinogen altered differentiation in mouse epidermal cells, *Carcinogenesis,* 1, 995, 1980.

78. **Oberley, L. W., Ridnour, L. A., Sierra-Rivera, E., Oberley, T. D., and Guernsey, D. L.,** Superoxide dismutase activities of differentiating clones from an immortal cell line, *J. Cell. Physiol.,* 138, 50, 1989.

79. **Wilson, V. L. and Jones, P. A.,** Inhibition of DNA methylation by chemical carcinogens in vitro, *Cell,* 32, 239, 1983.

Chapter 7

THE EARLY REACTIONS AND FACTORS INVOLVED IN *IN VITRO* ERYTHROID DIFFERENTIATION OF MOUSE ERYTHROLEUKEMIA (MEL) CELLS

Michio Oishi and Toshio Watanabe

TABLE OF CONTENTS

I. INTRODUCTION

Mouse erythroleukemia (MEL) cells,[1] established by Friend in 1966, undergo dramatic changes in their morphological and biochemical characteristics following exposure to various agents, such as dimethyl sulfoxide (DMSO),[2] hexamethylene *bis*-acetamide (HMBA),[3] and sodium butyrate.[4,5] These changes include synthesis of globin mRNA, globin, and heme, appearance of erythrocyte-specific membrane antigens, and cessation of cell proliferation.[6] Since the cellular changes induced are similar to those of erythroid cells, this experimental system has been a useful model for erythropoiesis. The entire differentiation process, which takes over 100 h to complete, may be separated into two stages. The first (early) stage consists presumably of a cascade of reactions which is initiated by the contact of the cells with inducing agents and is completed by commitment of the cells to differentiate. In the second (late) stage, genes specific to erythroid cells are sequentially activated. In this review, we will deal with only the first (early) stage of differentiation, with special emphasis on the nature of the reactions and intracellular factors involved in this stage of differentiation.

II. DNA METABOLISM AND MEL CELL DIFFERENTIATION

To date, most of the experiments examining the mechanism of MEL cell differentiation have been conducted by measuring changes in various biochemical parameters following exposure of the cells to inducing agents such as DMSO or HMBA. The changes frequently observed include alteration of membrane structure,[7-9] changes in signal transduction,[10,11] oncogene expression,[12-17] polyamine metabolism,[18-21] Ca ion influx,[22-27] changes in the methylation of DNA,[28-30] and arrest of the cells at certain stages of the cell cycle.[31-34] These experiments, however, often failed to provide concrete evidence whether the observed change is the cause or the effect (result) of differentiation. In fact, despite these numerous broad and extensive studies, it must be emphasized that the molecular mechanism by which MEL cells transmit signals from outside inducers and eventually commit themselves to differentiate still remains unknown.

One of the technical problems in analyzing the intracellular reactions at the early stage of differentiation is that MEL cells must be exposed to inducing agents for a considerable length of time (often over 24 h), as well as at relatively high inducer concentrations, to obtain an efficient level of differentiation. Under such conditions, inducing agents should affect a large number of intracellular reactions, thus making it difficult to pinpoint whether an observed change is actually responsible for differentiation. To circumvent this problem, the early reactions have been analyzed by cell fusion of two genetically marked MEL cells (Tk$^-$ Hgprt$^+$ and Tk$^+$ Hgprt$^-$).[35] Nomura and Oishi[35] treated one of these two types of mutant cells with a suboptimal concentration (at which no erythroid differentiation was observed) of a known inducing agent and fused them to another mutant cell which had been treated with a different inducing agent (also at a suboptimal level). The fused cells were then incubated in HAT medium, and differentiated cells were counted among the surviving cells. The cell fusion experiments were able to dissociate the early stage of differentiation into two different components, and the results suggested that the early reaction consisted of at least two different inducible reactions which act synergistically to induce MEL cell commitment. The first reaction is induced by agents which affect DNA replication, and the second reaction is induced by typical inducing agents such as DMSO or HMBA (even at suboptimal concentrations).

The effect of agents which affect DNA metabolism or damage DNA on MEL cell differentiation and its synergism with conventional inducing agents, such as DMSO or HMBA, have been described previously. In earlier studies, Terada et al.[36] and Scher and Friend[37] demonstrated that following ultraviolet light (UV) irradiation, MEL cells became very sensitive to the induction of differentiation by DMSO or HMBA. To explain this phenomenon, they proposed that changes in chromatin structure mediated by differentiation-inducing agents were an important, although not exclusive, factor in MEL cell differentiation. In fact, they also showed that DMSO and HMBA induced alterations in DNA structure. Involvement of DNA nicking in MEL cell differentiation was also suggested by the experiments of McMahon et al.,[38] but challenged by Pulito et al.[39] who obtained different experimental results. In this connection, Terada et al.[40] showed that inhibitors of poly(ADP-ribose) polymerase, such as nicotinamide, also induce MEL cell differentiation. Since it is well established that poly(ADP-ribose) polymerase is associated with the repair of UV-damaged DNA in mammalian cells, the effect of inhibitors of this enzyme on MEL cell differentiation is also likely to be related to the impairment of repair or changes in DNA structure. In this respect, it is interesting to note that when MEL cells were arrested in the G1 period of the cell cycle by differentiation-inducing agents, poly(ADP-ribose) polymerase activity in the cells was greatly increased.[41] The relationship between the induction of MEL cell differentiation by poly(ADP-ribose) polymerase inhibitors and the synergistic effect of UV treatment, however, has to be examined with caution, since poly(ADP-ribose) inhibitors induce MEL cell differentiation by themselves. In addition, poly(ADP-ribose) inhibitors also induce *in vitro* differentiation of other cell types, such as embryonal carcinoma cells, in which DNA damaging agents apparently have little effect on differentiation, with the exception of tissue plasminogen activator production by MMC in F9 cells.[42]

DNA methylation (and demethylation) has been the focus of extensive research for its role in the regulation of gene expression, although the generality of its effect on gene regulation in eukaryotic cells has been the subject of much debate. An interesting observation relative to the possible involvement of DNA methylation (demethylation) in MEL cell differentiation has come from work by Razin et al.[28] They provided evidence for temporal demethylation of genomic DNA following exposure of MEL cells to a differentiation-inducing agent (HMBA).[28] This demethylation, triggered within 12 to 18 h after HMBA treatment, appeared to be caused by a novel biochemical reaction, replacement of a cytosine moiety by 5'-methylcytosine. A possible cause-effect relationship between demethylation and the induction of differentiation was suggested by subsequent studies by Razin et al.[29] They demonstrated that inhibition of demethylation by deazaadenosine and homocysteine treatment also inhibited MEL cell differentiation, suggesting that the demethylation process was directly associated with differentiation.[29] However, since cycloheximide and dexamethasone inhibited the induction of differentiation, but not demethylation, it is unlikely that the demethylation reaction alone is sufficient to trigger differentiation. The importance of demethylation in the induction of MEL cell differentiation was also supported by the studies of Creusot et al.,[30] who showed that an inhibitor of 5'-methyltransferase, 5'-aza-2'-deoxycytidine, induced MEL cell differentiation. These interesting results also implicate DNA metabolism in MEL cell differentiation, but the relationship to the other effects caused by DNA damaging agents or poly(ADP-ribose) inhibitors is unclear.

The experiments described above on the effect of agents which modify DNA metabolism on MEL cell differentiation, nevertheless, have provided several clues to the nature of the early reactions. Since agents such as cytosine arabinoside or hydroxyurea have the same effect on MEL cell differentiation as UV or mitomycin (MMC),[76] inhibition of DNA replication (or cessation of cell division as a possible consequence) rather than simply damage to DNA structure seems to be an important factor in differentiation. Furthermore, the reaction responsible for the induction by these agents is likely to be indispensable for cell survival, since among several dozen mutant MEL cells which were unable to differentiate, none were defective in the UV (or MMC)-induced reaction.[77] The DNA damage-induced reaction studied by cell fusion had the following characteristics:[35] (1) the reaction triggered by these agents apparently accompanied *de novo* protein synthesis and reached a maximum activity at approximately 24 h incubation after the initial treatment and (2) the induced reaction is not specific to MEL cells, since UV-irradiated baby hamster kidney (BHK) cells or other cell lines also induced the reaction. These experiments implicate a factor, which is not necessarily specific to MEL cells, which is induced in these cells when DNA replication is inhibited. The factor seems to act cooperatively (synergistically) with an additional factor (see below) to induce MEL cell differentiation.

III. CELL PROLIFERATION AND CELL CYCLE

A close relationship between cell differentiation and cell proliferation is suggested because DNA replication is tightly coupled to cell proliferation and the cell cycle. In fact, cell proliferation has been considered to be a primary cellular function involved in cell differentiation. It is well established that during differentiation MEL cells, as well as other cell types, undergo dramatic alterations in their ability to proliferate, frequently losing the capacity to divide. Furthermore, there are numerous studies supporting an association between cell cycle and MEL cell differentiation. For example, Terada et al.[31] showed that, after the addition of differentiation-inducing agents, the G1 period in the MEL cell cycle was temporarily elongated. They suggested that this change in the cell cycle mediated by inducing agents was essential for differentiation. Friedman et al.,[43] however, questioned this interpretation after demonstrating that the extension of the G1 period was not necessarily followed by differentiation and therefore, was not a prerequisite for differentiation.

In any case, since most of the MEL cell differentiation-inducing agents do cause changes in the pattern of the cell cycle, the presence of a certain period in the cell cycle where the cells are particularly sensitive to inducing agents is a possibility. A typical experiment to illustrate this point was conducted by Gambari et al.,[44] who analyzed the induction of differentiation markers at different periods of the cell cycle. They showed that following exposure to differentiation-inducing agents, transcription of globin mRNA was initiated in the G1 period after having passed one S period. This is in agreement with the results of Mclintock[45] and Levy et al.,[46] who demonstrated, using synchronized cells, the necessity for the cells to undergo one cell cycle following treatment with inducing agents to differentiate. Furthermore, temperature-sensitive MEL cell mutants that were arrested in the G1 period at nonpermissive temperature could not be induced to differentiate when they were exposed to differentiation-inducing agents at the G1 period, while they were normally inducible in the other stages of the cell cycle.[47]

Aphidicolin is a specific inhibitor of DNA polymerase alpha and blocks the cell cycle at the S period. Beckman et al.[32] showed that aphidicolin by itself could induce MEL cell differentiation, but they also found that the combination of aphidicolin and HMBA reduced the time required for the appearance of differentiation markers. They argued that the inhibition of cell cycle progression is an initiation reaction for the induction of MEL cell differentiation.[32] Yoshida et al.[33] reported that trichostatin, originally isolated as an antifungal agent, is an extremely effective inducer of MEL cell differentiation. Trichostain inhibits cell proliferation by blocking the entry of cells into the G1 and G2 periods of the cell cycle.[34] These experiments suggest that inhibition of the cell cycle at a specific stage is important for the induction of MEL cell differentiation and implicate an intracellular differentiation-inducing factor(s) that is induced or accumulated at a certain stage of the cell cycle.

How are we able to reconcile these experiments with the fact that typical inducing agents such as DMSO or HMBA, when employed alone, can efficiently induce MEL cells to differentiate, although at relatively high concentrations and after a prolonged exposure? One possible answer to this question would be that most of the known inducing agents have dual effects on MEL cells. One effect is to generate a specific signal(s) through the cell membrane and the other is to affect DNA metabolism. If this is the case, then agents capable of causing both effects may now be classified as the inducing agents for erythroid differentiation in MEL cells.

IV. REACTIONS INDUCED BY TRANSMEMBRANE SIGNALS

By employing cell fusion, Kaneko et al.[48] characterized the transmembrane signal reaction, which is induced by a brief exposure of MEL cells to DMSO or HMBA and acts synergistically with the reaction induced by the interference of DNA metabolism (or cell division). They found that the transmembrane signal reaction was fully induced by exposure of MEL cells to DMsO (HMBA), even for a short period (2 h with 2% DMSO), a situation which normally does not induce erythroid differentiation. The induced reaction (resulting from transmembrane signaling) was detected only after cell fusion with UV (or MMC)-treated cells. This inducible activity remained active only transiently, since it reached a maximum at 6 to 8 h after treatment with DMSO (2 h with 2% DMSO) and declined to the pretreatment level after 10 to 16 h incubation. Besides the transient nature of this inducible activity, several interesting features were found to be associated with it. First, the differentiation-inducing activity was specific to MEL cells, since it was not induced in other cell lines in which the transmembrane signal reaction was induced. Second, the inducible factor(s) was completely inhibited in the presence of cycloheximide, suggesting that *de novo* protein synthesis was involved in this reaction. Interestingly, addition of cycloheximide at the peak of the induction blocked the decline (turnover?) of this differentiation-inducing activity. Furthermore, inhibitors of mitochondrial protein synthesis, such as chloramphenicol or tetracycline, also inhibited the induction as well as the turnover of this activity.[49] The effect of inhibitors of mitochondrial protein synthesis on MEL cell differentiation was quite surprising and difficult to understand, based on our current understanding of the functions associated with mitochondria.

By employing cytoplast fusion rather than cell fusion, Watanabe et al.[50] showed that cytoplasts prepared from DMSO (or HMBA)-treated cells were also able to trigger MEL cell differentiation after fusion with UV-irradiated cells. The inducing activity in the cytoplasts appears to have the same characteristics observed with the cell fusion experiments described above. The differentiation-inducing activity, however, was not induced when cytoplasts were directly treated with DMSO (or HMBA). These results indicate that the intracellular erythroid-inducing activity (1) is located in cytoplasts, (2) acts in *trans* and induces erythroid differentiation as a dominant factor, and (3) requires nuclei for its production.

Several agents are known to inhibit MEL cell differentiation in a specific manner. In their classical works, Rovera et al.[51] and Yamasaki et al.[52] showed that TPA *(12-0-tetradecanoylphorbol-13-acetate)* specifically inhibits the induction of MEL cell differentiation. Since TPA is now known to be an activator of protein kinase C in place of its natural activator, diacylglycerol,[53] it is reasonable to assume that signal transduction involving protein kinase C plays a central role in the differentiation cascade.[54] A number of reports have dealt with the possible relationship between the signal transduction and the early cascade leading to differentiation. TPA inhibited the DMSO-inducible transient induction detected by cell and cytoplast fusion experiments, but had no effect on the induction of the differentiation-inducing activity by DNA damaging agents.[48,50] It seems that TPA inhibits MEL cell differentiation by blocking the induction of one of the two inducible activities required for differentiation.

Faletto et al.[10] measured the metabolic turnover of phosphatidylinositol in MEL cells following exposure to inducing agents and found that there was a decrease in the levels of both inositol triphosphate and diacylglycerol within 2 h after exposure to the inducer. Furthermore, diacylglycerol analogues L-α-1-oleoyl-2-acetyl-*sn*-3-glycerol and sn-1,2-dioctanoylglycerol were found to inhibit the induction of differentiation, suggesting that the decrease of diacylglycerol by inducing agents is critical in the induction of MEL cell differentiation. The importance of a decreased diacylglycerol level in MEL cell differentiation was also suggested by studies conducted by Pincus et al.,[11] who showed that addition of diacylglycerol to the medium or phospholipase C treatment of the cells inhibited the induction of MEL differentiation. The inhibition by diacylglycerol, however, was only partial, implying that factors in addition to the decreased diacylglycerol level are required in triggering MEL cell differentiation. In related work, Melloni et al.[55] reported that C-kinase activity in membrane fractions was reduced shortly (within 5 h) after the addition of HMBA, whereas C-kinase activity in the cytoplasm increased. Since this change was inhibited by a protease inhibitor, leupeptin, M-kinase produced from C-kinase, which was mediated by a protease reaction, was implicated in this reaction. Leupeptin also inhibited the induction of MEL cell differentiation by HMBA, suggesting that the transfer of C-kinase from membrane to cytoplasm is important for the transmembrane signaling event in MEL cell differentiation. Furthermore, addition of TPA reduced the total as well as membrane C-kinase activity, which suggested C-kinase itself may be a member of the induction cascade. This was also supported by the observations that (1) induction of MEL cell differentiation by HMBA did not occur in MEL cells in which C-kinase activity was reduced by growth arrest and (2) HMBA-induced differentiation was restored following recovery of C-kinase activity.[55] It is interesting that bryostatin, a functional analogue of TPA and an activator of protein kinase C, reversed the inhibition by TPA of MEL cell differentiation.[56]

In addition to the examples described above, there have been an increasing number of reports concerning the possible role of signal transduction in inducing MEL cell differentiation. Most of the experimental results, however, are quite circumstantial and it is still too early to tell whether or not signal transduction is, in fact, a part of the early induction cascade. Recent experiments indicate that biological reactions regulated by protein kinases are quite diversified and interrelated with each other. This diversity of effects probably contributes to some of the conflicting experimental results and their interpretations in studies on the role of signal transduction in MEL cell differentiation.

V. ONCOGENE EXPRESSION AND MEL CELL DIFFERENTIATION

A number of reports have focused on changes in oncogene expression, as well as possible functions executed by various oncogenes, in mediating cell differentiation. Because most of the established cell lines which undergo *in vitro* differentiation, including MEL cells, are transformed, the expression of oncogenes was expected to affect induction of cell differentiation. In earlier work, Graf et al.[57] demonstrated that chicken red blood cell differentiation was inhibited by the expression of *erb*A and *erb*B, providing evidence that oncogene expression affects, in this situation adversely, induction of erythroid differentiation. In fact, the level of expression of some of the oncogenes change significantly during cell differentiation. In MEL cell differentiation, Lachman et al.[12] showed that the c-*myc* RNA level fluctuated in a complex manner. The level decreased immediately after the addition of inducing agents, but was restored quickly to the pretreatment level in the next 24 h. The c-*myc* mRNA level then decreased again, remaining at low levels thereafter. The complex two-phase change in the expression of the c-*myc* gene was also investigated by employing inhibitors of MEL cell differentiation, such as dexamethasone and TPA, and mutant MEL cells defective in differentiation.[13] Whereas the decrease in c-*myc* RNA level immediately after the addition of inducing agents was also observed in MEL cells which failed to differentiate (due to the presence of inhibitors or the mutation), the second decrease occurred only in MEL cells where the differentiation was actually induced. These results suggest a close association of the second, not the first, c-*myc* RNA decrease with MEL cell differentiation.

The direct effect of c-*myc* gene expression on MEL cell differentiation was demonstrated by introducing an exogenous c-*myc* gene into MEL cells. In their elegant studies, Coppola and Cole,[58] and Dmitrovsky et al.[59] showed that MEL cells, in which a foreign c-*myc* gene was constitutively expressed, could not be induced to differentiate. This was the first experimental evidence that oncogene expression actually inhibits cellular differentiation. A similar conclusion was reached by experiments employing MEL cells in which c-*myc* gene expression was regulated by outside inducers.[60,61] Kume et al.[61] showed that, of the two phases in c-*myc* gene expression reported by Lachman et al.,[12] the second decrease in c-*myc* gene expression was essential for induction of MEL cell differentiation. These findings were consistent with the conclusion reached by Sasaki et al.[13] Prochownik et al.[62] analyzed the effect of c-*myc* expression on the differentiation of MEL cells by employing an antisense c-*myc* gene. They found that a decrease in the c-*myc* RNA level due to the expression of the c-*myc* antisense RNA by itself was not sufficient to induce MEL cell differentiation. However, the combination of the decrease in c-*myc* and exposure of the cells to differentiation-inducing agents was sufficient for MEL cell differentiation. These results strongly suggest that expression of the c-*myc* gene suppresses MEL cell differentiation but the decrease in the c-*myc* RNA level is only one of the essential components required for completing the induction process. In more recent studies employing cell fusion between MEL cells, in which an artificially introduced c-*myc* gene had been placed under the control of human metallothionein promoter, it was demonstrated that the expression of the c-*myc* gene blocked only one of the two early inducible reactions (the DMSO- or HMBA-induced reaction) required for initiating differentiation.[63] The effect of c-*myc* gene expression on the induction of the transmembrane signaling activity was very rapid (within less than 1 h after its expression), suggesting a close association of the effect of c-*myc* gene expression with the early stage of MEL cell differentiation.

The expression of another oncogene, c-*myb*, has also been reported to fluctuate in a manner similar to that of c-*myc* during the induction process.[14,15] The expression of an exogenous c-*myb* gene also inhibited the induction of differentiation in a manner similar to the expression of the c-*myc* gene.[64] A transient increase in c-*fos* mRNA, as well as c-K*ras* mRNA, during MEL cell differentiation has also been reported.[15] The level of p53 protein, which is believed to play a role in the proliferation-related cellular functions, decreased during the initial stage of induction of differentiation.[16] However, p53 mRNA levels decreased in the later stage of differentiation.[17] The biological implication of the rather complex behavior of p53, however, is not clearly understood at the present time. In any event, when the target proteins of each oncogene product are identified, the mechanism by which changes in oncogene expression modulate the induction of MEL cell differentiation will become clearer.

A possible involvement of c-*myc* and other oncogenes in the early differentiation cascade is suggested by the observation that inhibitors of tyrosine phosphorylation by *src* and other oncogenes effectively induce MEL cell differentiation. In recent work, Kondo et al.[65] reported that herbimycin A, one of the benzenoid ansamycin antibiotics and a specific inhibitor for tyrosine kinase associated with the *src* gene product, induced MEL cell differentiation quite effectively. In related work, another inhibitor of tyrosine phosphorylation, genestein (an antibiotic), as well as a synthetic structural analogue of tyrosine (ST638, α-cyano-3-ethoxy-4-hydroxy-5-phenyl-thiomethylcinnamamide) were also found to be inducers of MEL cell differentiation.[66] These compounds acted only synergistically, however, with agents which damage DNA, such as UV or MMC. An interesting observation is that these agents are not limited to the induction of MEL cell differentiation. Herbimycin A also induced mouse embryonal carcinoma (F9) cell differentiation,[65] which is normally induced by retinoic acid, and human K562 differentiation.[67] These cell types are believed to be induced through mechanisms different from those functioning in the MEL cell system because none of the typical inducing agents for MEL cell differentiation (DMSO or HMBA) induce differentiation of F9 and K562 cells. In contrast, inhibitors of tyrosine phosphorylation implicate a common mechanism which underlies differentiation in a variety of cell systems. At present, we do not know the precise molecular mechanism by which the inhibition of protein phosphorylation at tyrosine residues induced differentiation, but these observations suggest a possible link between cell differentiation and cell proliferation.

Potentially important recent insights into the mechanism of MEL cell differentiation have come from the studies of Marks, Rifkind, and associates.[68] They demonstrated that MEL cells resistant to vincristine, an alkaloid antitumor agent, were very sensitive to induction by HMBA.[68] The cells were induced to differentiate not only at very low HMBA concentrations, but also reached commitment in a shorter period, suggesting that a factor induced by HMBA was constitutively expressed in the vincristine-resistant cells. The interesting behavior of vincristine-resistant MEL cells is not easily explained at the molecular level, but a clue to the mechanism may lie in the fact that vincristine inhibits cell division by interfering with the assembly and functions of tubulin and related intracellular cytoskeletal proteins.

VI. DETECTION OF DIFFERENTIATION-INDUCING FACTORS (DIF) IN CELL-FREE EXTRACTS

One of the important conclusions derived from the experiments described above is that, although the early sequence of events leading to MEL cell differentiation is quite complex and consists of an array of time-consuming events, there seems to be *trans*-acting intracellular factors involved in these reactions. Cell and cytoplast fusion experiments described previously strongly implicate such *trans*-acting factors. Nomura et al.[69] attempted to identify the putative intracellular factors responsible for erythroid differentiation in MEL cells. MEL cells which had been briefly exposed to DMSO and permeabilized to proteins were exposed to cell-free extracts from MEL cells treated with UV or MMC. When such cells were cultured for 5 d, the number of cells that accumulated hemoglobin increased severalfold over the untreated cells. The erythroid-inducing activity in the extracts was first apparent at 10 to 15 h after UV (or MMC) treatment and reached a maximal level at approximately 24 h incubation following UV (or MMC) exposure. Although the inducing activity was low, these experiments indicated the presence of an erythroid-inducing factor which appeared in an inducible manner after UV (or MMC) treatment in cell extracts. They showed that the induction process was inhibited by cycloheximide, suggesting that *de novo* protein synthesis was involved in the induction process. The inducing activity was not detected unless the recipient MEL cells were permeabilized to proteins. A similar differentiation-inducing activity was also found in other UV- or MMC-treated nonerythroid cells, such as mouse FM3A or human HeLa cells. Most of the differentiation-inducing activity was located in the cytoplasm, although recent experiments suggested that a minor portion of this inducing activity was also present in nuclei. These characteristics are consistent with a putative *trans*-acting differentiation-inducing factor resulting from inhibition of DNA replication (or metabolism), which was suggested by the previous cell fusion experiments.

This *trans*-acting differentiation-inducing factor was apparently proteinaceous, since it was nondialyzable, sensitive to proteases (trypsin and proteinase K), and lost after heat treatment (60°C for 15 min). The molecular weight of this *trans*-acting factor was approximately 80,000 Da. The partially purified protein induced erythroid differentiation at a level (60 to 70%, benzidine-positive cells) almost equivalent to that attained after DMSO or HMBA treatment. The apparently induced cells lost their proliferative capacity, as was observed with cells induced to differentiate by DMSO or HMBA.[70,71] This factor was termed "differentiation-inducing factor I" (DIF-I).

By using essentially the same strategy employed in the detection of DIF-I, Watanabe and Oishi[72] detected a second factor (DIF-II)[72] which was also suggested by the previous cell and cytoplast fusion experiments.[48,50] Cell-free extracts were prepared from MEL cells which had been exposed to DMSO (1.8%) for 6 h, which should induce a maximal level of the putative DMSO/HMBA-inducible second factor in these cells. The extracts were then introduced into MEL cells which had been exposed to UV treatment, thereby maximizing the production of DIF-I, which would be predicted to act synergistically with the DMSO-inducible factor (DIF-II). Although no erythroid-inducing activity was detected in the crude extract, a further purification utilizing DEAE chromatography revealed an active component in the eluate (at 50 mM NaCl). No similar activity was detected in the same eluate from nontreated control cells. DIF-I activity is usually eluted with 250 mM NaCl from a DEAE column under the same conditions. The DIF-II activity was not detected without UV pretreatment or permeabilization of the recipient cells, suggesting that the induction of DIF-I (by UV treatment) in the recipient cells and incorporation of this factor into the cells are required for detection of DIF-II activity in cell extracts.

Watanabe and Oishi[72] examined the erythroid-inducing activity in the extracts prepared from MEL cells which had been treated with various inducing agents and inhibitors and also from nonerythroid cells (FM3A). In addition to DMSO, DIF-II activity was also induced by HMBA. On the other hand, no DIF-II activity was detected in either the eluate prepared from MEL cells which had been treated with DMSO plus TPA or DMSO plus cycloheximide. Similarly, no DIF-II activity was induced in mouse nonerythroid cells (FM3A) by DMSO. Intracellularly, more than 93% of the DIF-II activity was located in cytoplasts (cytosol).

The DIF-II activity (in the extract) began to appear soon after adding DMSO, reached a maximum at 6 h incubation, but decreased after 10 h, after which it remained at low or nondetectable levels. These characteristics and the induction kinetics, especially with regard to the transient behavior, were very similar to those suggested by the previous cell and cytoplast fusion experiments.[48-50]

DIF-II activity was examined in a mutant MEL cell (DR-I) defective in erythroid differentiation.[72] The level of erythroid induction in DR-I was less than 0.1% (at DMSO, 1.8%; HMBA, 5 mM; or sodium butyate, 1 mM). Cell fusion experiments suggested that DR-1 was normal for DIF-I induction, but defective in the induction of DIF-II. No DIF-II activity was detected in the 250 mM (NaCl) eluate of DR-I cells, whereas a normal level of DIF-I activity was detected in these cells. Thus, the mutant is apparently impaired in the process leading to the induction of DIF-II; these data being consistent with the cell fusion experiments provide further evidence that DIF-II is directly involved in differentiation induced *in vitro* by DMSO or HMBA.

VII. MECHANISM OF THE COMMITMENT TO DIFFERENTIATE IN MEL CELLS

In the preceding sections, we have described recent progress concerning the early molecular events leading to MEL cell differentiation. It is widely believed that during MEL cell differentiation, the cells pass a point at which they commit themselves to differentiate.[70,71,73] In other words, before this point, the induced intracellular reactions in MEL cells do not result in differentiation, and the cells revert to the original uncommitted state when the inducing agents are withdrawn from the medium. In contrast, once the cells pass a specific point, the induction process proceeds independently of inducing agents. Levenson et al.[74] investigated the nature of the commitment process in detail. They showed that when differentiation-inducing agents were added in combination with cordycepin, an analogue of deoxyadenosine and an inhibitor of DNA precursor metabolism, commitment was inhibited. When cordycepin was removed, however, the cells reached a commitment state within only 2 h, provided that inducing agents were continuously present. This experiment suggested that cordycepin blocked the process to commitment immediately before completion, but differentiation-inducing agents were necessary for the cells to progress beyond the point of inhibition by cordycepin. This experiment also indicated that the process to commitment comprises several distinct steps. In related work, Murate et al.[75] showed that an effect similar to the one produced by cordycepin could be produced by dexamethasone, an inhibitor of MEL cell differentiation. In this case, a brief incubation with HMBA after removal of dexamethasone was also necessary for commitment to occur. This HMBA-dependent process for commitment was inhibited by cycloheximide, but not by actinomycin D, implying that mRNA for a gene required for commitment was already accumulated (but not translated) intracellularly in the presence of the inhibitor (dexamethasone).

Despite the importance of elucidating the biochemical nature of commitment, information regarding this process is relatively scarce. Among several questions which may be asked, the most interesting one would be whether there is any commitment factor(s). In this connection, it may be worthwhile discussing the possible relationship between cellular commitment and the intracellular differentiation-inducing factors (DIFs) described previously. Since MEL cell commitment generally occurs after 24 to 36 h incubation with DMSO or HMBA, well after the induction of DIFs, DIFs are probably not the commitment factors. Rather, it is possible that DIFs regulate a reaction(s) which interacts with a yet-to-be-identified commitment factor. Alternatively, it is possible that DIFs may interact with each other to activate one of these two factors to become a commitment factor. Recent experiments suggest the presence of such an interaction between DIF-I and DIF-II, producing another DIF (DIF-III) which alone is capable of inducing MEL cells to differentiate.[78] Considering the complexity of the early events leading to differentiation, however, it is necessary to conduct more careful biochemical and molecular and biological studies on the molecular and biochemical nature of commitment in MEL cell differentiation.

VIII. CONCLUSION

In this review, we have discussed the early events leading to erythroid differentiation *in vitro*. Although MEL cell differentiation is one of the most extensively studied *in vitro* differentiation systems, the experiments conducted in the last 2 decades, after the establishment of MEL cells by Friend, have now made us realize that the mechanism of commitment and induction of differentiation is much more complex than originally thought. Because of the large amount of sometimes conflicting data, we have discussed the recent progress in MEL cell differentiation in a rather descriptive manner. More research is required to define the critical biochemical reactions involved in differentiation and to completely understand the molecular mechanism(s) mediating MEL cell differentiation *in vitro*. An obstacle in defining the pivotal reactions involved in the early events in differentiation is the fact that this cascade of events, at least in the beginning, is not a result of a single biochemical reaction, but, rather, is a consequence of multiple induced intracellular reactions which are closely associated with fundamental cellular functions, such as DNA metabolism (replication) and/or cell proliferation. Recent progress in molecular biology now permits an analysis of complex cellular reactions. This approach can now be applied to defining the biochemical and molecular basis of MEL cell differentiation.

ACKNOWLEDGMENTS

The authors wish to thank our colleagues at the University of Tokyo for their helpful discussions and Ms. Takako Kobayashi and Yoko Okamoto for preparation of the manuscript. We also thank Mr. Masaki Oishi for his help in editing and translating the manuscript, which was originally written in Japanese, and gratefully acknowledge the editorial assistance of Gary M. Graham. This work was supported by a Special Grant for Cancer Research from the Japanese Ministry of Education.

REFERENCES

1. **Friend, C., Patuleia, M. C., and deHaven, E.,** Erythrocytic maturation *in vitro* of murine (Friend) virus-induced leukemia cells, *Natl. Cancer Inst. Monogr.,* 228, 505, 1966.
2. **Friend, C., Scher, W., Holland, J. G., and Sato, T.,** Hemoglobin synthesis in murine virus-induced leukemic cells *in vitro*: stimulation of erythroid differentiation by dimethyl sulfoxide, *Proc. Natl. Acad. Sci. U.S.A.,* 68, 378, 1971.
3. **Reuben, R. C., Wife, R. L., Breslow, R., Rifkind, R. A., and Marks, P. A.,** A new group of potent inducers of differentiation in murine erythroleukemia cells, *Proc. Natl. Acad. Sci. U.S.A.,* 73, 862, 1976.
4. **Leder, A. and Leder, P.,** Butyric acid, a potent inducer of erythroid differentiation in cultured erythroleukemic cells, *Cell,* 5, 319, 1975.
5. **Takahashi, E., Yamada, M., Saito, M., Kuboyama, M., and Ogasa, K.,** Differentiation of cultured Friend leukemia cells induced by short-chain fatty acids, *Gann,* 66, 577, 1975.
6. **Marks, P. A. and Rifkind, R. A.,** Erythroleukemic differentiation, *Annu. Rev. Biochem.,* 47, 419, 1978.
7. **Lyman, G. H., Preisler, H. D., and Papahadjopoulos, D.,** Membrane action of DMSO and other chemical inducers of Friend leukaemic cell differentiation, *Nature,* 262, 360, 1976.
8. **Rittmann, L. S., Jelsema, C. L., Schwartz, E. L., Tsiftsoglou, A. S., and Sartorelli, A. C.,** Lipid composition of Friend leukemia cells following induction of erythroid differentiation by dimethyl sulfoxide, *J. Cell. Physiol.,* 110, 50, 1982.
9. **Zwingelstein, G., Tapiero, H., Portoukalian, J., and Fourcade, A.,** Changes in phospholipid and fatty acid composition in differentiated Friend leukemic cells, *Biochem. Biophys. Res. Commun.,* 98, 349, 1981.
10. **Faletto, D. L., Arrow, A. S., and Macara, I. G.,** An early decrease in phosphatidylinositol turnover occurs on induction of Friend cell differentiation and precedes the decrease in c-myc expression, *Cell,* 43, 315, 1985.
11. **Pincus, S. M., Beckman, B. S., and George, W. J.,** Inhibition of dimethylsulfoxide-induced differentiation in Friend erythroleukemic cells by diacylglycerols and phospholipase C, *Biochem. Biopys. Res. Commun.,* 125, 491, 1984.
12. **Lachman, H. M. and Skoultchi, A. I.,** Expression of c-myc changes during differentiation of mouse erythroleukemia cells, *Nature,* 310, 592, 1984.
13. **Sasaki, H., Watanabe, T., Nomura, S., and Oishi, M.,** The level of c-myc transcript in differentiation-defective mouse erythroleukemia cells, *Jpn. J. Cancer Res.* (Gann), 78, 776, 1987.
14. **Ramsay, R. G., Ikeda, K., Rifkind, R. A., and Marks, P. A.,** Changes in gene expression associated with induced differentiation of erythroleukemia: protooncogenes, globin genes, and cell division, *Proc. Natl. Acad. Sci. U.S.A.,* 83, 6849, 1986.
15. **Todokoro, K. and Ikawa, Y.,** Sequential expression of proto-oncogenes during a mouse erythroleukemia cell differentiation, *Biochem. Biophys. Res. Commun.,* 135, 1112, 1986.
16. **Shen, D.-W., Real, F. X., Deleo, A. B., Old, L. J., Marks, P. A., and Rifkind, R. A.,** Protein p53 and inducer-mediated erythroleukemia cell commitment to terminal cell division, *Proc. Natl. Acad. Sci. U.S.A.,* 80, 5919, 1983.
17. **Khochbin, S., Principaud, E., Chabanas, A., and Lawrence, J.-J.,** Early events in murine erythroleukemia cells induced to differentiate: accumulation and gene expression of the transformation-associated cellular protein p53, *J. Mol. Biol.,* 200, 55, 1988.
18. **Gazitt, Y. and Friend, C.,** Polyamine biosynthesis enzymes in the induction and inhibition of differentiation in Friend erythroleukemia cells, *Cancer Res.,* 40, 1727, 1980.
19. **Sugiura, M., Shafman, T., and Kufe, D.,** Effects of polyamine depletion on proliferation and differentiation of murine erythroleukemia cells, *Cancer Res.,* 44, 1440, 1984.
20. **Watanabe, T., Shafman, R., and Kufe, D.,** Requirement of spermidine for induction of both heme synthesis and globin transcription in murine erythroleukemia cells, *J. Cell. Physiol.,* 122, 435, 1985.
21. **Klinken, S. P., Billelo, J., Bauer, S., Morse, H. C., III, and Thorgeirsson, S. S.,** Altered expression of β-globin, transferrin receptor, and ornithine decarboxylase in Friend murine erythroleukemia cells inhibited by α-difluoromethylonithine, *Cancer Res.,* 47, 2638, 1987.
22. **Mager, D. and Bernstein, A.,** Early transport changes during erythroid differentiation of Friend leukemic cells, *J. Cell. Physiol.,* 94, 275, 1978.
23. **Levenson, R., Housman, D., and Cantley, L.,** Amiloride inhibits murine erythroleukemia cell differentiation: Evidence for a Ca^{2+} requirement for commitment, *Proc. Natl. Acad. Sci. U.S.A.,* 77, 5945, 1980.
24. **Bridges, K., Levenson, R., Housman, D., and Contley, L.,** Calcium regulates the commitment of murine erythroleukemia cells to terminal erythroid differentiation, *J. Cell Biol.,* 90, 542, 1981.
25. **Levenson, R., Macara, I. G., Smith, R. L., Cantley, L., and Housman, D.,** Role of mitochondrial membrane potential in the regulation of murine erythroleukemia cell differentiation, *Cell,* 28, 855, 1982.
26. **Smith, R. L., Macara, I. G., Levenson, R., Housman, D., and Cantley, L.,** Evidence that a Na^+/Ca^{2+} antiport system regulates murine erythroleukemia cell differenitiation, *J. Biol. Chem.,* 257, 773, 1982.

27. **Faletto, D. L. and Macara, I. G.,** The role of Ca^{2+} in dimethyl sulfoxide-induced differentiation of Friend erythroleukemia cells, *J. Biol. Chem.,* 260, 4884, 1985.

28. **Razin, A., Szyf, M., Kafri, T., Roll, M., Giloh, H., Scarpa, S., Carotti, D., and Cantoni, G. L.,** Replacement of 5-methylcytosine by cytosine: a possible mechanism for transient DNA demethylation during differentiation, *Proc. Natl. Acad. Sci. U.S.A.,* 83, 2827, 1986.

29. **Razin, A., Levine, A., Kafri, T., Agostini, S., Gomi, T., and Cantoni, G. L.,** Relationship between transient DNA hypomethylation and erythroid differentiation of murine erythroleukemia cells, *Proc. Natl. Acad. Sci. U.S.A.,* 85, 9003, 1988.

30. **Creusot, F., Acs, G., and Christman, J. K.,** Inhibition of DNA methyltransferase and induction of Friend erythroleukemia cell differentiation by 5-azacytidine and 5-aza-2'-deoxycytidine, *J. Biol. Chem.,* 257, 2041, 1982.

31. **Terada, M., Fried, J., Nudel, U., Rifkind, R. A., and Marks, P. A.,** Transient inhibition of initiation of S-phase associated with dimethyl sulfoxide induction of murine erythroleukemia cells to erythroid differentiation, *Proc. Natl. Acad. Sci. U.S.A.,* 74, 248, 1977.

32. **Beckman, B. S., Kopfler, W., Koury, P., and Jeter, J. R., Jr.,** Effect of aphidicolin on Friend erythroleukemia cell maturation, *Exp. Cell Res.,* 169, 223, 1987.

33. **Yoshida, M., Nomura, S., and Beppu, T.,** Effects of trichostatins on differentiation of murine erythroleukemia cells, *Cancer Res.,* 47, 3688, 1987.

34. **Yoshida, M. and Beppu, T.,** Reversible arrest of proliferation of rat 3Y1 fibroblasts in both the G_1 and G_2 phase by trichostain A, *Exp. Cell Res.,* 177, 122, 1988.

35. **Nomura, S. and Oishi, M.,** Indirect induction of erythroid differentiation in mouse Friend cells: evidence for two intracellular reactions involved in the differentiation, *Proc. Natl. Acad. Sci. U.S.A. ,* 80, 210, 1983.

36. **Terada, M., Nudel, U., Fibach, E., Rifkind, R. A., and Marks, P. A.,** Changes in DNA associated with induction of erythroid differentiation by dimethyl sulfoxide in murine erythroleukemia cells, *Cancer Res.,* 38, 835, 1978.

37. **Scher, W. and Friend, C.,** Breakage of DNA and alterations in folded genomes by inducers of differentiation in Friend erythroleukemic cells, *Cancer Res.,* 38, 841, 1978.

38. **McMahon, G., Alsina, J. L., and Levy, S. B.** Induction of a Ca^{2+}, Mg^{2+}-dependent endonuclease activity during the early stages of murine erythroleukemic cell differentiation, *Proc. Natl. Acad. Sci. U.S.A.,* 81, 7461, 1984.

39. **Pulito, V. L., Miller, D. L., Sassa, S., and Yamane, T.,** DNA fragments in Friend erythroleukemia cells induced by dimethyl sulfoxide, *Proc. Natl. Acad. Sci. U.S.A.,* 80, 5912, 1983.

40. **Terada, M., Fujiki, H., Marks, P. A., and Sugimura, T.,** Induction of erythroid differentiation of murine erythroleukemia cells by nicotinamide and related compounds, *Proc. Natl. Acad. Sci. U.S.A.,* 76, 6411, 1979.

41. **Rastl, E. and Swetly, P.,** Expression of poly(adenosine diphosphate-ribose) polymerase activity in erythroleukemic mouse cells during cell cycle and erythropoietic differentiation, *J. Biol. Chem.,* 253, 4333, 1978.

42. **Nishimune, Y., Kume, A., Ogiso, Y., and Matsushiro, A.,** Induction of teratocarcinoma cell differentiation, *Exp. Cell Res.,* 146, 439, 1983.

43. **Friedman, E. and Schildkraut, C. L.,** Lengthening of G_1 phase is not strictly correlated with differentiation in Friend erythroleukemia cells, *Proc. Natl. Acad. Sci. U.S.A.,* 75, 3813, 1978.

44. **Gambari, R., Marks, P. A., and Rifkind, R. A.,** Murine erythroleukemia cell differentiation: relationship of globin gene expression and of prolongation of G_1 to inducer effects during G early S, *Proc. Natl. Acad. Sci. U.S.A.,* 76, 4511, 1979.

45. **McClintock, P. R. and Papaconstantinou, J.,** Regulation of hemoglobin synthesis in a murine erythroblastic leukemic cell: the requirement for replication to induce hemoglobin synthesis, *Proc. Natl. Acad. Sci. U.S.A.,* 71, 4551, 1974.

46. **Levy, J., Terada, M., Rifkind, R. A., and Marks, P. A.,** Induction of erythroid differentiation by dimethylsulfoxide in cells infected with Friend virus: relationship to the cell cycle, *Proc. Natl. Acad. Sci. U.S.A.,* 72, 28, 1975.

47. **Conkie, D., Harrison, P. R., and Paul, J.,** Cell-cycle dependence of induced hemoglobin synthesis in Friend erythroleukemia cells temperature-sensitive for growth, *Proc. Natl. Acad. Sci. U.S.A.,* 78, 3644, 1981.

48. **Kaneko, T., Nomura, S., and Oishi, M.,** Early events leading to erythroid differentiation in mouse Friend cells revealed by cell fusion experiments, *Cancer Res.,* 44, 1756, 1984.

49. **Kaneko, T., Watanabe, T., and Oishi, M.,** Effect of mitochondrial protein synthesis inhibitors on erythroid differentiation of mouse erythroleukemia (Friend) cells, *Mol. Cell. Biol.,* 8, 3311, 1988.

50. **Watanabe, T., Nomura, S., and Oishi, M.,** Induction of erythroid differentiation by cytoplast fusion in mouse erythroleukemia (Friend) cells, *Exp. Cell Res.,* 159, 224, 1985.

51. **Rovera, G., O'Brien, T. G., and Diamond, L.,** Tumor promoters inhibit spontaneous differentiation of Friend erythroleukemia cells in culture, *Proc. Natl. Acad. Sci. U.S.A.,* 74, 2894, 1977.

52. **Yamasaki, H., Fibach, E., Nudel, U., Weinstein, I. B., Rifkind, R. A., and Marks, P. A.,** Tumor promoters inhibit spontaneous and induced differentiation of murine erythroleukemia cells in culture, *Proc. Natl. Acad. Sci. U.S.A.,* 74, 3451, 1977.

53. **Castagna, M., Takai, Y., Kaibuchi, K., Sano, K., Kikkawa, U., and Nishizuka, Y.,** Direct activation of calcium-activated phospholipid-dependent protein kinase by tumor-promoting phorbol esters, *J. Biol. Chem.,* 257, 7847, 1982.

54. **Nishizuka, Y.,** Studies and perspectives of protein kinase C, *Science,* 233, 305, 1986.

55. **Melloni, E., Pontremoli, S., Michetti, M., Sacco, O., Cakiloglu, A. G., Jackson, J. F., Rifkind, R. A., and Marks, P. A.,** Protein kinase C activity and hexamethylene-bisacetamide-induced erythroleukemia cell differentiation, *Proc. Natl. Acad. Sci. U.S.A.,* 84, 5282, 1987.

56. **Dell'Aquila, M. L., Nguyen, H. T., Herald, C. L., Pettit, G. R., and Blumberg, P. M.,** Inhibition by bryostatin 1 of the phorbol ester-induced blockage of differentiation in hexamethylene bisacetamide-treated Friend erythroleukemia cells, *Cancer Res.,* 47, 6006, 1987.

57. **Graf, T. and Beug, H.,** Role of the v-erbA and v-erbB oncogenes of avian erythroblastosis virus in erythroid cell transformation, *Cell,* 34, 7, 1983.

58. **Coppola, J. A. and Cole, M. D.,** Constitutive c-myc oncogene expression blocks mouse erythroleukemia cell differentiation but not commitment, *Nature,* 320, 760, 1986.

59. **Dmitrovsky, E., Kuehl, W. M., Hollis, G. F., Kirsch, I. R., Bender, T. P., and Segal, S.,** Expression of a transfected human c-myc oncogene inhibits differentiation of a mouse erythroleukemia cell line, *Nature,* 322, 748, 1986.

60. **Prochownik, E. V. and Kukowska, J.,** Deregulated expression of c-myc by murine erythroleukemia cells prevents differentiation, *Nature,* 322, 848, 1986.

61. **Kume, T. U., Takada, S., and Obinata, M.,** Probability that the commitment of murine erythroleukemia cell differentiation is determined by the c-myc level, *J. Mol. Biol.,* 202, 779, 1988.

62. **Prochownik, E. V., Kukowska, J., and Rodgers, C.,** C-myc antisense transcripts accelerate differentiation and inhibit G_1 progression in murine erythroleukemia cells, *Mol. Cell. Biol.,* 8, 3683, 1988.

63. **Kaneko-Ishino, T., Kume, T. U., Sasaki, H., Obinata, M., and Oishi, M.,** Effect of c-myc gene expression on early inducible reactions required for erythroid differentiation in vitro, *Mol. Cell. Biol.,* 8, 5545, 1988.

64. **Clarke, M. F., Kukowska-Latallo, J. F., Westin, E., Smith, M., and Prochownik, E.,** Constitutive expression of a c-myb cDNA blocks Friend murine erythroleukemia cell differentiation, *Mol. Cell. Biol.,* 8, 884, 1988.

65. **Kondo, K., Watanabe, T., Sasaki, H., Uehara, Y., and Oishi, M.,** Induction of *in vitro* differentiation of mouse embryonal carcinoma (F9) and erythroleukemia (MEL) cells by herbimycin A, an inhibitor of protein phosphorylation, *J. Cell Biol.,* 109, 285, 1989.

66. **Watanabe, T., Shiraishi, T., Sasaki, H., and Oishi, M.,** Inhibitors for protein-tyrosine kinases, ST638 and genistein, induce differentiation of mouse erythroleukemia cells in a synergistic manner, *Exp. Cell Res.,* 183, 335, 1989.

67. **Honma, Y., Okabe-Kado, J., Hozumi, M., Uehara, Y., and Mizuno, S.,** Induction of erythroid differentiation of K562 human leukemic cells by herbimycin A, an inhibitor of tyrosine kinase activity, *Cancer Res.,* 49, 331, 1989.

68. **Melloni, E., Pontremoli, S., Damiani, G., Viotti, P., Weich, N., Rifkind, R. A., and Marks, P. A.,** Vincristine-resistant erythroleukemia cells have marked increased sensitivity to hexamethylene bisacetamide induced differentiation, *Proc. Natl. Acad. Sci. U.S.A.,* 85, 3835, 1988.

69. **Nomura, S., Yamagoe, S., Kamiya, T., and Oishi, M.,** An intracellular factor that induces erythroid differentiation in mouse erythroleukemia (Friend) cells, *Cell,* 44, 663, 1986.

70. **Gusella, J., Geller, R., Clarke, B., Weeks, V., and Housman, D.,** Commitment to erythroid differentiation by Friend erythroleukemia cells: a stochastic analysis, *Cell,* 9, 221, 1976.

71. **Friedman, E. A. and Schildkraut, C. L.,** Terminal differentiation in cultured Friend erythroleukemia cells, *Cell,* 12, 901, 1977.

72. **Watanabe, T. and Oishi, M.,** Dimethyl sulfoxide-inducible cytoplasmic factor involved in erythroid differentiation in mouse erythroleukemia (Friend) cells, *Proc. Natl. Acad. Sci. U.S.A.,* 84, 6481, 1987.

73. **Fibach, E., Reuben, R. C., Rifkind, R. A., and Marks, P. A.,** Effect of hexamethylene bisacetamide on the commitment to differentiation of murine erythroleukemia cells, *Cancer Res.,* 37, 440, 1977.

74. **Levenson, R., Kernen, J., and Housman, D.,** Synchronization of MEL cell commitment with cordycepin, *Cell,* 18, 1073, 1979.

75. **Murate, T., Kaneda, T., Rifkind, R. A., and Marks, P. A.,** Inducer-mediated commitment of murine erythroleukemia cells to terminal cell division: expression of commitment, *Proc. Natl. Acad. Sci. U.S.A.,* 81, 3394, 1984.

76. **Nomura, S. and Oishi, M.,** unpublished results.

77. **Watanabe, T., Nomura, S., and Oishi, M.,** unpublished results.

78. **Watanabe, T. and Oishi, M.,** unpublished results.

Chapter 8

REGULATION OF CELL CONTACTS, CELL CONFIGURATION, AND CYTOSKELETAL GENE EXPRESSION IN DIFFERENTIATING SYSTEMS

Avri Ben-Ze'ev, José Luis Rodríguez Fernández, Gideon Baum, and Barbara Gorodecki

TABLE OF CONTENTS

I. INTRODUCTION

Cellular and tissue morphogenesis are determined, to a large extent, by cell-cell and cell-substrate contacts. In the cytoplasmic face, at the point of these contacts, cytoskeletal filaments organize to form a coordinated interaction with transmembrane receptors that mediate cell-surface contact with components of the extracellular matrix, or with similar transmembrane receptors on the surface of neighboring cells. These structural assemblies are believed to transmit signals that regulate primary cellular processes such as cell proliferation, morphogenetic cell locomotion, and gene expression that are related to growth and differentiation. The molecular details of such structural interactions are under intensive investigation. The regulation of development and cytodifferentiation requires, in addition, an understanding of the control of gene expression by environmental conditions that affect cell contacts and cell shape. As a model for gene regulation by environmental factors that modulate cell and tissue morphogenesis, we are studying the regulation of cytoskeletal gene expression in response to changes induced in cell shape and cell contacts. Since in the cell cytoskeletal proteins are found in either a soluble form or in various organized filamentous structures, we asked: (1) is there a relationship between the mode or organization and the level of expression of the respective cytoskeletal elements? and (2) if the answer is yes, are such changes in cytoskeletal gene expression integral parts of programs that regulate cell growth and cytodifferentiation? In this chapter, we summarize our studies with respect to the questions in various sytems of established cell lines, primary cell cultures, and whole tissues. We present data supporting the hypothesis that conditions in the environment that affect cell-cell and cell-substrate contacts influence cytoskeletal gene expression, and that such changes are integral parts of various programs of differentiation.

II. REGULATION OF CYTOSKELETAL PROTEIN SYNTHESIS IN ESTABLISHED CELL LINES IN RESPONSE TO CHANGES IN CELL SHAPE AND CELL CONTACTS

In earlier studies, we have shown that in established cell lines the synthesis of the major cytoskeletal filament proteins, i.e., tubulin, actin, and intermediate filament proteins, is responsive to the mode of organization of the respective cytoskeletal elements in the cell.[1-5] Disruption of the microtubular network by colchicine is followed by a rapid decrease in tubulin synthesis,[6,7] due to the loss of cytoplasmic tubulin mRNA.[8] On the other hand, sequestration of the cellular tubulin into artificial paracrystals by treatment with vinblastine, or the formation of "superstable" microtubules with taxol, causes an increase in tubulin synthesis.[6,7]

The analysis of actin synthesis in fibroblasts in response to changes in cell configuration revealed a decrease in actin synthesis in fibroblasts placed in suspension culture.[9,10] Furthermore, a dramatic increase in actin synthesis and mRNA content was detected in reattaching fibroblasts following suspension culture, when extensive mobilization of actin into stress fibers occurs.[11]

Alterations in the expression of intermediate filament (IF) proteins are most probably the best example of differentiation-related cytoskeletal gene regulation, since the expression of IF proteins is tissue-type specific.[12] We found that the synthesis of vimentin-type IF, which is mainly characteristic to mesenchymal cells, is reduced when the extent of cell substrate contact is dramatically affected, as in fibroblasts cultured in suspension.[13] This cell shape-related regulation of vimentin synthesis is also maintained in cancer cells which are anchorage independent,[13,14] and therefore points to a basic mode of vimentin regulation in response to changes in the level of cell-matrix contacts. Moreover, when suspension-cultured cells are allowed to reattach, there is a rapid recovery in the synthesis and mRNA levels of vimentin.[13,14]

Epithelial cells express *in vivo* a cytokeratin type IF network, but when placed in culture, many types of epithelial cells start to synthesize vimentin. In these cells, the two IF networks are organized independently, with the cytokeratin network often looping into the desmosomal intercellular plaque, while the vimentin filaments are organized closer to the cell's center.[15] The expression of vimentin and cytokeratin protein and mRNA is differentially regulated in these established epithelial cell lines, in response to changes in cell-substrate and cell-cell contacts[16,17] (reviewed in References 18 and 19). We have shown that the synthesis of cytokeratins and the desmosomal plaque protein, desmoplakin, is maximal when optimal conditions of cell-cell contact are established, as in dense cultures either in monolayer or in suspension, when extensive desmosomal intercellular contacts are formed.[20,21] In contrast, at low plating density, these cells acquire an extended morphology on the substrate, have very few intercellular contacts, and synthesize high levels of vimentin and minimal amounts of cytokeratin.[16] The splitting of intercellular junctions by the tumor promoter, TPA (tetradecanoyl phorbol acetate), causes a dramatic change in cell morphology, including the formation of vimentin-rich long cellular processes and a decrease in the expression of cytokeratins, without affecting the synthesis of vimentin or that of other cytoskeletal proteins.[21]

The various very different functions fulfilled by the major cytoskeletal networks are determined by a large number of cytoskeleton-associated proteins. An example of a cytoskeletal protein which is involved directly in defining cell-cell and cell-substrate contacts of the adherens type is vinculin.[22-25] We have shown that the synthesis of vinculin in cultured fibroblasts is responsive to the extent of cell-cell and cell-substrate contacts.[26] Fibroblasts cultured at low density (Figure 1B) form large vinculin-containing adhesion plaques (Figure 1C), but synthesize vinculin at a much lower level (Figure 1A) than fibroblasts cultured at high plating density (Figure 1D). In dense cultures (Figure 1E), cell substrate contacts are numerous (Figure 1F) and, in addition, many cell-cell contacts containing vinculin (Figure 1F, arrowheads), characteristic of adherens-type junctions,[26] are evident. When the contribution by cell-cell contact to vinculin organization is reduced, as in cells cultured sparsely on increasing concentrations of a nonadhesive polymer (poly HEMA[26]), the synthesis of vinculin correlates with the extent of cell-substrate contact. Thus, cells synthesize vinculin at maximal levels when they are fully spread, but vinculin synthesis is drastically reduced in poorly adherent spherical cells.[26] These alterations in vinculin synthesis in response to changes in cell substrate contacts correlate with the level of vinculin mRNA in these cells.[27]

III. REGULATION OF CYTOSKELETAL PROTEIN EXPRESSION IN DIFFERENTIATING SYSTEMS

A. THE ADIPOCYTE SYSTEM
1. Cytoskeletal, Extracellular Matrix (ECM), and β-Integrin Regulation in Differentiating Adipocytes

We investigated the functional significance of these shifts in cytoskeleton expression in response to changes in cell contacts in cells that develop and maintain the differentiated phenotype. An example of fibroblasts changing shape from a fusiform to a rounded shape in a differentiating system (by analogy to the system described above) is the adipose conversion of 3T3 preadipocytes. Subclones of 3T3 isolated by Green and Kehinde[28] differentiate into adipocytes at very high frequency when stimulated with an appropriate medium, and maintain this property in an inheritable fashion.[29] Moreover, upon subcutaneous injection into animals, cells of the 3T3-F442A subclone form fat pads.[30] The conversion of these cells into adipocytes is characterized by cell rounding, the accumulation of lipids (Figure 2D through F), and the induction of key enzymes involved in lipid metabolism.[31] The dramatic changes in cell shape that occur during adipogenesis are accompanied by a decrease in the synthesis of the major cytoskeletal proteins, actin, tubulin, and vimentin[32,33] (Figure 2A

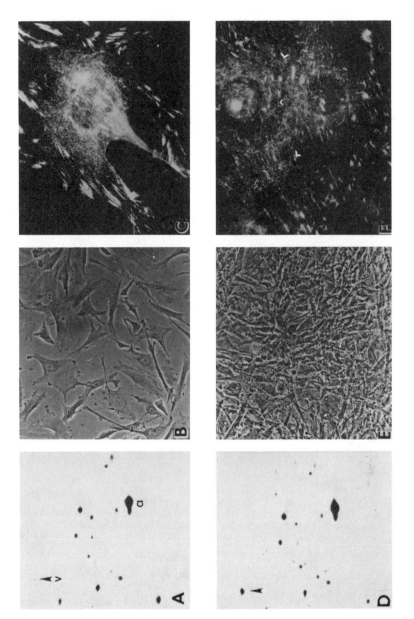

FIGURE 1.   Regulation of vinculin synthesis by cell-culture density. Chick embryo fibroblasts seeded at 10^5 cells/35-mm diameter dish (A—C) and 10^6 cells/35-mm dish were pulse labeled with ^{35}S-methionine 2 d after plating and the pattern of labeled proteins was determined by two dimensional gel electrophoresis (A and D). The morphology of these cultures under the phase contrast microscope (B and E) and the organization of vinculin by a monoclonal antivinculin antibody (C and F) were determined. a, actin; v, vinculin; arrowheads in F mark vinculin at areas of cell-cell contact.

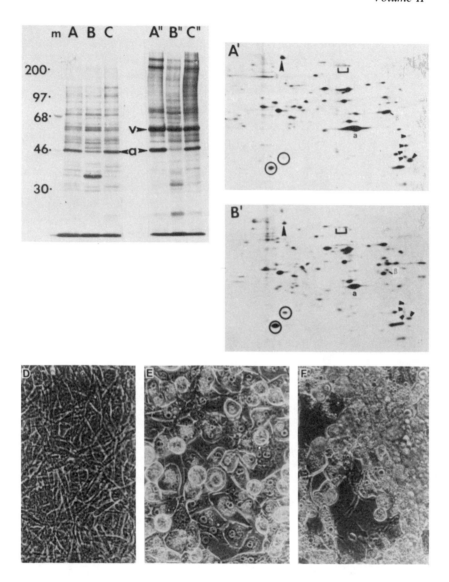

FIGURE 2.   The adipose conversion of 3T3 preadipocytes is associated with a decreased synthesis of cytoskeletal proteins. 3T3 preadipocytes were grown to confluence in the presence of calf serum (A,A′,A″, and D) and were then stimulated for 3 d (E) or 5 d (B,B′,B″, and F) with adipogenic medium. Cells were also kept in nonadipogenic medium for 5 d (C and C″). (A—C); ^{35}S-methionine-labeled cytoplasmic protein pattern; (A″—C″) intermediate filament-enriched cytoskeletal fraction. a, actin; v, vimentin; β, β-tubulin. Large arrowhead points to vinculin; bracket, α-actinin; small arrowheads, tropomyosin isoforms. The circles mark proteins whose synthesis is induced in adipogenic cells. (Modified from Rodríguez Fernández, J. L., and Ben-Ze'ev, A., *Differentiation*, 42, 65, 1989.)

through C, A″ through C″, A′, and B′). In addition, the synthesis of the microfilament-associated proteins, vinculin and α-actinin, and of the non-muscle tropomyosin isoforms is also reduced (Figure 2B′; compare to Figure 2A′). Analysis of mRNAs coding for fibronectin (FN), the β-subunit of the fibronectin receptor integrin (INT), and for various cytoskeletal elements revealed a decrease in their level following stimulation with adipogenic medium (Figure 3). The expression of a key enzyme in lipid metabolism (glycerophosphate dehydrogenase, GPD) increases drastically during this time (Figure 3, GPD).

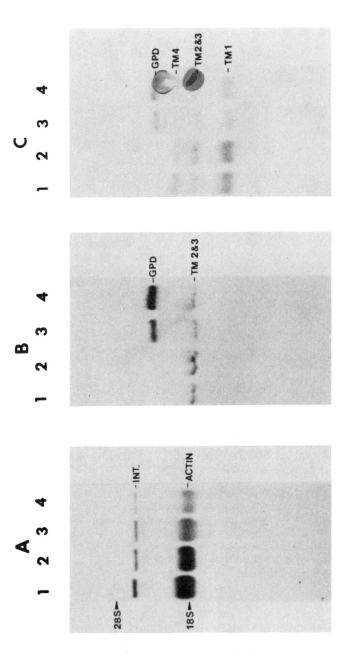

FIGURE 3.   Decrease in mRNA levels of cytoskeletal proteins, fibronectin, and β-integrin in differentiating adipocytes. RNA was isolated from confluent preadipocytes (lanes 1) and after various periods of challenging the cells with adipogenic medium. Lanes 2, 2 d; lanes 3, 4 d; lanes 4, 7 d; and lanes 5, 14 d after the shift to adipogenic medium. The Northern blots were hybridized with cDNAs to various cytoskeletal proteins and to a fat cell-specific probe (GPD). INT., β-integrin; TM, tropomyosin isoform-specific cDNAs; VIM, vimentin; FN, fibronectin; β TUB, β-tubulin; GPD, glycerophosphate dehydrogenase. (Modified from Rodríguez Fernández, J. L., and Ben-Ze'ev, A., *Differentiation*, 42, 65, 1989.)

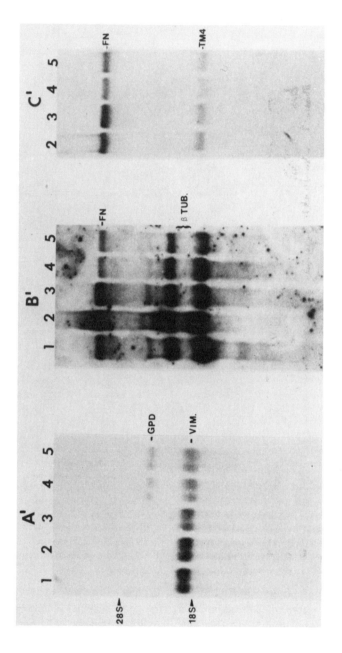

FIGURE 3 (continued)

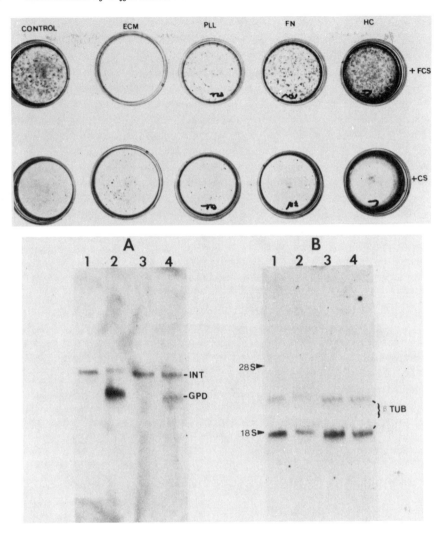

FIGURE 4.   The adipose conversion of preadipocytes is delayed under conditions that inhibit the change in cell shape. Preadipocytes were grown to confluency on various substrata and then either challenged with adipogenic medium (FCS) or with nonadipogenic calf serum (CS). ECM, cells grown on an extracellular matrix deposited on the culture dish by corneal endothelial cells; PLL, cells grown on poly L-lysine; FN, fibronectin-coated substrate; HC, substrate-coated with hydrated rat tail collagen. (A and B) Northern blots with RNA isolated from cells cultured on plastic (lanes 1 and 2) or on ECM (lanes 3 and 4). Lanes 1 and 3 represent RNA from preadipocytes; lanes 2 and 4, RNA from cells challenged for 4 d with adipogenic medium. (A) Hybridization with cDNAs to β-integrin (INT) and glycerophosphate dehydrogenase (GPD); (B) hybridization with β-tubulin (β TUB) cDNA. (Modified from Rodríguez Fernández, J. L., and Ben-Ze'ev, A., *Differentiation*, 42, 65, 1989.)

The change in cell morphology accompanying adipogenesis does not appear to result from lipid accumulation, since it occurs when the expression of lipogenic enzymes increases, even in the absence of lipid accumulation.[34] We tested the hypothesis that the cell shape changes occurring during adipose conversion are part of programmed events related to preadipocyte differentiation, and that interference with these morphological changes will inhibit adipogenesis. For this purpose, we plated preadipocytes on various substrates and at confluence challenged the cells with adipogenic medium (Figure 4). The accumulation of lipids was determined by the intensity of cell staining with the lipophylic stain Oil Red O. The differentiation of preadipocytes into adipocytes was drastically inhibited when the cells were cultured on an extracellular matrix (ECM) which was deposited on the culture dish by corneal endothelial cells. Some of the molecular changes associated with adipogenesis, such as the induction of GPD synthesis and the decrease in β-integrin and tubulin synthesis, were also blocked in cells cultured on ECM (Figure 4A and B). On this substrate and on poly L-lysine (PLL),the cells maintained a fibroblastic morphology, while on hydrated collagen, as on plastic, the cells acquired a round morphology and accumulated lipid droplets. The common property that allowed both PLL and the endothelial ECM to inhibit adipose conversion was the ability of these matrices to "hold down" the cells tightly adherent to the substrate and, therefore, prevent the change in cell shape which follows the challenge with adipogenic medium. This hypothesis was further suported by our finding that the morphological changes and adipose conversion proceeded at normal rates in cells cultured on a maleable, hydrated ECM, such as the one secreted by the EHS tumor.[34a] When plated and treated with adipogenic medium on the EHS matrix, the cells rounded up and induced GPD expression, as on plastic culture dishes. Using a different protocol for inducing differentiation, Spiegelman and Ginty[35] demonstrated that in cells cultured on fibronectin-coated dishes, the morphological and biochemical changes characteristic of adipogenesis were prevented. Furthermore, by treating cells growing on fibronectin with cytochalasin D to disrupt the organization of actin filaments, the inhibitory action of fibronectin was reversed.[35]

The involvement of fibronectin in adipogenesis regulation is also suggested by studies with TGF-β. TGF-β was shown to inhibit adipocyte differentiation, most probably by elevating the synthesis of mRNAs coding for fibronectin and integrin.[36-38] Moreover, the processing and appearance of mature fibronectin receptors on the cell surface, as well as the secretion of fibronectin, were both enhanced by TGF-β.[37,38] Taken together, these results strongly support the notion that changes in the extracellular matrix and its receptor may alter not only cell morphology and cytoskeletal organization and expression, but that these changes also represent programmed alterations that are necessary for new gene expression to bring about the differentiated phenotype of adipocytes.

B. THE GRANULOSA CELL SYSTEM
1. Regulation of Microfilament Protein Organization and Expression During Granulosa Cell Differentiation

We studied the importance of cell-ECM contacts in the induction, acquisition, and maintenance of differentiated functions in ovarian granulosa cells. These are a major cell type of the ovarian follicle, the functional unit of the ovary.[39,40a] In the ovary, granulosa cells are separated from other cell types and blood vessels by a basement membrane. Gonadotropic hormones secreted from the pituitary initiate the cycle of maturation of granulosa cells to produce the ovum and to form and secrete high amounts of progesterone. In primary cultures of granulosa cells, many of the morphological and biochemical changes associated with granulosa cell differentiation can be reproduced. When cultured in the presence of gonadotropins, granulosa cells undergo massive aggregation, maintain a more rounded shape, and develop numerous lipid droplets (Figure 5B), while in the absence of hormone, the cells are flat and devoid of these organelles (Figure 5A). The increase in the production and accumulation of the mitochondrial key steroidogenic enzyme cytochrome P-450 cholesterol side chain cleavage (P-450$_{scc}$) is evident in hormone-treated cells (Figure 5D), while cells cultured in its absence display a more faint immunostaining of the mitochondria, which have a filamentous appearance (Figure 5C; also see References 41 through 43).

The morphological changes associated with granulosa cell differentiation are characterized by a disorganized microfilament system, with cytoplasmic subcortical actin (Figure 6E), very few vinculin (Figure 6F)- and α-actinin-containing (Figure 6G) adhesion plaques, and diffuse tropomyosin staining (Figure 6H). Cells cultured in the absence of the hormone develop an extensive array of stress fibers (Figure 6A) and large vinculin (Figure 6B)- and α-actinin-containing (Figure 6C) adhesion plaques, and organize tropomyosin (Figure 6D) and α-actinin (Figure 6C) along stress fibers. The organization of microtubules and vimentin-

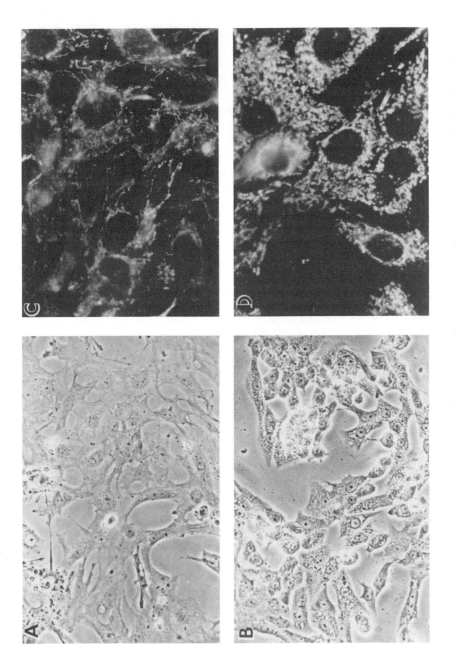

FIGURE 5.   Changes in cell shape, cell aggregation, and cytochrome P-450$_{scc}$ expression in differentiating granulosa cells. Granulosa cells obtained from rats were cultured for 48 h in the absence (A and C) and in the presence of gonadotropin (FSH) (B and D). Cell morphology (A and B) and cytochrome P-450$_{scc}$ distribution (C and D) indicative of the luteinization of granulosa cells are presented.

containing IF remains largely unaffected in differentiating rat and human granulosa cells.[44,45] In correlation with these changes in the organization of the actin-cytoskeleton system, in hormone-treated cells the synthesis of the respective cytoskeletal proteins is markedly reduced (Figure 7B), compared to cells cultured without the hormone (Figure 7A). The synthesis of tubulin and vimentin,[44,45] however, remain unchanged under these conditions. In addition to an induction in the expression of key steroidogenic enzymes, the synthesis of 30- and 220-kDa proteins and a major heat shock protein, hsp90,[46] is elevated in differentiating granulosa cells (Figure 7B). These biochemical and morphological changes induced by gonadotropins are also obtained when granulosa cells are treated with agents that elevate cellular cAMP levels, which are believed to act as second messengers in gonadotropin action.[40a,43,45]

Granulosa cells within the same follicle express a gradient of differentiation, with the cells adjacent to the follicular basement membrane containing the highest density of receptors for gonadotropin and the highest intracellular level of key steroidogenic enzymes.[47-50] The positive role of the basement membrane in developing and maintaining the differentiated phenotype of granulosa cells was demonstrated by showing similar changes in the organization and expression of the actin cytoskeleton of granulosa cells cultured on an endothelial cell-type ECM in the absence of gonadotropins.[51] These cells, when plated on ECM, produce progesterone at levels similar to that of cells cultured in the presence of gonadotropin, or by mature granulosa-lutein cells in the ovary.[51] These results, therefore, emphasize the central role of granulosa cell-ECM interactions in the regulation of the morphological and biochemical changes occurring during the development of the steroidogenic potential.[40a]

The involvement of the microfilament network in the process of steroidogenesis is suggested by showing that primary granulosa cells cultured in the presence of cytochalasin B display a disorganized microfilament network and decreased synthesis of vinculin, α-actinin, actin, and tropomyosin, but increased production of progesterone.[43,44,54a] Moreover, if granulosa cells induced to differentiate with cytochalasin B are washed and incubated in drug-free medium, they evidence a recovery in the organization and expression of microfilament proteins and a decrease in the synthesis of proteins associated with granulosa cell luteinization.[46,54a] It appears, therefore, that the changes in the microfilament protein organization and synthesis are important programmed events in the pathway of steroidogenesis mobilization. These events can be partially prevented by culturing primary granulosa cells on fibronectin-coated substrate in serum-free medium. Under these conditions, the cells remain well spread on the substrate, even in the presence of gonodotropins, and they express high levels of the microfilament proteins and a much reduced level of progesterone.[44]

2. Cytoskeleton Expression in Established Granulosa Cell Lines and in the Ovary During Granulosa Cell Maturation

We determined the expression of cytoskeletal proteins in several recently isolated, established granulosa cell lines which maintain stimulable steroidogenesis.[52] Established granulosa cell lines obtained by transfection with SV40 DNA form colonies of epitheloid cells (Figure 8E) resembling undifferentiated primary granulosa cells cultured in the absence of gonadotropins (Figure 5A). These SV40-transfected cells lost their ability to produce progesterone in response to tropic hormones and elevated levels of cellular cAMP.[52] Granulosa cell lines obtained by the co-transfection of primary cell cultures with SV40 and Ha-*ras* DNA have a very different morphology (Figure 8F). They are rounded to spindly, form aggregates, and are reminiscent of differentiating primary granulosa cell cultures (Figure 5B). These SV40 and Ha-*ras*-transfected cell lines maintain the ability to produce progesterone in response to cAMP elevation.[52] In accordance with the morphological characteristics of cells transfected with SV40, the synthesis of the actin cytoskeleton proteins in these cells is somewhat reduced (Figure 8C), compared to primary cultures of granulosa cells (Figure 8A). Transformation with SV40 also elevates the synthesis of the low-molecular-weight tropomyosin isoform no. 5, in agreement with other studies on tropomyosin expression in

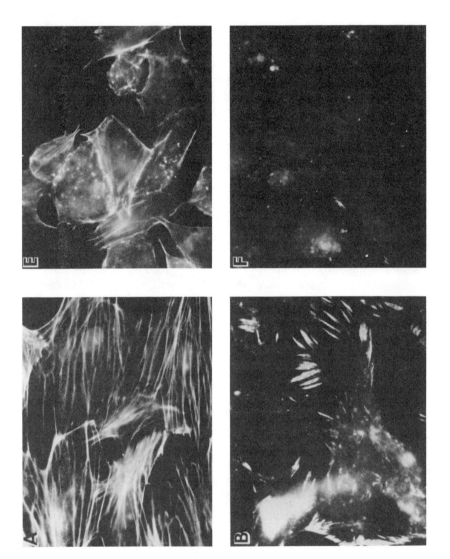

FIGURE 6.   Reorganization of the actin cytoskeleton in differentiating granulosa cells. Granulosa cells cultured for 48 h in the absence (A—D) or presence (E—H) of gonadotropin (FSH) were treated with antibodies against various cytoskeletal proteins, followed by rhodamine-labeled secondary antibodies. (A and E) Staining with rhodamine-phalloidin to visualize actin; (B and F) staining with antivinculin; (C and G) anti-α-actinin; (D and H) antitropomyosin.

FIGURE 6 (continued)

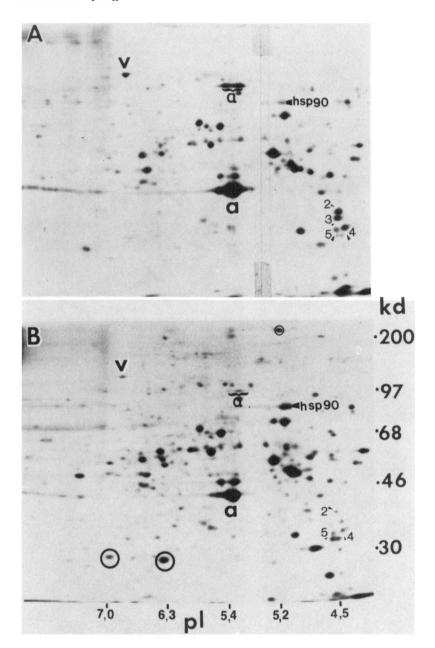

FIGURE 7.   Decreased synthesis of actin and actin-associated proteins in differentiating granulosa cells. Cells cultured for 48 h in the absence (A) or presence (B) of gonadotropin (FSH) were pulse labeled with ^{35}S-methionine and the proteins were analyzed by two dimensional gel electrophoresis. a, actin; v, vinculin; a, α-actinin; 2, 3, 4, 5, tropomyosin isoforms; hsp90, heat shock protein of 90-kDa mol wt. Circles mark proteins whose synthesis is consistently increased in differentiating granulosa cells. (From Ben-Ze'ev, A. and Amsterdam, A., *J. Biol. Chem.*, 262, 5366, 1987. With permission.)

SV40-transformed rodent cells.[53] In the co-transfected (SV40 + Ha-*ras*) granulosa cell lines (Figure 8D), the synthesis of tropomyosin as well as that of actin, α-actinin, and vinculin are markedly decreased, similarly to differentiating primary granulosa cells treated with 8-Br-cAMP (Figure 8B). These changes in cytoskeletal protein synthesis result from a decrease in mRNA translation *in vitro* (Figure 9A and B) and mRNA content (Figure 9C through L).

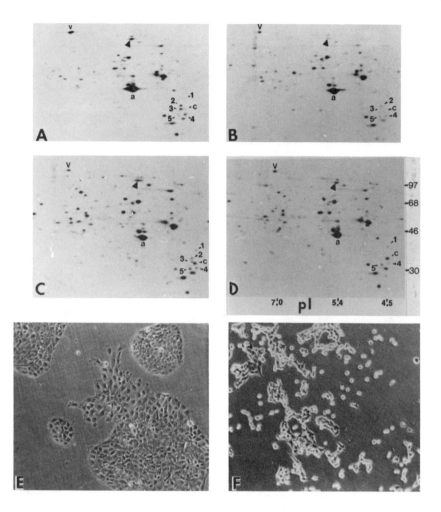

FIGURE 8. Expression of cytoskeletal proteins in primary granulosa cell cultures and in established granulosa cell lines. Primary granulosa cells cultured for 48 h in the absence (A) or in the presence (B) of cAMP; (C and E) granulosa cell lines obtained by transfection with SV40; (D and F) granulosa cell lines obtained by transfection with SV40 plus activated Ha-*ras*. A-D cells cultured for 48 h were pulse labeled with ^{35}S-methionine and the proteins were analyzed by 2-D gel electrophoresis. Arrowheads mark the position of α-actinin; c, cyclin — a protein whose synthesis is elevated in proliferating and transformed cells, a, actin; 1—5, tropomyosin isoforms; v, vinculin. (From Baum, G., Suh, B. S., and Amsterdam, A., Ben-Ze'ev, A., unpublished observations.)

Thus, in differentiating primary granulosa cells, the translation *in vitro* of mRNAs for vinculin, actin, and tropomyosin (Figure 9B) is reduced, when compared to unstimulated cells (Figure 9A). Similarly, the level of actin mRNA and the mRNAs coding for the various tropomyosin isoforms is much lower in gonadotropin-treated cells (Figure 9D, G, and I) than in control cells (Figure 9C, H, and J). In contrast, the level of the IF protein mRNA (vimentin) remained unchanged in differentiating cells (Figure 9E and F). In the SV40 and Ha-*ras* co-transfected cell lines, the level of a 1.8-kb mRNA coding for tropomyosins 2 and 3[54] (Figure 9L) is drastically reduced, compared to SV40-transfected cells (Figure 9K), or to primary unstimulated granulosa cells (Figure 9J). These SV40- and Ha-*ras*-transfected cell lines, which maintain the ability to undergo stimulable steroidogenesis, do not signif- icantly change the synthesis of their cytoskeletal proteins upon stimulation with cAMP. It appears, therefore, that in the SV40 + Ha *ras*-transfected cells, the cytoskeleton is already expressed in a conformation that allows inducible steroidogenesis to occur. This is achieved

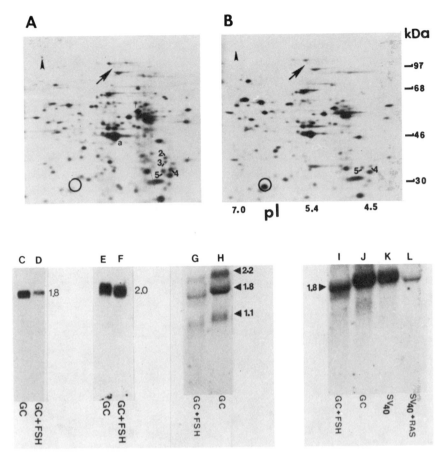

FIGURE 9.   The translation and mRNA content of the actin-cytoskeleton proteins is decreased in differentiating primary granulosa cell cultures and in established granulosa cell lines. *In vitro* translation of poly(A)-containing RNA from untreated (A) and gonadotropin (FSH)-treated primary (B) granulosa cells (GC). (C,E,H, and J) Northern blots of RNA from control cells; (D,F,G, and I) from gonadotropin-treated (FSH) cells; (K) from SV40-transfected cell lines; (L) from SV40 + Ha-*ras* transfected cell lines. (C and D) Hybridization with actin cDNA; (E and F) vimentin cDNA; (G—L) tropomyosin cDNAs; (I—L) hybridization with a cDNA recognizing tropomyosins 2 and 3 (1.8-kb mRNA); followed by (G and H) hybridization with cDNAs to tropomyosin 4 (2.2-kb mRNA) and tropomyosin 1 (1.1-kb mRNA). The small arrowheads in A and B point to vinculin, while the large arrows to α-actinin; a, actin, 2—5; tropomyosin isoforms; circles mark proteins as in Figure 7. (From Baum, G., Suh, B. S., Amsterdam, A., and Ben-Ze'ev, A., unpublished observations.)

by down regulating the organization and expression of the actin cytoskeleton and adopting a morphology more compatible with mobilization of steroidogenesis.

The physiological relevance of the cytoskeletal changes observed in primary cultures and in established granulosa cell lines was determined by isolating RNA from ovaries at various stages of follicular development and luteinization and hybridizing with tropomyosin isoform-specific cDNAs.[54a] This is a relevant approach, since the great majority of the ovarian tissue of immature rats after hormonal synchronization consists of granulosa cells of preantral or preovulatory follicles.[39,40a] As shown in Figure 10, the level of the various tropomyosin isoform RNA is reduced in the highly steroidogenic corpora lutea (Figure 10A and B, lanes 1), when compared to RNA isolated from earlier stages of granulosa cell maturation (Figure 10A and B, lanes 2 and 3). In corpora lutea and in the preovulatory follicles (Figure 10C, lanes 1 and 2) there is, at the same time, a high level of RNA coding for the key steroidogenic enzyme cytochrome P-450$_{xcc}$. Therefore, regulation of the actin-

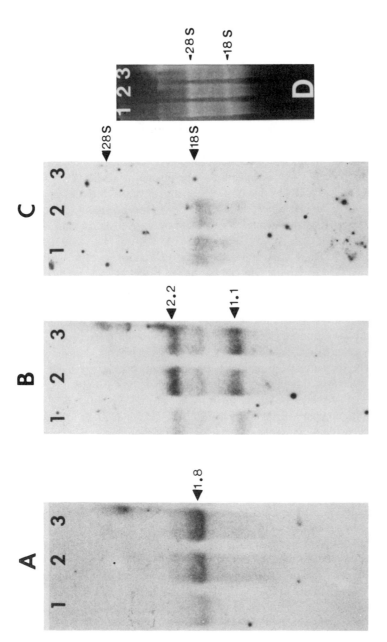

FIGURE 10. Decrease in tropomyosin expression during ovarian folliculogenesis and luteinization. RNA was isolated from ovaries at the preovulatory stage (lanes 3), the preantral stage (lanes 2), and from corpora lutea (lanes 1). The RNA was hybridized with cDNAs to tropomyosin 2 and 3 (A), followed by hybridization with cDNAs to tropomyosin 1 and 4 (B). The blot was rehybridized with a cDNA specific to cytochrome P-450$_{scc}$ (C); ethidium bromide pattern of the blot used (D). (From Ben-Ze'ev, A., Baum, G., and Amsterdam, A., Regulation of tropomyosin expression in the maturing ovary and in primary granulosa cell cultures, *Dev. Biol.*, 135, 191, 1989.)

cytoskeleton expression is an integral part of the modulation of steroidogenesis in granulosa cells in primary culture, in established cell lines, and in the ovary during differentiation of granulosa-lutein cells.

C. THE HEPATOCYTE SYSTEM

1. Cytoskeletal and Tissue-Specific Gene Expression in Growing Hepatocytes *In Vitro* and in Regenerating Liver

The studies discussed above suggest that signals for induction and maintenance of the differentiated phenotype involve changes in the organization and expression of various cytoskeletal elements.

Signals regulating the initiation of DNA synthesis are also transmitted through an "organized cytoskeleton" since interference with the structure of the cytoskeleton drastically affects the level of DNA synthesis.[55,56] Moreover, the organization and gene expression of components of the adhesion plaques, of β-integrin, and of various cytoskeletal elements are dramatically affected when cells are stimulated to proliferate with purified growth factors.[57-61] The importance of changes in cell shape and cell substrate contacts in the regulation of growth was supported by studies in various experimental systems[62-69] (reviewed in Reference 65). In a recent study, it was demonstrated that cells arrested in the G_o stage of the cell cycle by suspension culture rapidly induce the expression of growth-associated genes such as c-*fos* and c-*myc* upon reattachment to the substrate, even in the absence of growth factors.[66]

A very useful system where the relationship between cytoskeletal gene expression and the regulation of both growth and differentiation can be studied is that of liver hepatocytes. In the liver, hepatocytes constitute over 90% of the total cell population,[67] and a very large number of tissue-specific genes were characterized in liver cells.[68] In addition, in adult liver, hepatocytes are arrested in the G_o stage of the cycle, but exit synchronously to initiate DNA synthesis upon hepatectomy to restore normal tissue size.[69] In culture, primary hepatocytes, when incubated in the presence of EGF and insulin, induce DNA synthesis after an 18- to 24-h lag (Figure 11 H). The induction of growth in cultured hepatocytes is associated with a dramatic increase in the synthesis of the major cytoskeletal filament proteins and of several cytoskeletal-associated proteins, while the synthesis of tissue-specific proteins, such as albumin, decreases severalfold (compare Figure 11 D to 11C, and 11G to 11E). Northern blot analysis showed that these shifts in cytoskeletal and tissue-specific protein synthesis result from similar changes in the abundance of the respective mRNAs.[70-73] A kinetic comparison, at the level of mRNA coding for cytoskeletal and several liver-specific proteins, between hepatocytes incubated *in vitro* in the presence of EGF and regenerating liver at various times following hepatectomy revealed parallel changes in the two systems; the levels of cytoskeletal mRNA increased dramatically, while the level of tissue-specific mRNAs decreased severalfold.[74] Nuclear run-on experiments in both systems demonstrated a decrease in tissue-specific gene transcription, but no change in cytoskeletal gene transcription.[74] Hence, under appropriate *in vitro* conditions and in regenerating liver, similar shifts in gene regulation occur; cytoskeletal gene expression is enhanced by a posttranscriptional regulation, while tissue-specific gene expression is reduced mainly by a decrease in the transcription of the gene. It appears, therefore, that this inverse relationship between cytoskeletal and tissue-specific gene expression is of some physiological significance.

2. Cell-Cell and Cell-Matrix Contacts as Regulators of Cytoskeletal and Tissue-Specific Gene Expression in Differentiated Hepatocytes

Previous studies suggested that the culturing of hepatocytes on collagenous substrata and extracellular matrices and the extent of cell-cell contacts are all important factors in maintaining differentiated hepatocyte functions *in vitro*.[75-78] We cultured hepatocytes on

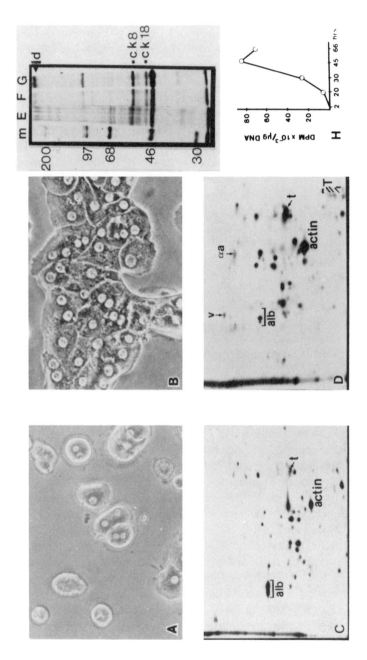

FIGURE 11.   Pattern of proteins synthesized by growing hepatocytes. Primary cultures of rat hepatocytes prepared by collagenase perfusion were plated in the presence of insulin and EGF and after 5 h (A,C, and E), 24 h (F), and 48 h (B,D, and G), the pattern of cytoplasmic proteins (C and D) and intermediate filament proteins (E—G) synthesized by the cells was determined by ^{35}S-methionine pulse labeling and gel electrophoresis. The level of DNA synthesis (H) was determined by pulse labeling the cells with ^{3}H-thymidine. ck 8 and 18, cytokeratins; d, desmoplakin; m, molecular weight markers; alb, albumin; t, β-tubulin; αa, α-actinin; v, vinculin; T, tropomyosin isoforms.

EHS-type hydrated ECM and compared the synthesis and secretion of liver-specific proteins and the synthesis of cytoskeletal proteins to that in cells cultured on dried rat tail collagen (Figure 12). Cells cultured on the EHS matrix adhere tenaciously to the matrix, but remain largely spherical, whether in groups of several cells or as single cells (Figure 12B). Cells plated on dried rat tail collagen attach and spread out to form the characteristic epitheloid morphology which is obtained on all solid substrata with these cells. The pattern of proteins synthesized (Figure 12C through F) and secreted (Figure 12G and H) by cells cultured on these two different matrices is also very different. Cells on the EHS matrix synthesize (Figure 12D) and secrete (Figure 12H) high levels of albumin and other liver-specific proteins, but synthesize low levels of cytoskeletal proteins (Figure 12D and F). In contrast, cells grown on dried rat tail collagen express high levels of cytoskeletal proteins (Figure 12C and E), but synthesize (Figure 12C) and secrete (Figure 12G) little liver-specific proteins. In these short-term cultures on the EHS matrix, the cells maintain a pattern of tissue-specific and cytoskeletal gene expression very similar to that of adult liver cells in the rat, as revealed by Northern blot analysis (Figure 13, lanes 3; compare to lanes 1) and nuclear run-on assays.[79] In contrast, cells cultured on dried rat tail collagen proliferate, as shown by the expression of histone mRNA (Figure 13, lanes 2) and DNA synthesis (Figure 11H), and express high levels of cytoskeletal mRNA and reduced levels of liver-specific mRNAs (Figure 13, lanes 2).

The nature of the individual component(s) of the EHS matrix that can allow maintenance of differentiated functions in cultured hepatocytes remains to be determined.[5,73,79] It is suggested, however, that glycosaminoglycans and heparan sulfate proteoglycans can preserve functional gap-junction formation and gap-junction protein expression in cultured hepatocytes.[80] In short-term cultures, hepatocytes plated on hydrated rather than on dried rat tail collagen can express, to a high degree, the characteristics of differentiated hepatocytes (Figure 14). On hydrated collagen, the cells form a compact trabecular pattern (Figure 14E), which is different from the more extended flat morphology on dry rat tail collagen (Figure 14D). The pattern of proteins synthesized on hydrated collagen (Figure 14B and H) resembles more the one obtained on the EHS matrix (Figure 14C and I) than that obtained on dried rat tail collagen (Figure 14A and G). Thus, a low level of actin, tubulin, cytokeratin, vinculin, and desmoplakin synthesis (Figure 14B, C, H, and I), but a high level of albumin synthesis and secretion (Figure 14K and L), is characteristic of hepatocytes cultured on both the EHS matrix and hydrated collagen. The ability of hydrated collagen gels to support the expression of differentiated functions is obtained only in cells cultured at high density. Cells plated very sparsely spread out equally well on hydrated and dried-rat tail collagen, and do not support the expression of tissue-specific genes.[5,73] Hence, cell-cell in addition to cell-matrix contacts are important for maintaining the differentiated phenotype of hepatocytes on hydrated collagen gels.

The synthesis of DNA in cultured hepatocytes requires the presence of growth factors; it is inhibited in dense cultures on all matrices, and in sparse cultures where a rounded cell shape is maintained, as on the EHS matrix.[73] The conditions allowing hepatocyte proliferation *in vitro* and during liver regeneration in the animal are characterized by the low levels of differentiation-function expression. Inhibition of DNA synthesis by the omission of growth factors, or by cell culturing at high density is not sufficient, however, for the maintenance of high levels of tissue-specific protein expression (see Figure 14). The more general conclusion that can be drawn from these results is that the expression of the differentiated phenotype in cultured hepatocytes is enhanced on collagenous matrices, and it depends on cell-cell as well as on cell-matrix contact.

3. Regulation of Albumin Synthesis and Secretion in SV40-Immortalized Hepatocytes Cultured on Different Matrices

The recently isolated immortalized hepatocytes that express high levels of albumin[81] are

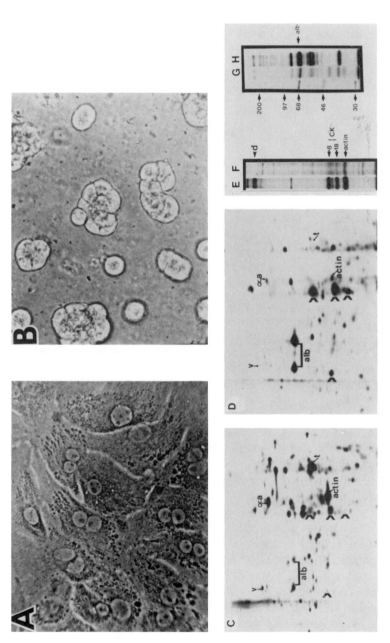

FIGURE 12.   Maintenance of the differentiated phenotype in hepatocytes cultured on an extracellular matrix. Hepatocytes were cultured for 48 h on dried rat tail collagen (A,C,E, and G) or on a hydrated extracellular matrix secreted by the EHS tumor (B,D,F, and H). Cell morphology (A and B), the pattern of cytoplasmic proteins (C and D), intermediate filament proteins (E and F), and proteins secreted into the medium (G and H) were determined. Arrowheads in C and D point to liver-specific proteins whose synthesis is elevated on the EHS matrix. ck 8 and 18, cytokeratins; d, desmoplakin; alb, albumin; t, β-tubulin; αa, α-actinin; v, vinculin.

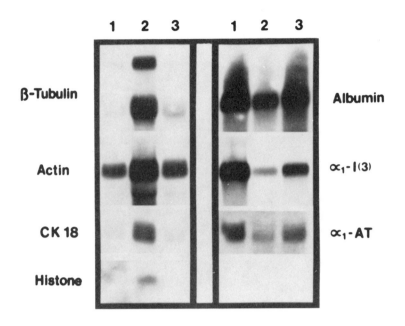

FIGURE 13. High levels of tissue-specific RNAs are maintained on EHS matrix, as in adult liver, but not on dried rat tail collagen. RNA was isolated from adult rat liver (lanes 1), from hepatocytes cultured for 48 h on rat tail collagen (lanes 2) and on EHS matrix (lanes 3). Northern blots were hybridized to tissue-specific and cytoskeletal protein cDNAs. α_1-I(3), α_1-inhibitor 3; ;α_1-AT, α_1-antitrypsin; CK 18, cytokeratin number 18. (From Ben-Ze'ev, A., Robinson, G. S., Bucher, N. L. R., and Farmer, S. R., *Proc. Natl. Acad. Sci. U.S.A.*, 85, 2161, 1988. With permission.)

very useful for the study of tissue-specific versus cytoskeletal protein expression, since in these cells, which proliferate continuously, the expression of several tissue-specific functions is maintained at levels similar to those of nonproliferating, well-differentiated hepatocytes in the adult liver.[81,82]

The results summarized in Figure 15 show that SV40-immortalized hepatocytes synthesize (Figure 15A) and secrete (Figure 15F, lane 1) albumin when cultured in serum-free medium on plastic plates. The synthesis and secretion of albumin by these cells is much higher when the cells are cultured on hydrated collagenous substrata (Figure 15D, E, and F, lanes 4 through 6). Dried rat tail collagen (Figure 15B), fibronectin (Figure 15C), and endothelial ECM (Figure 15F, lane 7) are not efficient in inducing alubumin synthesis and secretion over the levels obtained on plastic (Figure 15A and F, lane 1). It appears, therefore, that the ability of these hepatocytes to alter these hydrated collagenous matrices (floating collagen, hydrated collagen, and EHS-hydrated gel), by contracting them and thereby changing cell shape, is an essential feature of maintaining a high level of tissue-specific gene expression. More rigid matrices such as dried rat tail collagen, fibronectin, and the endothelial ECM, all of which support adhesion but do not allow matrix deformation and a change in cell morphology, are not more efficient than the tissue culture plastic in maintaining the differentiated phenotype. The expression of cytoskeletal proteins in these cells on the various matrices did not change substantially (Figure 15B-E; compare to Figure 15A), suggestive, as in SV40-immortalized granulosa cells (Figure 8), of an uncoupling between the regulation of expression of the cytoskeleton and that of differentiated functions in oncogene-transfected cells.

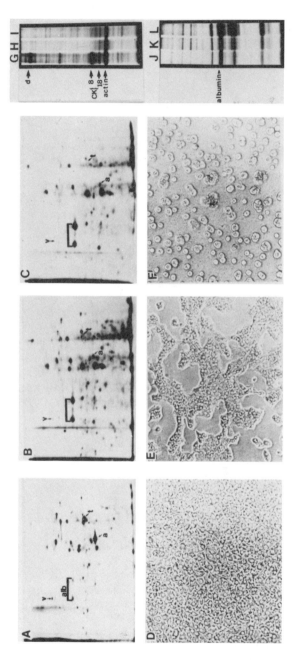

FIGURE 14. Hepatocytes cultured on hydrated rat tail collagen at high density maintain the differentiated phenotype. Hepatocytes were cultured for 48 h at high plating density on dried rat tail collagen (A,D,G, and J), on hydrated rat tail collagen (B,E,H, and K), and on the EHS-type extracellular matrix (C,F,I, and L). Cell morphology (D—F), the pattern of cytoplasmic proteins (A—C), intermediate filament proteins (G—I), and secreted proteins (J—L) were determined. Alb, albumin; t, β-tubulin; v, vinculin; a, actin; d, desmoplakin; CK 8 and 18, cytokeratins. (Modified from Ben-Ze'ev, A., *Cell Shape: Determinants, Regulation and Regulatory Role*, Stein, N. D. and Bronner, F., Eds., Academic Press, Orlando, 1989, 95.)

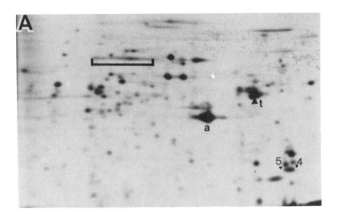

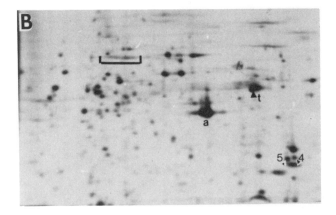

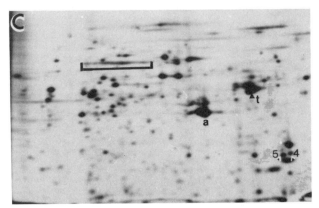

FIGURE 15.   Synthesis of cytoskeletal and tissue-specific proteins in SV40-immortalized hepatocytes cultured on various matrices. Hepatocytes immortalized by transfection with SV40 and selected for high albumin production[81] were cultured in serum-free medium for 48 h on plastic (A), dried rat tail collagen (B), fibronectin (C), hydrated rat tail collagen (D), and EHS-type ECM (E). The cells were pulse labeled with ^{35}S-methionine and the pattern of cytoplasmic proteins was determined (A-E).

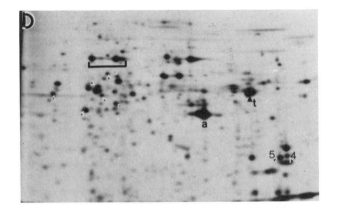

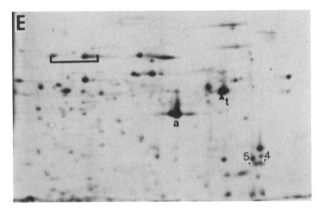

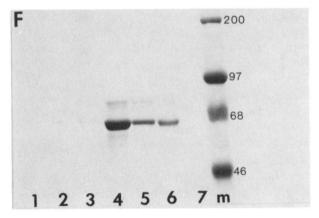

FIGURE 15 (continued). The pattern of proteins secreted (F) during a period of 3 h from these cells after 48 h culture on plastic (lane 1), fibronectin (lane 2) dried rat tail collagen (lane 3), hydrated rat tail collagen gel (lane 4), floating rat tail collagen gel (lane 5), EHS matrix, (lane 6), and endothelial ECM (lane 7). a, actin; t, β-tubulin; 4,5, tropomyosin isoforms. The bracket marks the position of albumin. The small arrowheads in D mark the positions of proteins whose synthesis is high on hydrated collagen but is reduced in cells grown on EHS matrix (E).

IV. MECHANISMS THAT LINK CYTOSKELETAL PROTEIN ORGANIZATION TO THE REGULATION OF CYTOSKELETAL GENE EXPRESSION

In the studies presented above, the regulation of cytoskeletal protein synthesis correlated with changes in cell morphology and cell contacts. In general, the alterations in cytoskeletal protein synthesis are followed by similar changes in the level of the corresponding cytoskeletal mRNA. In systems where nuclear run-on experiments were performed, the changes in cytoskeletal protein synthesis and mRNA levels were suggested to occur mainly by post-transcriptional mechanisms, since nuclear transcription of these genes remained largely unchanged.[74,79,83-85]

A high degree of specificity in regulating the level of an mRNA coding for a cytoskeletal element, by a mechanism that is linked to the mode of organization of the respective protein, was elegantly demonstrated in the case of tubulin autoregulation.[8] Complete depolymerization of microtubules leads to a rapid decrease in tubulin synthesis,[6,7] while the artificial raising of polymerized tubulin levels by drugs[6,7] or by tubulin microinjection[8] enhances tubulin synthesis. These changes in tubulin synthesis and mRNA content, in response to changes in the level of free tubulin, operate in enucleated cells[87,88] and require an association between tubulin mRNA and polyribosomes.[89] Using site-directed mutagenesis of the first four N-terminal codons, it was demonstrated that the nascent β-tubulin chain is recognized by the unpolymerized tubulin subunits,[90] and that this protein-protein interaction most probably induces, by an as yet unidentified mechanism, a rapid degradation of the tubulin mRNA. This type of autoregulation of cytoskeletal mRNA stability by the level of the corresponding unpolymerized protein subunits provides the specificity for a very basic mechanism for cytoskeletal protein synthesis regulation in response to changes in the organization of the respective cytoskeletal elements.

When considering the regulation of cytoskeletal protein organization *in vivo*, one should note that the binding of cytoskeletal elements to areas of cell-cell and cell-matrix contacts is mediated by specific transmembrane receptors.[24,25,91,92] The organization, expression, and phosphorylation of these receptors were suggested as key regulatory events in cytoskeletal organization and cell differentiation and transformation.[38,93-96]

Furthermore, in the cytoplasm, the translation of mRNA into protein requires an obligatory association of translation and mRNA factors with the cytoskeleton.[97-107] This is, therefore, an additional level of control that could operate in the mechanisms that link changes in cytostructure with changes in gene expression.

Signals transmitted by changes in the composition of the extracellular matrix, alterations in the expression or phosphorylation of receptors for cell-matrix and cell-cell contacts, which alter the interaction of the cytoskeleton with the cell membrane or affect the organization of the cytoskeleton directly, or signals transmitted by various second messengers of growth and differentiation factors could all converge on regulatory pathways that affect the levels of mRNAs by mechanisms such as the one described above for β-tubulin.

It remains to be determined what are the signals that convey the specificity for regulating tissue-specific mRNA levels during alterations in cell contacts and cytoskeletal organization. The systems and the probes are available, and we should hope in the near future for progress in elucidating the molecular details of the mechanisms that link cell structure to the expression of the cells' differentiated functions.

ACKNOWLEDGMENTS

We thank R. Reiss for technical assistance and C. Banks for typing the manuscript. Studies from the author's laboratory were supported by grants from the U.S.A.-Israel Binational Foundation (BSF), Jerusalem, Israel; the Fund for Basic Research administered by the Israel Academy of Sciences and Humanities; and the Forchheimer Center for Molecular Genetics at the Weizmann Institute. Avri Ben-Ze'ev holds the Lunenfeld-Kunin Professorial Chair.

REFERENCES

1. **Ben-Ze'ev, A.,** Cell shape, the complex cellular networks and gene expression: cytoskeletal protein genes as a model system, in *Cell and Muscle Motility,* Vol. 6, Shay, J. W., Ed., Plenum Press, New York, 1985, 23.
2. **Ben-Ze'ev, A.,** The relationship between cytoplasmic organization, gene expression and morphogenesis, *Trends Biochem. Sci.,* 11, 478, 1986.
3. **Ben-Ze'ev, A.,** Regulation of cytoskeletal protein synthesis in normal and cancer cells, *Cancer Rev.,* 4, 91, 1986.
4. **Ben-Ze'ev, A.,** The role of changes in cell shape and contacts in the regulation of cytoskeleton expression during differentiation, *J. Cell Sci. Suppl.,* 8, 293, 1987.
5. **Ben-Ze'ev, A.,** Cell shape and cell contacts: molecular approaches to cytoskeleton expression, in *Cell Shape: Determinants, Regulation and Regulatory Role,* Stein, W. D. and Bronner, F., Eds., Academic Press, Orlando, 1989, 95.
6. **Ben-Ze'ev, A., Farmer, S. R., and Penman, S.,** Mechanisms of regulating tubulin synthesis in cultured mammalian cells, *Cell,* 17, 319, 1979.
7. **Cleveland, D. W., Lopata, M. A., Sherline, P., and Kirschner, M. W.,** Unpolymerized tubulin modulates the level of tubulin mRNAs, *Cell,* 24, 537, 1981.
8. **Cleveland, D. W.,** Autoregulated instability of tubulin mRNAs: a novel eukaryotic regulatory mechanism, *Trends Biochem. Sci.,* 13, 339, 1988.
9. **Benecke, B.-J., Ben-Ze'ev, A., and Penman, S.,** The control of mRNA production, translation and turnover in suspended and reattached anchorage-dependent fibroblasts, *Cell,* 14, 931, 1978.
10. **Farmer, S., Ben-Ze'ev, A., Benecke, B. J., and Penman, S.,** Altered translatability of messenger RNA from suspended anchorage-dependent fibroblasts: reversal upon cell attachment to a surface, *Cell,* 15, 627, 1978.
11. **Farmer, S., Wan, K., Ben-Ze'ev, A., and Penman, S.,** The regulation of actin mRNA levels and translation responds to changes in cell configuration, *Mol. Cell. Biol.,* 3, 182, 1983.
12. **Osborn, M. and Weber, K.,** Intermediate filaments: cell-type-specific markers in differentiation and pathology, *Cell,* 31, 303, 1982.
13. **Ben-Ze'ev, A.,** Cell configuration related control of vimentin biosynthesis and phosphorylation in cultured mammalian cells, *J. Cell Biol.,* 97, 858, 1983.
14. **Ben-Ze'ev, A., Zoller, M., and Raz, A.,** Differential expression of intermediate filament proteins in metastatic and non metastatic variants of the BSp73 tumor, *Cancer Res.,* 46, 785, 1986.
15. **Franke, W. W., Schmid, E., Winter, S., Osborn, M., and Weber, K.,** Widespread occurrence of intermediate-sized filaments of the vimentin-type in cultured cells from diverse vertebrates, *Exp. Cell Res.,* 123, 25, 1979.
16. **Ben-Ze'ev, A.,** Differential control of cytokeratins and vimentin synthesis by cell-cell contact and cell shape in cultured epithelial cells, *J. Cell Biol.,* 99, 1424, 1984.
17. **Ben-Ze'ev, A.,** Control of intermediate filament protein synthesis by cell-cell interaction and cell configuration, *FEBS Lett.,* 171, 107, 1984.
18. **Ben-Ze'ev, A.,** Cell-cell interaction and cell shape related control of intermediate filaments protein synthesis, in *Molecular Biology of the Cytoskeleton,* Borisy, G., Cleveland, D., and Murphy, D., Eds., Cold Spring Harbor Laboratory, Cold Spring Harbor, NY, 1985, 435.
19. **Ben-Ze'ev, A.,** Cell-cell interaction and cell configuration related control of cytokeratins and vimentin expression in epithelial cells and in fibroblasts, in *Intermediate Filaments,* Wang, E., Fischman, D., Liem, R. K. H., and Sun, T.-T., Eds., Ann. N.Y. Acad. Sci., 1985, 455, 597.
20. **Ben-Ze'ev, A.,** Cell density and cell shape related regulation of vimentin and cytokeratin synthesis: inhibition of vimentin and induction of a new cytokeratin in dense epithelial cell cultures, *Exp. Cell Res.,* 157, 520, 1985.
21. **Ben-Ze'ev, A.,** Tumor promoter-induced disruption of junctional complexes in cultured epithelial cells is followed by the inhibition of cytokeratin and desmoplakin synthesis, *Exp. Cell Res.,* 164, 335, 1986.

22. **Geiger, B.,** A 130-K protein from chicken gizzard: its localization at the termini of microfilament bundles in cultured chicken cells, *Cell*, 18, 193, 1979.
23. **Burridge, K. and Feramisco, J. R.,** Microinjection and localization of a 130K protein in living fibroblasts: a relationship to actin and fibronectin, *Cell*, 19, 587, 1980.
24. **Geiger, B., Volk, T., Volberg, T., and Bendori, R.,** Molecular interactions in adherens-type contacts, *J. Cell Sci. Suppl.*, 8, 251, 1987.
25. **Burridge, K., Fath, K., Kelly, T., Nuckolls, G., and Turner, C.,** Focal adhesions; transmembrane junctions between the extracellular matrix and the cytoskeleton, *Ann. Rev. Cell. Biol.*, 4, 487, 1988.
26. **Ungar, F., Geiger, B., and Ben-Ze'ev, A.,** Cell contact- and shape-dependent regulation of vinculin synthesis in cultured fibroblasts, *Nature*, 319, 787, 1986.
27. **Bendori, R., Salomon, D. and Geiger, B.,** Contact-dependent regulation of vinculin expression in cultured fibroblasts: a study with vinculin-specific cDNA probes, *EMBO J.*, 6, 2897, 1987.
28. **Green, H. and Kehinde, O.,** Sublines of mouse 3T3 cells that accumulate lipid, *Cell*, 1, 113, 1974.
29. **Green, H. and Kehinde, O.,** Spontaneous heritable changes leading to increased adipose conversion in 3T3 cells, *Cell*, 7, 105, 1976.
30. **Green, H. and Kehinde, O.,** Formation of normally differentiated subcutaneous fat pads by an established preadipocyte cell line, *J. Cell. Physiol.*, 101, 169, 1979.
31. **Spiegelman, B., Frank, M., and Green, H.,** Molecular cloning of mRNA from 3T3 adipocytes: regulation of mRNA content for glycerophosphate dehydrogenase and other differentiation-dependent proteins during adipocyte development, *J. Biol. Chem.*, 258, 10083, 1983.
32. **Spiegelman, B. M. and Farmer, S. R.,** Decrease in tubulin and actin gene expression prior to morphological differentiation of 3T3-adipocytes, *Cell*, 29, 53, 1982.
33. **Sidhu, R.,** Two-dimensional electrophoresis analysis of proteins synthesized during differentiation of 3T3-L1 preadipocytes, *J. Biol. Chem.*, 254, 11111, 1979.
34. **Huri-Harcuch, W., Wise, L. S., and Green, H.,** Interruption of the adipose conversion of 3T3 cells by biotin deficiency: differentiation without triglyceride accumulation, *Cell*, 14, 53, 1978.
34a. **Rodríguez Fernández, J. L. and Ben-Ze'ev, A.,** Regulation of fibronectin, integrin, and cytoskeleton expression in differentiating adipocytes: inhibition by extracellular matrix and polylysine, *Differentiation*, 42, 65, 1989.
35. **Spiegelman, B. M. and Ginty, C. A.,** Fibronectin modulation of cell shape and lipogenic gene expression in 3T3-adipocytes, *Cell*, 35, 657, 1983.
36. **Roberts, C. J., Birkenmeier, T. M., McQuillan, J. J., Akiyama, S. K., Yamada, S. S., Chen, W.-T., Yamada, K. M., and McDonald, J. A.,** Transforming growth factor β stimulates the expression of fibronectin and of both subunits of the human fibronectin receptor by cultured human lung fibroblasts, *J. Biol. Chem.*, 263, 4586, 1988.
37. **Ignotz, R. and Massague, J.,** Type β transforming growth factor controls the adipogenic differentiation of 3T3 fibroblasts, *proc. Natl. Acad. Sci. U.S.A.* 82, 8530, 1985.
38. **Ignotz, R. A. and Massague, J.,** Cell adhesion protein receptors as targets for transforming growth-factor-β action, *Cell*, 51, 1899, 1987.
39. **Hsueh, A. J. W., Adashi, E. Y., Jones, P. B. C., and Welsh, T. H.,** Hormonal regulation of the differentiation of cultured ovarian granulosa cells, *Endocr. Rev.*, 5, 76, 1984.
40. **Amsterdam, A. and Rotmensch, S.,** Structure-function relationships during granulosa cell differentiation. *Endocr. Rev.*, 8, 309, 1987.
40a. **Amsterdam, A., Rotmensch, S., and Ben-Ze'ev, A.,** Coordinated regulation of morphologial and biochemical differentiation in a steroidogenic cell: the granulosa cell model, *Trends Biochem. Sci.*, 14, 377, 1989.
41. **Goldring, N. B., Farkash, Y., Goldschmit, D., and Orly, J.,** Immunofluorescent probing of the mitochondrial cholesterol side-chain cleavage cytochrome P-450 expressed in differentiating granulosa cells in culture, *Endocrinology*, 119, 28211, 1986.
42. **Farkas, Y., Timberg, R., and Orly, J.,** Preparation of antiserum to rat cytochrome P-450 cholesterol side chain cleavage, and its use for ultrastructural localization of the immunoreactive enzyme by the protein A-gold technique, *Endocrinology*, 118, 1353, 1986.
43. **Ben-Ze'ev, A., Kohen, F., and Amsterdam, A.,** Gonadotropin-induced differentiation of granulosa cells is associated with the coordinated regulation of cytoskeletal proteins involved in cell-contact formation, *Differentiation*, 34, 222, 1987.
44. **Ben-Ze'ev, A., and Amsterdam, A.,** *In vitro* regulation of granulosa cell differentiation: involvement of cytoskeletal protein expression, *J. Biol. Chem.*, 262, 5366, 1987.
45. **Ben-Ze'ev, A. and Amsterdam, A.,** Regulation of cytoskeletal protein organization and expression in human granulosa cells in response to gonadotropin treatment, *Endocrinology*, 124, 1033, 1989.
46. **Ben-Ze'ev, A. and Amsterdam, A.,** Regulation of heat shock protein synthesis by gonadotropins in cultured granulosa cells, *Endocrinology*, 124, 2584, 1989.

47. **Albertini, D. F. and Anderson, E.,** The appearance and structure of intercellular connection during the ontogeny of the rabbit ovarian follicle with particular reference to gap junctions, *J. Cell Biol.,* 63, 234, 1974.

48. **Amsterdam, A., Koch, Y., Lieberman, M. E., and Lindner, H. R.,** Distribution of binding sites for human chorionic gonadotropin in the preovulatory follicle of the rat, *J. Cell Biol.,* 67, 894, 1975.

49. **Zoller, L. C. and Weisz, J.,** Identification of cytochrome P-450, and its distribution in the membrana granulosa of the preovulatory follicle using quantitative cytochemistry, *Endocrinology,* 103, 310, 1979.

50. **Zoller, L. C. and Weisz, J.,** A quantitative cytochemical study of glucose-6-0-phosphate dehydrogenase and 5-3β-hydroxysteroid dehydrogenase activity in the membrana granulosa of the ovulable type of follicle of the rat, *Histochemistry,* 62, 125, 1979.

51. **Ben-Ze'ev, A. and Amsterdam, A.,** Regulation of cytoskeletal proteins involved in cell contact formation during differentiation of granulosa cells on extracellular matrix, *Proc. Natl. Acad. Sci. U.S.A.,* 83, 2894, 1986.

52. **Amsterdam, A., Zauberman, A., Meir, G., Pinhasi-Kimhi, O., Suh, B. S., and Oren, M.,** Cotransfection of granulosa cells with simian virus 40 and Ha-RAS oncogene generates stable lines capable of induced steroidogenesis, *Proc. Natl. Acad. Sci. U.S.A.,* 85, 7582, 1988.

53. **Matsumura, F. and Yamashiro-Matsumura, S.,** Tropomyosin in cell transformation, *Cancer Rev.,* 6, 21, 1986.

54. **Yamawaki-Kataoka, Y. and Helfman, D. M.,** Isolation and characterization of cDNA clones encoding a low molecular weight nonmuscle tropomyosin isoform, *J. Biol.Chem.,* 262, 10791, 1987.

54a. **Ben-Ze'ev, A., Baum, G., and Amsterdam, A.,** Regulation of tropomyosin expression in the maturing ovary and in primary granulosa cell cultures, *Dev. Biol.,* 735, 191, 1989.

55. **Crossin, K. L. and Carney, D. H.,** Evidence that microtubule depolymerization early in the cell cycle is sufficient to initate DNA synthesis, *Cell,* 23, 61, 1981.

56. **Maness, P. F. and Walsh, R. C.,** Dihydrocytochalasin B disorganizes actin cytoarchitecture and inhibits initiation of DNA synthesis in 3T3 cells, *Cell,* 30, 252, 1982.

57. **Bockus, B. J. and Stiles, C. D.,** Regulation of cytoskeletal architecture by platelet-derived growth factor, insulin and epidermal growth factor, *Exp. Cell Res.,* 153, 186, 1984.

58. **Herman, B. and Pledger, W. J.,** Platelet-derived growth factor-induced alterations in vinculin and actin distribution in Balb/c-3T3 cells, *J. Cell Biol.,* 100, 1031, 1985.

59. **Bellas, R. E., Dike, L. E., Bendori, R., and Farmer, S. R.,** Response of components of the adhesion apparatus to growth activation: transcriptional activation of the vinculin gene following serum stimulation of quiescent Swiss 3T3 cells, *J. Cell Biol.,* 107 (Abstr.), 428, 1988.

60. **Dhawan, J., Dike, L. E., and Farmer, S. R.,** Cell adhesion regulates expression of the cytomatrix, *J. Cell Biol.,* 107 (Abstr.), 3315, 1988.

61. **Rysek, R.-P., MacDonald-Bravo, H., Zerial, M., and Bravo, R.,** Coordinate induction of fibronectin, fibronectin receptor, tropomyosin, and actin genes in serum-stimulated fibroblasts, *Exp. Cell Res.,* 180, 537, 1989.

62. **Folkman, J. and Moscona, A.,** Role of cell shape in growth control, *Nature,* 273, 345, 1978.

63. **Ben-Ze'ev, A., Farmer, S. R., and Penman, S.,** Protein synthesis requires cell-surface contact while nuclear events respond to cell shape in anchorage-dependent fibroblasts, *Cell,* 21, 365, 1980.

64. **Ingber, D. E., Madri, J. A., and Folkman, J.,** Endothelial growth factors and extracellular matrix regulate DNA synthesis through modulation of cell and nuclear expansion, *In Vitro Cell. Dev. Biol.,* 23, 387, 1987.

65. **Ingber, D. E. and Folkman, J.,** Tension and compression as basic determinants of cell form and function: utilization of a cellular tensegrity mechanism, in *Cell Shape: Determinants, Regulation and Regulatory Role* Stein, W. D., and Bronner, F., Eds., Academic Press, Orlando, 1989, 3.

66. **Dike, L. E. and Farmer, S. R.,** Cell adhesion induces expression of growth-associated genes in suspension-arrested fibroblasts, *Proc. Natl. Acad. Sci. U.S.A.,* 85, 6792, 1988.

67. **Wachstein, M.,** Cyto- and histochemistry of the liver, in *The Liver, Morphology, Biochemistry, Physiology,* Vol. 1, Rouiller, C. H., Ed., Academic Press, New York, 1963, 13.

68. **Derman, E., Drauter, K., Walling, L., Weinberger, C., Ray, M., and Darnell, J. E., Jr.,** Transcriptional control in the production of liver-specific mRNAs, *Cell,* 23, 731, 1981.

69. **Bucher, N. L. R. and Malt, R. A.,** *Regeneration of Liver and Kidney,* Little, Brown, Boston, 1971, 17.

70. **Clayton, D. F. and Darnell, J. E., Jr.,** Changes in liver-specific compared to common gene transcription during primary culture of mouse hepatocytes, *Mol. Cell. Biol.,* 3, 1552, 1983.

71. **Jefferson, D. M., Clayton, D. F., Darnell, J. E., Jr., and Reid, L. M.,** Post-transcriptional modulation of gene expression in cultured rat hepatocytes, *Mol. Cell. Biol.,* 4, 1929, 1984.

72. **Clayton, D. F., Harrelson, A. L., and Darnell, J. E., Jr.,** Dependence of liver-specific transcription on tissue organization. *Mol. Cell. Biol.,* 5, 2623, 1985.

73. **Ben-Ze'ev, A., Robinson, G. S., Bucher, N. L. R., and Farmer, S. R.,** Cell-cell and cell-matrix interactions differentially regulate the expression of hepatic and cytoskeletal genes in primary cultures of rat hepatocytes, *Proc. Natl. Acad. Sci. U.S.A.,* 85, 2161, 1988.

74. **Freidman, J. M., Chung, E. Y., and Darnell, J. E., Jr.,** Gene expression during liver regeneration, *J. Mol. Biol.,* 179, 37, 1984.

75. **Michalopoulos, G. and Pitot, H. C.,** Primary culture of parenchymal liver cells on collagen membranes, *Exp. Cell Res.,* 94, 70, 1975.

76. **Enat, R., Jefferson, D. M., Ruiz-Opazo, N., Gatmaitan, Z., Leinwand, L. A., and Reid, L. M.,** Hepatocyte proliferation in vitro: its dependence on the use of serum-free hormonally defined medium and substrata of extracellular matrix, *Proc. Natl. Acad. Sci. U.S.A.,* 81, 1411, 1984.

77. **Fraslin, J. M., Kneip, B., Vaulont, S., Glaise, D., Munnich, A., and Guguen-Guillouzo, C.,** Dependence of hepatocyte-specific gene expression on cell-cell interactions in primary culture, *EMBO J.,* 4, 2487, 1985.

78. **Bissell, M. D., Arenson, D. M., Maher, J. J., and Roll, F. J.,** Support of cultured hepatocytes by a laminin-rich gel, *J. Clin. Invest.,* 79, 801, 1987.

79. **Robinson, G. S., Bucher, N. L. R., and Farmer, S. R.,** Regulation of hepatic gene expression by cell-extracellular matrix (ECM) interactions in primary cultures of rat hepatocytes, *J. Cell Biol.,* 107, 4512, 1988 (Abstr).

80. **Spray, D. C., Fujita, M., Saez, J. C., Choi, H., Watanabe, T., Hertzberg, E., Rosenberg, L. C., and Reid, L. M.,** Proteoglycans and glycosaminoglycans induce gap junction synthesis and function in primary liver cultures, *J. Cell Biol.,* 105, 541, 1987.

81. **Woodworth, C. D. and Isom, H.,** Regulation of albumin gene expression in a series of rat hepatocyte cell lines immortalized by simian virus 40 and maintained in chemically defined medium, *Mol. Cell. Biol.* 7, 3740, 1987.

82. **Woodworth, C. D., Kreider, J. W., Mengel, L., Miller, T., Yunlian, M., and Isom, H. C.,** Tumorigenicity of simian virus 40-hepatocyte cell lines: effect of in vitro and in vivo passage on expression of liver-specific genes and oncogenes, *Mol. Cell. Biol.,* 8, 4492, 1988.

83. **Cleveland, D. W. and Havercroft, J. C.,** Is apparent autoregulatory control of tubulin synthesis non-transcriptionally controlled?, *J. Cell Biol.,* 97, 919, 1983.

84. **Cook, K. S., Hunt, C. R., and Spiegelman, B. M.,** Developmentally regulated mRNAs in 3T3-adipocytes: analysis of transcriptional control, *J. Cell Biol.,* 100, 514, 1985.

85. **Djain, P., Phillips, M., and Green, H.,** The activation of specific gene transcription in the adipose conversion of 3T3 cells, *J. Cell. Physiol.,* 124, 554, 1985.

86. **Cleveland, D. W., Pittenger, M. F., and Feramisco, J. R.,** Elevation of tubulin levels by microinjection suppresses new tubulin synthesis, *Nature (London),* 305, 738, 1983.

87. **Pittenger, M. F. and Cleveland, D. W.,** Retention of autoregulatory control of tubulin synthesis in cytoplasts: demonstration of a cytoplasmic mechanism that regulates the level of tubulin expression, *J. Cell Biol.,* 101, 1941, 1985.

88. **Caron, J. M., Jones, A. L., Rall, L. B., and Kirschner, M. W.,** Autoregulation of tubulin synthesis in enucleated cells, *Nature,* 317, 648, 1985.

89. **Pachter, J. S., Yen, T. J., and Cleveland, D. W.,** Autoregulation of tubulin expression is achieved through specific degradation of polysomal tubulin mRNAs, *Cell,* 51, 283, 1987.

90. **Yen, T. J., Machlin, P., and Cleveland, D. W.,** Autoregulated instability of β-tubulin mRNAs by recognition of the nascent amino terminus of β-tubulin, *Nature,* 334, 580, 1988.

91. **Hynes, R. O.,** Integrins: a family of cell surface receptors, *Cell,* 48, 549, 1987.

92. **Buck, C. A. and Horwitz, A. F.,** Cell surface receptors for extracellular matrix molecules, *Annu. Rev. Cell Biol.,* 3, 179, 1987.

93. **Menko, S. A. and Boettiger, D.,** Occupation of the extracellular matrix receptor, integrin, is a control point for myogenic differentiation, *Cell,* 51, 51, 1987.

94. **Hirst, R., Horwitz, A., Buck, C., and Rohrschneider, L.,** Phosphorylation of the fibronectin receptor complex in cells transformed by oncogenes that encode tyrosine kinases, *Proc. Natl. Acad. Sci. U.S.A.,* 70, 3170, 1986.

95. **Plantefaber, L. C. and Hynes, R. O.,** Changes in integrin receptors on oncogenically transformed cells, *Cell,* 56, 281, 1989.

96. **Dahl, S. C. and Grabel, L. B.,** Integrin phosphorylation is modulated during the differentiation of F-9 teratocarcinoma stem cells, *J. Cell Biol.,* 108, 183, 1989.

97. **Ben-Ze'ev, A., Horowitz, M., Skolnik, H., Abulafia, R., Laub, O., and Aloni, Y.,** The metabolism of SV40 RNA is associated with the cytoskeletal framework, *Virology,* 111, 475, 1981.

98. **Bonneau, A. M., Darveau, A., and Sonenberg, N.,** Effect of viral infection on host protein synthesis and mRNA association with the cytoplasmic cytoskeletal structure, *J. Cell Biol.,* 100, 1209, 1985.

99. **Cervera, M., Dreyfuss, G., and Penman, S.,** Messenger RNA is translated when associated with the cytoskeletal framework in normal and VSV-infected HeLa cells, *Cell,* 23, 113, 1981.

100. **Fulton, A. B., Wan, K. W., and Penman, S.,** The spatial distribution of polyribosomes in 3T3 cells and the associated assembly of proteins into the skeletal framework, *Cell,* 20, 849, 1980.

101. **Howe, J. G. and Hershey, J. W. B.,** Translational initiation factor and ribosome association with the cytoskeletal framework fraction from HeLa cells, *Cell,* 37, 85, 1984.
102. **Jeffrey, W. R.,** Spatial distribution of messenger RNA in the cytoskeletal framework of Ascidian eggs, *Dev. Biol.* 103, 482, 1984.
103. **Lawrence, J. B. and Singer, R. H.,** Intracellular localization of messenger RNA for cytoskeletal proteins, *Cell,* 45, 407, 1986.
104. **Lenk, R., Ransom, L., Kaufmann, Y., and Penman, S.,** A cytoskeletal structure with associated polyribosomes obtained from HeLa cells, *Cell,* 10, 67, 1977.
105. **Lenk, R. and Penman, S.,** The cytoskeletal framework and poliovirus metabolism, *Cell,* 16, 289, 1979.
106. **Van Venrooij, J. J., Sillekens, P. T. G., van Ekelen, C. A. G., and Reinders, R. T.,** On the association of mRNA with the cytoskeleton in uninfected and adenovirus-infected human KB cells, *Exp. Cell Res.,* 135, 79, 1981.
107. **Ornelles, D. A., Fey, E. G., and Penman, S.,** Cytochalasin releases mRNA from the cytoskeletal framework and inhibits protein synthesis, *Mol. Cell. Biol.,* 6, 1650, 1986.

Chapter 9

MACROPHAGE (M) AND GRANULOCYTE MACROPHAGE-COLONY STIMULATING FACTOR (GM-CSF) EFFECTS ON MONOCYTE DIFFERENTIATION

R. Allan Mufson

TABLE OF CONTENTS

I. M AND GM-CSF EFFECTS ON HUMAN MONOCYTE DIFFERENTIATION

Understanding the regulation of human monocyte differentiation has recently become extremely important from both a therapeutic and a pathologic perspective. The application of recombinant DNA techniques to hematopoiesis has resulted in the availability of purified factors which can influence monocyte differentiation. These factors are termed colony-stimulating factors (CSF) and the two which most profoundly influence monocyte differentiation are granulocyte-macrophage (GM) and macrophage (M) CSF.[1,2] Both of these factors can influence the expression of monocyte and macrophage effector functions which may be therapeutically useful in destroying pathogen-infected cells or tumor cells through anitbody-independent or dependent mechanisms. Clinical trials of both M-CSF and GM-CSF are currently in progress.[1,3,4] Monocytes, however, are also known to be reservoirs of human immunodeficiency virus-1 (HIV-1), and alterations in the differentiation of monocytic cells can lead to activation of viral replication.[5-8] It is thus important to understand how GM-CSF and M-CSF modulate monocyte differentiation and macrophage function, and to define differences in their mechanisms of action which may be responsible for differential gene activation.

The molecular cloning of the cDNAs for GM-CSF and M-CSF has advanced our understanding of their biochemistry and biosynthesis. GM-CSF is a 14- to 35-kDa glycoprotein coded for by a 1.0-kb mRNA found in T-cells, monocytes, endothelial cells, and fibroblasts.[9] The biosynthesis of M-CSF is more complex and the factor occurs as two protein species, a 70- to 90-kDa form and a 36- to 52-kDa form, which are coded for by mRNAs of 4.0 and 1.8 kb, respectively.[10,11] M-CSF is also produced by monocytes, endothelial cells, and fibroblasts. Each CSF has a unique receptor on the monocyte cell membrane. GM-CSF binds to a receptor with a K_d of 20 to 1000 pM, with 100 to 500 receptors per cell.[2,12] The binding of GM-CSF to this receptor is saturable and specific, and cross-linking studies in the murine system have shown two polypeptides of about 50 and 130 kDa mol wt which bind GM-CSF. M-CSF binds to a receptor which is the product of the c-*fms* protooncogene. The receptor has a molecular weight of 165 kDa and K_d of 30 mM. The M-CSF receptor is very abundant, with approximately 10^4 receptors per monocyte. The receptor protein also has a tyrosine kinase activity associated with it, which is activated after interaction of M-CSF with the binding domain of the protein.[2,12]

Murine monocytes and macrophages require M-CSF for maintenance of adherence, proliferation, and colony formation *in vitro*.[13] M-CSF also enhances murine macrophage differentiation and function, as evidenced by increased production of prostaglandins, interferon, plasminogen activator, G-CSF, the intracellular killing of candida, H_2O_2 release in response to phorbol ester, and Fc receptor expression.[14,15] Thus, M-CSF, in addition to its ability to support macrophage colony formation from monocyte progenitors *in vitro*, can also enhance monocyte/macrophage differentiation and effector function.

GM-CSF, in contrast to M-CSF, supports the formation of granulocyte and macrophage colonies *in vitro* from hematopoietic progenitor cells, and enhances neutrophil effector function.[2] Additionally, it is becoming apparent that GM-CSF can also affect monocyte/macrophage differentiation and effector function *in vitro*, although the effects are less well studied. GM-CSF in murine macrophages has been reported to increase the release of H_2O_2 in response to phorbol ester, increase Fc receptor-mediated phagocytosis, enhance adherence, and increase their ability to kill the intracellular parasites *Trypanasoma cruzii* and *Leishmanii donovanii*.[14-16]

Studies in which the differentiation of murine monocytes or macrophages cultured in GM-CSF have been compared to cells cultured in M-CSF have revealed striking differences in the resulting functional phenotype. Mouse bone marrow-derived monocytes/macrophages cultured in GM-CSF for 7 d were capable of lysing P815 mastocytoma target cells without additional activation; however, addition of the macrophage activator γ-interferon, alone or in combination with lipopolysaccharide, significantly enhanced tumorcidial capacity.[19] In contrast, CSF-1-derived macrophages were incapable of directly lysing the P815 cell, and addition of even high concentrations of γ-interferon did not enhance their lytic capacity. Significant tumor cell cytotoxicity was observed only when CSF-1-derived macrophages were exposed to high doses of both γ-interferon and lipopolysaccharide. In addition to differences in their tumor cytolytic capacity, bone marrow macrophages derived in either M-CSF or GM-CSF exhibit differences in phenotype associated with antigen presentation and their interaction with T-cells.[20] Murine bone marrow macrophages cultured in M-CSF or GM-CSF were examined for surface expression of the major histocompatibility class II antigen, Ia. Low numbers of cells derived in either cytokine were Ia antigen positive. Treatment of macrophages cultured in either cytokine with γ-interferon increased the percentage of cells which were Ia antigen positive in both populations. More M-CSF-treated cells than GM-CSF-treated cells were induced to become Ia positive. In contrast, however, GM-CSF-derived macrophages exhibited greater basal Ia antigen expression and Ia density per cell. RNA blot analysis confirmed that untreated GM-CSF-derived macrophages had fourfold more Ia mRNA than M-CSF-derived macrophages. In the presence of γ-interferon, M-CSF-derived macrophages increased their total Ia expression to levels found in GM-CSF-treated macrophages. Antigen-induced T-cell proliferation was also significantly greater with GM-CSF-derived macrophages than with M-CSF-derived macrophages.[20]

Other investigators have noted that if bone marrow macrophages maturing in M-CSF were subsequently switched to GM-CSF-containing medium, the cells would change from a state in which they were Ia antigen negative and poorly induced antigen-dependent T-cell proliferation to a state in which they were Ia antigen positive and much more efficient in inducing antigen-dependent T-cell proliferation.[21] Expression of membrane-bound IL-1 was also significantly augmented by the switch to GM-CSF, but not to γ-interferon-containing medium. The continued presence of GM-CSF was necessary for maintenance of expression of this phenotype, but the change in phenotype could be induced only early in macrophage maturation.

The possibility that macrophage phenotype and differentiation may be dependent on the CSF in which cells are incubated is further strengthened by studies of murine alveolar and peritoneal macrophages.[22] Culturing murine alveolar macrophages in the presence of GM-CSF maintains their alveolar macrophage phenotype, which is characterized by expression of the glycosphingolipid asialo GM-1, and no expression of Mac-1 antigen or capacity to bind fluorescein isothiocyanate (FITC)-labeled lipopolysaccharide. The cells cultured in GM-CSF also maintained the round alveolar macrophage morphology. Incubating these alveolar macrophages in M-CSF changed their cellular morphology to the markedly stretched and elongated form of the peritoneal macrophage. The alveolar macrophages cultured in M-CSF also changed their surface phenotype to that characteristic of the peritoneal macrophage, i.e., high Mac-1 antigen expression, staining with FITC-labeled lipopolysaccharide, and no staining with antibody against asialo GM-1. Lung tissue is a rich source of GM-CSF, and it is possible that this contributes to the maintenance of the alveolar macrophage phenotype. Selection of alveolar macrophage cells rather than induction of the alveolar phenotype is unlikely, since alveolar macrophages from incubations with GM-CSF could switch their phenotype after incubation in M-CSF.

Whether different cytokines can induce different monocyte/macrophage phenotypes in the human system is a less well-explored question. It cannot be assumed that human monocytes/macrophages are exactly equivalent to their murine counterparts, because they differ in several important physiological and biochemical respects. Human monocytes and macrophages do not proliferate in response to M-CSF, while the murine cells do proliferate in response to this factor.[13,29] M-CSF does not act as a colony-stimulating factor for human monocyte progenitors *in vitro*, while it does for murine monocyte progenitors.[1] Finally, murine macrophages can release reactive nitrogen intermediates after activation, while human cells do not produce these nitrogen intermediates,[24,25] and human monocytes require a second signal to release interferon α/β, while murine monocytes release low levels of these cytokines in response to M-CSF alone.[26,27] M-CSF has been shown to enhance human monocyte survival, differentiation, and release of TNF_α interferon and CSF.[21,23] Only two published studies compare the functional phenotypes induced by M-CSF vs. GM-CSF in cultured human monocytes. Culturing human peripheral blood monocytes in M-CSF greatly enhanced their ability to kill both TNF-resistant and -sensitive tumor target cells in response to a second signal, such as interferon γ, lipopolysaccharide, or a combination of these macrophage activators. Monocytes cultured in GM-CSF, however, did not show this enhanced tumor cell cytolysis in response to a second signal.[28a] It has also been shown that human monocytes cultured in M-CSF have an enhanced capacity to perform antibody dependent cell mediated cytotoxicity (ADCC) while human monocytes cultured in GM-CSF do not demonstrate this enhanced ADCC capacity.[29] These data would indicate that human M-CSF and GM-CSF may also induce different paths of monocyte differentiation, but the resulting phenotypes cannot be predicted from the murine model.

Studies of the cellular distribution of the human immunodeficiency virus HIV-1 have demonstrated that human monocytes are a major reservoir for this lentivirus, and it is believed that the differentiation state of the monocyte can alter virus expression.[5-8] For example, the U-1 clone of U-937 cells which is infected with HIV-1 produces only low levels of the virus constitutively; however, after treatment with phorbol ester, the production of virus was enhanced 20-fold.[5] Futher studies of the phorbol ester-treated cells indicated they had undergone macrophage differentiation, as judged by increased adherence, expression of CD11 antigen, and increased production of superoxide anion in response to phorbol ester. Most recently, experiments have been described in which normal human monocytes have been infected with HIV-1 and then cultured in either GM-CSF or M-CSF for up to 14 d.[29] During this period, HIV-1 virion production was monitored by enzyme-linked immunoabsorbent assay for HIV-1 p24 (core) antigen in culture supernatants. Although M-CSF and GM-CSF were both effective in significantly enhancing virion production at the optimal concentrations used for each factor, GM-CSF appeared to be three- to sevenfold more effective than M-CSF. The authors of this study ascribed the enhanced virus production to increased levels of DNA synthesis and cell proliferation in response to cytokines. This, however, may not be the complete explanation, because M-CSF enhanced virus production almost 20-fold, but it did not induce human monocytes to synthesize DNA or proliferate.[23,30] Another possible explanation is that the cytokines are altering the differentiation state of the monocytes and that the difference between the viral inductive activity of GM-CSF and M-CSF resides in their abilities to induce different states of cellular differentiation. This idea would be consistent with studies of ovine lentivirus replication. Normal undifferentiated sheep monocytes were permissive for the virus, but showed low levels of viral replication. Addition of interferon to infected cells induced cellular differentiation, which was accompanied by enhanced viral gene expression.[31]

Finally, it is necessary to suggest a molecular mechanism whereby GM-CSF and M-CSF may induce distinct macrophage differentiation patterns and viral replication. A possible explanation may lie in the set of immediately early response (IER) genes induced in quiescent serum-deprived fibroblasts stimulated by serum. In quiescent fibroblasts, a family of about 100 genes is induced rapidly after readdition of serum to these cells.[32-34] Some of these genes are inducible even in the presence of cycloheximide. Among these are the nuclear protooncogenes c-*fos* and c-*myc*, whose expression is associated with growth and differentiation. These nuclear protooncogenes are also induced by GM-CSF in the factor-dependent, virally transformed murine monocytic cell lines NSF 60.8 and BAC 1.2F5.[35,36] These murine monocytic cell lines require GM-CSF for continued proliferation. After depriving the cells of CSF, they enter a quiescent state and cease proliferation. In NSF 60.8, readdition of CSF to quiescent cells transiently induces c-*fos* and c-*myc* mRNA within 15 to 30 min of factor addition. This gene induction is followed by renewed cell proliferation. M-CSF also transiently induces the mRNA for the c-*fos* and c-*myc* oncogenes in factor-deprived normal murine monocytes and BAC 1.2F5 cells. Significantly, in BAC 1.2F5 cells, M-CSF also induces the IER genes JE and KC, but GM-CSF does not induce JE and KC, although c-*myc* and c*fos* are induced.[36] Thus, GM-CSF and M-CSF may not induce the same sets of immediate early response genes. No studies have been reported on the induction of IER genes in response to GM-CSF or M-CSF in normal human monocytic cells.

Studies of the induction of these genes in human monocytes could be important in understanding the regulation of cell differentiation and activation of HIV-1 replication, because some of these genes have a role in activating gene transcription and altering cellular differentiation. For example, the c-*fos* gene codes for a protein which is complexed with the c-*jun* protein to form transcription factor complex AP-1.[37] There are specific nucleotide motifs in promoter enhancer elements of mammalian genes which bind AP-1, and this interaction enhances gene transcription.[37-41] The induction of the AP-1 proteins in response to CSF could be important in triggering expression of a class of genes containing AP-1 enhancer binding sites. Moreover, transfection into embryonal carcinoma cells of plasmids containing the c-*fos* gene resulted in cells with a more differentiated phenotype,[42] and transfection of plasmids expressing c-*fos* antisense sequences blocked the ability of these cells to differentiate in response to dibutyryl cyclic AMP.[43] Although c-*fos* expression is often observed in association with myelomonocytic differentiation, correlational studies have led to conflicting conclusions as to its role in myelomonocytic differentiation.[44] The HIV LTR is also known to contain motifs which can bind the cellular transcription factors NF$_{KB}$, Sp-1, and AP-1.[45-47]

REFERENCES

1. **Clark, S. C. and Kamen, R.,** The human hematopoietic colony-stimulating factors, *Science,* 236, 1229, 1987.
2. **Metcalf, D.,** *The Molecular Control of Blood Cells,* Howard University Press, Washington, D.C., 1988.
3. **Groopman, J. E., Mitsuyasu, R. T., DeLeo, M. J., Oette, D. H., and Golde, D. W.,** Effect on recombinant human granulocyte-macrophage colony-stimulating factor on myelopoiesis in the acquired immunodeficiency syndrome, *N. Engl. J. Med.,* 317, 593, 1987.
4. **Andreesen, R., Scheibenhagen, C., Buegger, W., Kopf, S., Shumachen, C., and Lohr, G. W.,** Adpotive transport of autologous tumor cytotoxic macrophages grown from blood monocytes: a new approach to cancer immunotherapy, *Proc. Am. Assoc. Cancer Res.,* 30, 411, 1634, 1989.
5. **Folks, T. M., Justement, J., Kinter, A., Schnittman, S., Orenstein, J., Pali, G., and Janci, A.,** Characterization of a promonocyte clone chronically infected with HIV and inducible by 13-phorbol-12-myristate acetate, *J. Immunol.,* 140, 1117, 1988.
6. **Pauza, C. D., Galindo, J., and Richman, D.,** Human immunodeficiency virus infection of monoblastoid cells: cellular differentiation determines the pattern of virus replication, *J. Virol.,* 62, 3558, 1988.
7. **Gendelman, H. E., Orenstein, J. M., Martin, M. A., Ferrua, C., Mitra, R., Phipps, T., Wahl, L. A., Lane, H. C., Fauci, A. S., Burke, D. S., Skillman, D., and Meltzer, M. S.,** Efficient isolation and propagation of human immunodeficiency virus on recombinant colony-stimulating factor 1-treated monocytes, *J. Exp. Med.,* 167, 1428, 1988.
8. **Orenstein, J. M., Meltzer, M. S., Phipps, T., and Gendelman, H. E.,** Cytoplasmic assembly and accumulation of immunodeficiency virus types H2 in recombinant human colony stimulating factor treated human monocytes: an ultrastructural study, *J. Virol.,* 62, 2578, 1988.
9. **Wong, G. G., Witik, G. S., Temple, P. A., Wilkins, K. M., Leary, A. C., Luxemberg, D., Jones, S. S., Brown, E. L., Kay, R. M., Orr, E. C., Shoemaker, C., Golde, E. W., Kaufman, R. J.,**

Hewick, R. M., Wang, E. A., and Clark, S. C., Human GM-CSF: molecular cloning of complementary DNA and purification of the natural and recombinant proteins, *Science,* 228, 810, 1985.

10. **Kawaskai, E. S., Ladner, M. D., Wang, A. M., Van Arsdell, J., Warren, M. K., Coyne, M. W., Scheikert, W. L., Lee, M. T., Wilson, K. J., Boosman, A., Stanley, E. R., Ralph, P., and Mark, P. F.,** Molecular cloning of a complementary DNA encoding human macrophage specific colony stimulating factor (CSF-1), *Science,* 230, 291, 1985.

11. **Wong, G. G., Temple, P., Leary, A., Witek-Gianotti, J., Yang, Y. C., Carletta, A., Chung, M., Murtha, P., Kriz, R., Kaufman, R. J., Farenz, C. R., Sibley, B., Turner, K., Hewick, R., Clark, S., Yanai, N., Yokota, H., Yamada, M., Saito, M., Motoyoshi, K., and Takaku, F.,** Human CSF-1: molecular cloning and expression of a 4 Kb cDNA encoding the human primary protein, *Science,* 235, 1504, 1987.

12. **Nicola, N. A.,** The hematopoietic colony stimulating factors, *Immunol. Today,* 8, 134, 1987.

13. **Tushinski, R. J., Oliver, I. T., Guilbert, L. J., Tynan, P. W., Warner, J. R., and Stanley, E. R.,** Survival of mononuclear phagocytes depends on a lineage-specific growth factor that the differentiated cells selectively destroy, *Cell,* 28, 71, 1982.

14. **Ralph, P., Warren, M. K., Ladner, M. B., Kawasaki, E. S., Boosman, A., and White, T. J.,** Molecular and biological properties of human macrophage growth factor, CSF-1, in *Cold Spring Harbor Symposia on Quantitative Biology,* Vol. 51, 1986, 679.

15. **Wing, E. J., Ampel, N. M., Waheed, A., and Sadduck, R. K.,** Macrophage colony stimulating factor (M-CSF) enhances the capacity of murine macrophages to secrete oxygen reduction products, *J. Immunol.,* 135, 2052, 1985.

16. **Coleman, D. L., Chodakewitz, J. A., Bartiss, A. H., and Mellors, J.W.,** Granulocyte-macrophage colony-stimulating factor enhances selective effector functions of tissue-derived macrophages, *Blood,* 72, 573, 1988.

17. **Reed, S. G., Nathan, C. F., Pihl, D. L., Rodricks, P., Shanebeck, K., Conlon, P. J., and Grabstein, K. H.,** Recombinant granulocyte/macrophage colony-stimulating factor activates macrophages to inhibit *Trypanosoma cruzi* and release hydrogen peroxide, *J. Exp. Med.,* 166, 1436, 1987.

18. **Weiser, W. Y., Van Niel, A., Clark, S. C., David, J. R., Remold, H. G.,** Recombinant human granulocyte/macrophage colony-stimulating factor activates intracellular killing of *Leishmania donovani* by human monocyte-derived macrophages, *J. Exp. Med.,* 166, 1436, 1987.

19. **Falk, L. A., Hogan, M., and Vogel, S. N.,** Bone marrow progenitors cultured in the presence of granulocyte-macrophage colony-stimulating factor versus macrophage colony-stimulating factor differentiate into macrophages with distinct tumoricidal capacities, *J. Leuk. Biol.,* 43, 471, 1988.

20. **Falk, L. A., Wahl, L. M., and Vogel, S. N.,** Analysis of Ia antigen expression in macrophages derived from bone marrow cells cultured in granulocyte-macrophage colony-stimulating factor or macrophage colony-stimulating factor, *J. Immunol.,* 140, 2652, 1988.

21. **Fischer, H.-G., Frosch, S., Reske, K., and Reske-Kunz, A. B.,** Granulocyte-macrophage colony-stimulating factor activates macrophages derived from bone marrow cultures to synthesis of MHC class II molecules and to augmented antigen presentation function, *J. Immunol.,* 141, 3882, 1988.

22. **Akagawa, K. S., Kamoshita, K., and Tokunaga, T.,** Effects of granulocyte-macrophage colony-stimulating factor and colony-stimulating factor-1 on the proliferation and differentiation of murine alveolar macrophages, *J. Immunol.,* 141, 3383, 1988.

23. **Becker, S., Warren, M. K., and Haskill, S.,** Colony-stimulating factor-induced monocyte survival and differentiation into macrophages in serum-free cultures, *J. Immunol.,* 139, 3703, 1987.

24. **Meltzer, M.,** personal communication.

25. **Ding, A. H., Nathan, C. F., and Stuehr, D. J.,** Release of reactive nitrogen intermediates and reactive oxygen intermediates from mouse peritoneal macrophages, *J. Immunol.,* 141, 2407, 1988.

26. **Warren, M. K. and Ralph, P.,** Macrophage growth factor CSF-1 stimulates human monocyte production of interferon, tumor necrosis factor, and colony stimulating activity, *J. Immunol.,* 137, 2281, 1986.

27. **Warren, M. K. and Vogel, S. N.,** Bone marrow-derived macrophages: development and regulation of differentiation markers by colony-stimulating factor and interferons, *J. Immunol.,* 134, 982, 1985.

28. **Sampson-Johannes, A. and Carlino, J. A.,** Enhancement of human monocyte tumoricidal activity by recombinant M-CSF, *J. Immunol.,* 141, 3680, 1988.

28a. **Mufson, R. A., Aghajanian, J., Wong, G., Woodhouse, C., and Morgan, A. C.,** Macrophage colony-stimulating factor enhances monocyte and macrophage antibody dependent cell mediated cytotoxicity, *Cell. Immunol.,* 119, 182-192, 1989.

29. **Koyanagi, Y., O'Brien, W. A., Zhao, J. Q., Golde, D. W., Gasson, J. C., and Chen, I. S. Y.,** Cytokines alter production of HIV-1 from primary mononuclear phagocytes, *Science,* 241, 1673, 1988.

30. **Mufson, R. A. and Sobieski, D. A.,** unpublished data.

31. **Gendelman, H., Narayan, S., Kennedy-Stoskopf, S., Kennedy, P., Gholiti, J., Clements, J., Stanley, J., and Peyeskpaier, B.,** Tropism of sheep lentiviruses for monocytes: susceptibility to infection and virus gene expression increase during maturation of monocytes to macrophages, *J. Virol.,* 58, 67, 1986.

32. **Cochran, B. H., Reffel, A. C., and Stiles, C. D.,** Molecular cloning of gene sequences regulated by platelet derived growth factor, *Cell,* 33, 939, 1983.

33. **Lau, L. F. and Nathans, D.,** Expression of a set of growth-related immediate early genes in BALB/c 3T3 cells: coordinate regulations with c-fos or c-myc, *Proc. Natl. Acad. Sci. U.S.A.,* 84, 1182, 1987.

34. **Almendral, J. M., Sommer, D., MacDonald-Bravo, H., Burkhardt, J., Perera, J., and Bravo, R.,** Complexity of early genetic response to growth factor and mouse fibroblasts, *Mol. Cell. Biol.,* 8, 2140, 1988.

35. **Harel-Bellan, A. and Farrar, W. L.,** Modulation of proto-oncogene expression by colony stimulating factors, *Biochem. Biophys. Res. Commun.,* 148, 1001, 1987.

36. **Orlofsky, A. and Stanley, E. R.,** CSF-1-induced gene expression in macrophages: dissociation from the mitogenic response, *EMBO J.,* 67, 2947, 1987.

37. **Rauscher, F. J., Cohen, D. R., Curran, T., Bos, T. J., Vogt, P. K., Bohmann, D., Tjian, R., and Franza, B. R.,** Fos associated protein p39: the product of the jun protogene, *Science,* 240, 1010, 1988.

38. **Lee, W., Mitchell, P., and Tjian, R.,** Purified transcription factor AP-1 interacts with TPA inducible enhancer elements, *Cell,* 49, 741, 1987.

39. **Angel, P., Imayawa, M., Chin, R., Stein, B., Imbra, R. J., Rahmsdorf, H., Jonat, C., Herrlich, P., and Karin, M.,** Phorbol ester inducible genes contain a common cis element recognized by a TPA modulated trans acting factor, *Cell,* 49, 729, 1987.

40. **Lamph, W., Wansley, P., Sassone-Corsi, P., and Verma, I.,** Production of protooncogene Jun/AP1 by serum and TPA, *Nature,* 334, 629, 1988.

41. **Sharma, A., Bos, T., Pekkala-Hagan, A., Vogt, P., and Lee, A.,** Interaction of cellular factors related to the jun oncoprotein and the promoter of a replication-dependent hamster histone H3-2 gene, *Proc. Natl. Acad. Sci. U.S.A.,* 86, 491, 1989.

42. **Lockett, T. J. and Sleigh, J. M.,** Oncogene expression in differentiating F-9 mouse embryonal carcinoma cells, *Exp. Cell Res.,* 173, 370, 1987.

43. **Edwards, S. A., Rundell, A. Y. K., and Adamson, E. D.,** Expression of c-fos antisense RNA inhibits the differentiation of F9 cells to parietal endoderm, *Dev. Biol.,* 129, 91, 1988.

44. **Mitchell, R. L., Henning-Chubb, C., Huberman, E., and Verma, I. M.,** c-fos Expression is neither sufficient nor obligatory for differentiation of monomyelocytes to macrophages, *Cell,* 45, 497, 1986.

45. **Nabel, G. and Baltimore, D.,** An inducible transcription factor activates expression of human immuno-deficiency virus in T-cells, *Nature,* 326, 711, 1987.

46. **Jones, K. A., Kadonaga, J. T., Luciw, P. A., and Tjian, R.,** Activation of the AIDS retrovirus promoter by the cellular transcription factor, Sp1, *Science,* 232, 755, 1986.

47. **Franza, R. B., Jr., Rauscher, F. J., III, Josephs, S. F., and Curran, T.,** The fos complex and fos related antigens recognizing sequences that contain AP-1 binding sites, *Science,* 239, 1150, 1988.

Chapter 10

REGULATION OF PHENOTYPE BY TRANSFORMING GROWTH FACTOR-β: ROLE OF THE EXTRACELLULAR MATRIX

Ronald A. Ignotz and Joan Massague

TABLE OF CONTENTS

I. THE TGF-β SYSTEM

The development and integrity of pluricellular organisms is controlled by multiple genetic and biochemical devices. Prominent among these devices is an intercellular communication network established by secretory polypeptide factors that control cell proliferation and differentiation. These factors act hormonally by binding to target cell-surface receptors that become allosterically activated and transduce specific signals across the plasma membrane. By acting on target cells located at varying distances from their sources, these factors establish autocrine, paracrine, or endocrine modes of intercellular stimulation. Some growth factors are promoters of cell proliferation, and have been the subject of much attention for over two decades because exacerbation of their function can lead to oncogenesis. However, the concept of balance between growth regulatory stimuli of opposite activity has materialized in recent years with the finding of a complex collection of factors that inhibit cell growth and regulate terminal differentiation. An important example of this class of factors is transforming growth factor-β_x(TGF-β).

TGF-β is the prototype of a superfamily of polypeptides that regulate growth, differentiation, and morphogenesis in organisms from insects through humans.[1-3] This superfamily is divided into four families, according to the structural and functional relatedness of their members (Figure 1). The most extensively studied group is the TGF-β family itself. Five TGF-β genes (β1, β2, β3, β4, and β5) have been identified thus far. They share a high degree (70 to 80%) of amino acid sequence identity with each other, and are extremely well conserved phylogenically.[4-13] The amino acid sequence deduced from the corresponding cDNAs indicates that TGFs-β are synthesized as part of larger precursors that are proteolytically processed (Figure 2). The bioactive domain corresponding to each factor resides in the carboxyl-terminal 100 to 132 amino acids of its precursor. Although TGF-β1 was originally isolated as a homodimer,[14] identification of multiple TGF-β genes suggests that homodimers as well as heterodimers may exist. In fact, the homodimers TGF-β1 and TGF-β2, and the heterodimer TGF-β1.2, have been isolated from porcine platelets, presumably due to co-expression of TGF-β1 and TGF-β2 genes in porcine megakaryocytes.[6] Hence, co-expression of multiple TGF-β genes may generate an array of dimers with differing regulatory or bioactive properties. TGF-β3, TGF-β4, and TGF-β5 have been identified at the cDNA level just recently, and polypeptides corresponding to these genes have not been isolated from natural sources yet.

Dimerization of the TGF-β precursor and proteolytic cleavage of the propeptide occur intracellularly, at least in cells that overexpress a transfected TGF-β1 gene.[15,16] However, TGF-β1 and TGF-β2 are secreted in an inactive or latent form complexed with other polypeptides.[15,17-19] One of the components of the latent TGF-β complex is the propeptide fragment of the TGF-β precursor.[16,19] Association of the TGF-β dimer with the propeptide appears to be sufficient to generate the latent form. The TGF-β1 propeptide contains N-linked mannose-6-phosphate,[20] through which it can bind to the man-6-P/insulin-like growth factor-II receptor.[21] The physiological role of this interaction is not clear. The predicted propeptide sequences of TGF-β1, TGF-β3, and TGF-β4 contain the arginine-glycine-aspartate (RGD) sequence that in cell adhesion molecules functions as a primary recognition site for binding to cell-surface adhesion receptors of the integrin class,[22] but in TGF-β precursors this sequence may not have the same function. In addition to the propeptide, the latent TGF-β complex isolated from natural sources carries a 125- to 160-kDa glycoprotein.[18,19] The primary structure of this binding protein consists of multiple EGF-like repeats[18] like those found in membrane-bound growth factor precursors of the EGF family[23] and in coagulation factors that have serine protease activity.[24] In addition to the latent secretory TGF-β complex, TGF-β may be sequestered or carried in serum by α2-macroglobulin.[25] *In vitro*, latent TGF-β can be activated by incubation at extreme pH conditions, or by treatment

THE TGF-β SUPERFAMILY OF GROWTH

DIFFERENTIATION AND MORPHOGENESIS FACTORS

Genes	Bioactive dimers	Genes	Bioactive Dimers
TGF-β Family		*Decapentaplegic/Vg Family*	
TGF-ß1	TGF-ß1 homodimer	DPP-C**	(Decapentaplegic transcript)
TGF-ß2	TGF-ß2 homodimer	Vg1*	(cDNA only)
TGF-ß3	TGF-ß3 homodimer	VGR-1	(Vg1-related; cDNA only)
TGF-ß4#	(cDNA only)	BMP-2	BMP-2 homodimer
TGF-ß5*	TGF-ß5 homodimer	BMP-3 through -7	(cDNAs only)
	TGF-ß1.2 heterodimer		

Activin/Inhibin Family		*Müllerian Inhibing Substance Family*	
α	Inhibin A, (α,ßA)	MIS	MIS homodimer
	Inhibin B, (α,ßB)		
ßA	Activin A, ßA homodimer		
ßB	Activin AB, (ßA.ßB)		

FIGURE 1.   TGF-β superfamily of growth and differentiation factors. This grouping of the members of the TGF-β superfamily is according to closest structural or functional homologies. The subunit composition (homo- or heterodimeric) for those members for which the proteins have been isolated is indicated. These factors have been identified in humans, except some identified only in *chick,* xenopus laevis, or **Drosophila melanogaster.

with proteolytic or deglycosylating enzymes.[17,26,27] The physiological mechanism for activation of latent TGF-β has not been fully elucidated yet.

Although TGF-β was first identified in the culture medium from transformed cells, many normal tissues and cultured cells have been found to express TGF-β. TGF-β is particularly abundant in platelets and bone.[5,6,14] The abundance of TGF-β in centers of active tissue development and in platelets suggests a role in morphogenesis and tissue repair. In the mouse embryo, TGF-β has been localized by immunohistochemistry to mesenchymally derived tissues such as bone, cartilage, and teeth, in subepithelial cell layers, and in hematopoietic centers.[28-30] The most intense staining is observed during periods of rapid morphogenesis.[29]

In addition to the multiple forms of TGF-β, several other proteins have been identified that have an amino acid sequence and precursor structure similar to TGF-β[1] (Figure 1). These factors also exhibit functional relatedness in that they are all involved in morphogenesis and/or control of differentiation processes. Within this group are the mammalian Müllerian inhibiting substance (MIS) that induces the regression of the Müllerian duct in male embryos.[31] Activins and inhibins antagonize each other in the regulation of follicle-stimulating hormone production by pituitary cells and aromatase activity in ovarian granulosa cells, as well as in erythroblast differentiation.[32-36] In *Xenopus*, a maternal mRNA designated Vg1 has been shown to have homology to TGF-β.[37] Vg1 is thought to be critically involved in mesoderm development in frog embryops. A TGF-β-related gene, decapentaplegic (dpp), has also been identified in *Drosophila*.[38] The dpp transcript (DPP) is critical for proper dorsoventral specification in the fly embryo, and for imaginal disc development in the larva. Three bone morphogenesis proteins (BMPs) expressed in mammalian bone have a high degree (up to 75%) of amino acid sequence identity to dpp.[39] Thus, the two layers of

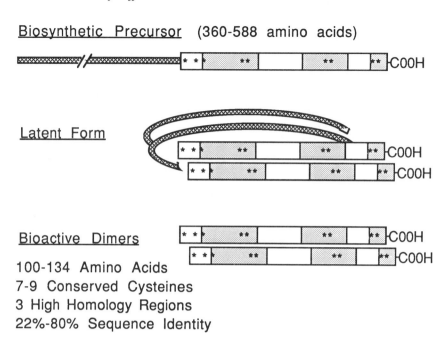

<u>Biosynthetic Precursor</u> (360-588 amino acids)

<u>Latent Form</u>

<u>Bioactive Dimers</u>
100-134 Amino Acids
7-9 Conserved Cysteines
3 High Homology Regions
22%-80% Sequence Identity

FIGURE 2. General structure and biosynthetic processing of TGF-β and related factors. Factors of this group are synthesized as larger precursors, with the bioactive domains located within the C-terminal portion of the precursors. In those cases examined, cleavage of the precursor and dimerization occur intracellularly. The propeptide remains associated with the bioactive domain, forming a latent complex that is activated by dissociation after secretion. The bioactive form of TGF-β and related factors is a disulfide-linked dimer. Many of the structural homologies between the TGF-β family members are concentrated in the bioactive domains.

complexity—structural and functional—that characterize the TGF-β group are also found in the other families of TGF-β-related factors.

II. MEMBRANE TGF-β RECEPTORS AND OTHER TGF-β BINDING PROTEINS

TGF-β elicits its biological effects presumably through binding to specific cell-surface receptors. The various forms of TGF-β (TGF-β1, TGF-β2, and TGF-β1.2) that have been isolated thus far display a complex pattern of interactions with at least four distinct types of high-affinity cell-surface binding components, or receptors.

TGF-β receptor types I and II are glycoproteins of 53 and 73 kDa, respectively, that bind TGF-β1 with higher affinity than TGF-β2.[6,40-42] The type III receptor is, remarkably, an integral membrane proteoglycan termed betaglycan. Betaglycan displays similar affinity for all forms of TGF-β tested.[6,40,41,43,44] Most mammalian cell types express at least one of these three receptor types, and in many cases all three receptors are expressed simultaneously.[45-47] Among more than 100 cell lines and tissues examined, only one rat pheochromocytoma cell line, and several human retinoblastoma cell lines, lack detectable TGF-β receptors.[48] A fourth TGF-β receptor type identified in pituitary cells is unique in its ability to bind inhibin and activin in addition to TGF-β1 and TGF-β2.[49]

The finding of this diversity of cell-surface TGF-β binding proteins has raised questions about which of these molecules may act as the mediator of TGF-β effects on the cell and what may be the function of the other TGF-β binding proteins.

A. RECEPTORS INVOLVED IN THE GROWTH-SUPPRESSIVE ACTION OF TGF-β

TGF-β acutely inhibits epithelial cell proliferation by acting at a level distal from the receptors for growth-stimulating factors.[50] Identification of the probable TGF-β receptor that mediates the growth inhibitory action has been achieved by isolating cell mutants that have become resistant to growth inhibition by this factor. Two classes of such mutants have been isolated by chemical mutagenesis of the Mv1Lu mink lung epithelial cell line.[51] Mv1Lu cells express simultaneously TGF-β receptor types, I, II, and III. TGF-β-resistant class R mutants derived from this parental cell line have selectively lost expression of functional type I TGF-β receptors. They have also lost the ability to respond to TGF-β with elevated expression of fibronectin and cell flattening. However, the receptor types II and III in these cells are expressed at normal levels, and display unaltered affinity for TGF-β1 and TGF-β2. In contrast, class S expresses receptor types I, II, and III, and binds TGF-β normally, but does not respond to this factor. Analysis of somatic cell hybrids indicates that both mutant phenotypes are recessive.[52] The results with R mutants identify the type I receptor as the one involved in mediating growth inhibition and other effects of TGF-β. S mutants might be defective in the TGF-β signal transduction mechanism, possibly due to mutations in the signaling domain of the TGF-β receptor.

The biological potency of various forms of TGF-β correlates well with their relative affinity for the type I receptor in various hematopoietic progenitor cell lines.[41,42] Paradoxically, however, many other cell lines, including Mv1Lu cells, respond equally well to TGF-β1 and TGF-β2 despite the lower affinity of TGF-β2 for the type I receptor. It is possible that one of the other two receptor types potentiates the actions of TGF-β2 mediated through the type I receptor. A fact potentially significant in this regard is that, unlike most other cell lines, the hematopoietic progenitor cells do not express detectable receptor types II or III.[41]

B. BETAGLYCAN, A POLYMORPHIC TGF-β BINDING PROTEOGLYCAN

In many cell lines, the most abundant cell-surface TGF-β binding component is a chondroitin/heparan sulfate proteoglycan now called betaglycan, but previously referred to as the type III TGF-β receptor. Betaglycan consists of a microheterogeneous core protein of 100 to 130 kDa linked to one or several glycosaminoglycan (GAG) chains.[43,44] Experiments with Chinese hamster ovary (CHO) cell mutants defective in GAG synthesis have shown that the TGF-β binding site resides in the core protein.[53] The GAG chains are not required for cell-surface expression or the ligand binding activity of betaglycan. Furthermore, the growth inhibitory response of CHO cells to TGF-β is not affected by the absence of GAGs. The GAG attachment site, the TGF-β binding site, and the membrane anchor are located in separate tryptic domains in betaglycan. Betaglycan purified to near homogeneity does not require interaction with other receptor types in order to bind TGF-β.

Forms of betaglycan that lack the membrane anchor are released by cultured cells into the medium and can also be found in extracellular matrices and serum.[54] The core protein of secretory betaglycan exhibits a molecular weight of 100 to 120 kDa. This core protein lacks the hydrophobic membrane anchor present in membrane betaglycan. The similarity in molecular weight between the core proteins of membrane-derived and secretory betaglycans suggests that the membrane anchor/cytoplasmic domain of betaglycan might be very small.

Several cell lines, including myoblasts and hematopoietic progenitors, are highly responsive to TGF-β1, yet they do not express detectable betaglycan. All the evidence available to date suggests that betaglycan may not be directly involved in the mediation of TGF-β action. However, betaglycan might be involved in the presentation of TGF-β to the type I signal transducing receptor. Given its structural features, relative abundance, and secretory nature, betaglycan may also be able to accumulate in extracellular matrices and act as a pericellular reservoir or clearance system for TGF-β.

III. CONTROL OF PHENOTYPE BY TGF-β

The biological effects of TGF-β are diverse. Originally described as a factor capable of inducing anchorage-independent growth of normal NRK-49F fibroblasts in semisolid medium,[55] TGF-β is now known to be a potent growth inhibitor for several cell types.[42,48,56-58] Certain fibroblast lines are growth stimulated by TGF-β, but it has been suggested that in some of these cell lines the growth stimulatory effect may be secondary to the induction of autocrine growth factors, such as platelet-derived growth factor.[59-61] The mechanism by which TGF-β negatively affects proliferation is not understood. Growth arrest of CCL39 fibroblasts or Mu1Lu lung epithelial cells by TGF-β does not correlate with a detectable block in any of various early growth-related responses, including activation of plasma membrane H^+/Na^+ antiport, induction of c-*fos* and c-*myc* expression, activation of protein kinase C, or phosphorylation of ribosomal protein S6, all of which are normally observed during the mitogenic stimulation of cells.[50,62]

TGF-β has profound effects in cell differentiation and expression of differentiated. As with the effects on proliferation, TGF-β can either inhibit or promote differentiation. Adipogenesis, myogenesis, and hematopoiesis, for example, are reversibly inhibited *in vitro* by TGF-β, whereas the differentiation of osteoblasts, chondroblasts, and epithelial cells can be promoted by this factor. The remainder of this article focuses on the control of cell differentiation by TGF-β and the biochemical effects of TGF-β that may mediate its actions on cell phenotype.

A. ADIPOGENIC DIFFERENTIATION

Several lines of mouse 3T3 fibroblasts have the capacity to spontaneously differentiate into adipocytes, a process that is stimulated by a set of pharmacologic agents and serum factors.[63] 3T3-L1 cell cultures induced with a mixture of insulin, dexamethasone, and methylisobutyl xanthine begin to exhibit the adipocyte phenotype within 3 to 4 d. Differentiation results in profound morphologic changes, high-level expression of several enzymes (glycerophosphate dehydrogenase, ATP-citrate lyase, and fatty acid synthetase) involved in lipid metabolism, and changes in receptors for hormones that control adipogenesis. The differentiating cells accumulate lipid droplets, assume a rounded morphology, and become postmitotic. The resulting adipocytes have the morphological and biochemical traits of the physiologically normal adipocyte phenotype.

When 3T3-L1 cells are induced to differentiate in the presence of picomolar concentrations of TGF-β, differentiation is completely blocked.[64] The cells retain a fibroblastic appearance and the enzymatic markers of differentiation are not induced. Interestingly, TGF-β can be added up to 30 h postinduction and still block differentiation. This indicates that the commitment to differentiate into adipocytes is not reached until about 30 h after induction. Furthermore, a transient 4-h exposure of the cells to TGF-β is sufficient to prevent expression of the adipocyte phenotype. Thus, relatively rapid events induced by TGF-β are dominant suppressors of adipocyte differentiation. If TGF-β is removed and the cultures are rechallenged with the induction mixture, differentiation occurs, indicating that the effects of TGF-β are reversible. Inhibition of adipogenic differentiation by TGF-β does not appear to be related to effects on cell proliferation, since TGF-β does not affect, reduce, or prolong the rounds of cell division that typically take place during the differentiation inductive phase. Once 3T3-L1 cells become committed to express the adipocyte phenotype, they are refractory to the inhibitory action of TGF-β. However, this is not due to a loss of TGF-β receptors, as 3T3-L1 adipocytes continue to express TGF-β receptor types I, II, and III, at levels similar to those of preadipocytes. TGF-β also blocks adipocyte differentiation of Balb/c 3T3 and TA1 preadipocytes.[65-66] In contrast to 3T3 adipocytes, differentiated TA1 cells can respond to TGF-β with inhibition of adipocyte-specific gene expression, albeit in a serum-dependent manner.[66]

B. MYOGENIC DIFFERENTIATION

Rat skeletal muscle myoblast cell lines L6, L_6E_9 and L8 can terminally differentiate into multinucleated, striated muscle cells when placed into medium of poor growth-promoting ability.[67-69] The expression of skeletal muscle-specific genes precedes cell fusion. In addition, there is a decrease in the expression of β/γ actin, fibronectin, and type I collagen, all of them markers of the fibroblastic phenotype. When TGF-β is included in the differentiation-promoting medium, the expression of skeletal muscle-specific genes is prevented and the myoblasts do not fuse into multinucleated myotubes.[70] However, there is a distinct reorganization of the cell monolayer. Cells cluster to form what appears as hills and valleys. These events correlate with a high increase in the expression of type I collagen, fibronectin, and the extracellular matrix proteoglycans decorin and biglycan.[71,72] Like preadipocytes, myoblasts also go through a critical temporal point, after which differentiation will proceed even in the presence of TGF-β. L_6E_9 cell monolayers commit to differentiation and become refractory to inhibition by TGF-β in a stochastic manner, the process being complete 3 d after the shift to the differentiation medium. As in differentiating preadipocytes, the resistance of myoblasts to the inhibitory action of TGF-β after the commitment point is not due to the lack of TGF-β receptors. Multinucleated L_6E_9 myotubes continue to express normal levels of TGF-β receptor types I and II, betaglycan being undetectable both before and after differentiation. Furthermore, the TGF-β receptors in L_6E_9 myotubes remain competent to mediate effects of TGF-β such as elevation of collagen expression. Other myoblast cell lines (L6-A1 and C2) whose differentiation is also blocked by TGF-β[73,74] appear to have a reduced capacity to bind TGF-β after differentiating into myotubes. However, this reduction is only partial, and is therefore unlikely to be the cause of complete resistance of the cells to TGF-β.

The differentiation of skeletal muscle satellite cells from mature rats as well as rat neonatal myoblasts, BC_3H1 myoblasts, and chick embryo myoblasts is also blocked by TGF-β.[70,75,76] In all cases tested, TGF-β does not affect the proliferation of myoblasts. Thus, the inhibition of myogenic differentiation by TGF-β is not likely to be based on retention of the cells in a proliferative state, as it may occur serum mitogens.

C. CHONDROGENESIS AND OSTEOGENESIS

Bone tissue undergoes a constant remodeling process whereby new skeletal components are made and deposited while old components are resorbed. These are necessarily finely regulated processes on which depend the normal development, maintenance, and repair of bone. The observation that bone can sustain its own regeneration led to the finding of bone matrix-derived polypeptide factors involved in the control of bone cell growth and differentiation. Searches based on assays designed to identify factors capable of promoting osteogenic cartilage formation led to the isolation of two factors, designated cartilage-inducing factors A and B (CIF-A, CIF-B), from extracts of bovine bone.[5] These factors induce the expression of a chondrocyte phenotype in embryonic rat muscle mesenchymal cells suspended in semisolid medium.[5] The cells start expressing cartilage-specific proteoglycan and type II collagen, and form suspended cell colonies. Chemical characterization of CIF-A and CIF-B has demonstrated that they correspond, respectively, to TGF-β1 and TGF-β2.[77,78]

In addition to its chondrogenic activity, TGF-β favors proliferation and expression of differentiation markers in cultured osteoblasts. TGF-β is abundant in bone matrix,[5,30] is synthesized by cultured bone cells, and is mitogenic for osteoblasts.[79-81] TGF-β also promotes the expression of several proteins found in bone matrix, including type I collagen, osteopontin, and osteonectin.[81-83] Rat osteosarcoma cells ROS 17/2.8 have been reported to express elevated levels of alkaline phosphatase enzymatic activity and mRNA in response to TGF-β.[84]

In the face of this evidence, however, TGF-β has been found to repress the expression of bone differentiation markers in various cell types that can undergo a certain degree of osteogenic differentiation *in vitro*. Examples of this type of response to TGF-β have been described in rat calvaria osteoblasts[85] and the nontransformed murine bone cell line, MC3T3L1.[86] Various potential explanations exist for this apparent paradox. First, it is possible that cell cultures maintained on plastic do not assemble an appropriate extracellular matrix support and tissue structure required for osteogenic differentiation. Another attractive possibility is that TGF-β might be just one factor among many whose concerted activity is required to induce osteogenesis differentiation. Recently, several additional factors have been identified that may contribute to the osteogenic process. Three of them are the TGF-β-related bone morphogenetic proteins (BMPs) mentioned above. It is likely that additional factors that collaborate with TGF-β to promote cartilage and bone formation exist. A detailed review of BMPs and their role in bone formation can be found in Reference 87.

D. HEMATOPOIESIS

Hematopoietic progenitor cells can be induced to proliferate into colonies of differentiated cells in response to colony-stimulating factors (CSFs). Mouse B6SUtA cells are a factor-dependent hematopoietic cell line that responds to multi-CSF (IL-3) with the formation of differentiated granulocyte-macrophage-mast cell colonies. These colonies also contain hemoglobinized erythroid cells if erythropoietin is present in the medium.[88] TGF-β1 at picomolar concentrations potently inhibits multilineage as well as hemoglobinized colony formation by B6SUtA cells.[42] A similar, albeit weaker, inhibitory effect of TGF-β is also observed in 32D-C13 cells, a bipotential cell line that differentiates into either mast cells or granulocytes in response to IL-3 or G-CSF, respectively.[42] The reduced effect of TGF-β1 on 32D-C13 cell colony formation may be related to the fact that these cells have progressed further along a differentiation pathway than B6SUtA cells. These effects of TGF-β1 are not restricted to established cell lines, but are also observed in freshly isolated as well as long-term bone marrow cultures, and are linked to inhibition of cell proliferation.

An important difference between the reponse to TGF-β in hematopoietic progenitor cells and other cell types is the differential sensitivity of the hematopoeitic progenitors to various forms of TGF-β. Thus, TGF-β2 is approximately 1% as potent as TGF-β1 in inhibiting the proliferation and differentiation of B6SUtA and 32D-C13 cells. The heterodimer, TGF-β1.2, has an intermediate potency.[41,42] The ability of these cell lines to distinguish between the various forms of TGF-β correlates with the presence of type I receptors as the only detectable TGF-β receptor type in these cells. The relationship that exists between the order of biological potency and receptor binding affinity of various TGFs-β in this cell system has provided one of the first clues implicating the type I TGF-β receptor in growth-inhibitory responses.

E. OTHER ACTIONS OF TGF-β ON CELL DIFFERENTIATION

Several other cell types in which TGF-β affects differentiation have been reported in addition to those reviewed in the preceding sections. Epithelial cells from skin, lung, intestine, and kidney are growth inhibited by TGF-β. In some cases, the effect of TGF-β on these cells is fully reversible, and the cells can resume the normal proliferative phenotype after removal of this factor. In certain cases, however, growth inhibition leads to or is accompanied by terminal differentiation. For example, normal human bronchial epithelial cells in culture assume a flattened squamous cell morphology, have elevated levels of plasminogen activator and cross-linked envelopes, and become postmitotic in response to TGF-β.[89] The rat intestinal crypt cell line IEC-6, when treated with TGF-β, becomes irreversibly growth arrested and differentiates, as characterized by the expression of sucrase activity.[90] Both human and mouse keratinocytes are reversibly growth inhibited by TGF-β acting alone,[91,92] but in combination with epidermal growth factor, TGF-β enhances the keratinization of these cells.[93]

TGF-β can also modulate the degree of expression of a differentiated phenotype. Examples include the elevation of follicle-stimulating hormone expression by pituitary cells, the modification of granulosa cell responsiveness to FSH, the inhibition of steroidogenesis in cultured adrenocortical cells and Leydig cells,[33,94,98] the immunosuppressive activity on B- and T-lymphocytes and natural killer cells,[9,95,96] and the deactivation of macrophages.[97]

IV. EARLY RESPONSE GENES IN TGF-β ACTION

How can a single factor like TGF-β exert such a wide range of biological effects? The answer to this question is unavailable, since there is no information yet on the nature of the intracellular signal(s) elicited by activated TGF-β receptors. One approach to this question has been to examine the potential involvement of known second-messenger systems in TGF-β action. The evidence obtained thus far shows that none of the parameters examined is modulated by TGF-β in a wide range of cell types, and if modulation by TGF-β occurs in certain cells it may be indirect. However, second messengers for other hormonally active agents may be involved in the pathway of TGF-β action several steps downstream from the immediate TGF-β postreceptor signal. It is also conceivable that TGF-β and the family of factors that it represents operate via a novel type of signaling pathway.

An alternative approach has been to identify biochemical events that are located downstream in the pathway of TGF-β action, but are important in mediating many of the ultimate proliferative or phenotypic responses. This approach has lead to the identification of genes whose level of expression changes early in response to TGF-β. Examination of genes encoding factors that control cell proliferation has led to the finding that c-*jun* and *jun*-B are both frequently activated in response to TGF-β, in some cases within 30 min of the addition of this factor.[99] Induction of these two genes by TGF-β is observed independently of the kind of proliferative response—positive, negative, or null—that the cells have to this factor. The c-*myc* expression is either stimulated or repressed by TGF-β in AKR-2B embryo fibroblasts,[59] endothelial cells,[100] and keratinocytes[92] correlating with the proliferative response of these cells to TGF-β. However, c-*myc* expression does not appear to be rapidly modified by TGF-β in CCl-39 lung fibroblasts, a cell line whose proliferation is arrested by TGF-β.[62] In addition, TGF-β can affect the expression of growth factors, including platelet-derived growth factor A and B chains, and TGFs-β themselves.[59-61,101] The physiological role of these effects remains to be clarified.

Another group of genes that has been identified as a major target for TGF-β action are the genes that encode the adhesion apparatus of the cell, i.e., the genes that encode extracellular matrix components and cell-surface adhesion receptors.

V. THE CELL ADHESION APPARATUS AS A BIOCHEMICAL TARGET

At the biochemical level, the actions of TGF-β on many of its target cells appear to have a central theme. Grossly, TGF-β amplifies the accumulation of extracellular matrix and cell adhesion receptors. This effect is accomplished by a concerted action at multiple levels, including enhanced synthesis and deposition of extracellular matrix components, induction of inhibitors of extracellular matrix degradation, and modification of the repertoire of cell-surface adhesion receptors, as summarized in Figure 3. The original impetus to examine the expression of these molecules in response to TGF-β derived from the observed effects of TGF-β on cell morphology and differentiation,[64] as well as its potential involvement in wound-healing responses that include accumulation of connective tissue and is suggested by the high level of TGF-β found in blood platelets.[14,102]

A. MODULATION OF THE EXTRACELLULAR MATRIX

TGF-β elevates fibronectin expression in many mesenchymal and epithelial cell types, both normal and transformed.[103,104] A two- to tenfold elevation in *de novo* fibronectin

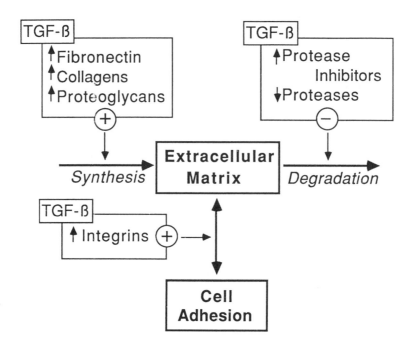

FIGURE 3.   Summary of TGF-β actions on the cell adhesion apparatus. TGF-β elevates the synthesis of several structural proteins of the extracellular matrix and prevents matrix degradation by stimulating the synthesis and release of specific protease inhibitors. TGF-β also alters (generally upregulates) the expression of cell-adhesion receptors, thus affecting cell-matrix interactions. These changes in cell-matrix interactions may in part be responsible for the effects of TGF-β on differentiation.

synthesis and a corresponding increase in extracellular matrix fibronectin accumulation are commonly observed in response to TGF-β. Elevated fibronectin synthesis is at least in part due to elevated fibronectin mRNA levels following TGF-β treatment.[71] Elevated transcription of fibronectin genes has been demonstrated with the use of expression plasmids containing the fibronectin gene promoter linked to a reporter gene.[105] Increased stability of fibronectin mRNA has also been observed in response to TGF-β, the relative contribution of each of these two mechanisms varying with the cell type.[106]

TGF-β also regulates the expression of several types of collagen, including type I collagen α1 and α2 chains, and collagen types III, VI, and X.[103,107,108] Expression of type II collagen is induced in mesenchymal muscle cells secondarily to their chondrogenic differentiation in response to TGF-β.[5] As with fibronectin, increased synthesis of collagen protein is correlated with higher mRNA levels.[71,108,109] Analysis of the promoter of the mouse α2(I) collagen gene has defined a binding site for the transcription factor NF-1 as the site involved in transcriptional activation of this gene by TGF-β.[110] Mutations in the NF-1 site abolish TGF-β activation of α2(I) promoter, as determined by expression of a reporter gene linked to it. Insertion of the NF-1 sequence 5′ to the SV40 promoter also confers TGF-β responsiveness to the SV40 promoter. Thus, TGF-β increases the rate of transcription driven by these two promoters apparently by regulating NF-1 activity. However, removal of NF-1 binding sites from the promoter region of other genes, such as the fibronectin gene or the TGF-β1 gene, results only in a partial decrease in the transcriptional activation of these genes by TGF-β.[105,111] At least two other sites in the fibronectin promoter are also responsive to TGF-β.

Other matrix glycoproteins whose synthesis is elevated in response to TGF-β include osteopontin, osteonectin, tenascin, and thrombospondin.[82,83,109,112] The expression of abundant extracellular matrix chondroitin/dermatan sulfate proteoglycans is also elevated in TGF-β-treated cultures of fibroblasts, myoblasts, and epithelial cells.[72] These proteoglycans probably include biglycan (PG I) and decorin (PG II). Not only does TGF-β elevate the expression of proteoglycan core proteins, but it also increases the size of the glycosaminoglycan chains attached to them.[72] The composition of GAG chains in the mouse epithelial cell membrane proteoglycan, syndecan, and the overall synthesis of glycosaminoglycans in human arterial smooth muscle cells are also regulated by TGF-β.[113,114] The mechanism by which TGF-β affects the elongation and termination of GAG is unknown at present.

Elevated synthesis of extracellular matrix components is not solely responsible for the net accumulation of extracellular matrix induced by TGF-β. Plasminogen activator inhibitor-1 (PAI-1) and the tissue inhibitor of metalloprotease (TIMP), two inhibitors of extracellular matrix degrading enzymes, are strongly upregulated by TGF-β.[115-117] In the case of PAI-1, the upregulation can be observed less than 30 min after the addition of TGF-β, sooner than the response of fibronectin, collagens, and proteoglycans, which ordinarily takes 3 to 4 h to appear and 12 to 24 h before it reaches its maximum. PAI-1 mRNA levels can increase up to 50-fold over the control in response to TGF-β. In some cell types, TGF-β also decreases the expression of collagenase,[117] but may increase collagenase activity in other cell types.[118] Thus, elevated synthesis as well as decreased degradation contribute to the accumulation of the extracellular matrix in response to TGF-β.

B. MODULATION OF CELL-ADHESION RECEPTORS

A large set of the cell-surface receptors that interact with the cell's environment is devoted to mediate cell adhesion to extracellular matrix components and to other cells. One of the best characterized families of cell-adhesion receptors are the integrins. Integrins are heterodimeric complexes of integral membrane α and β glycoprotein subunits that link the extracellular matrix with the cytoskeleton via specific binding interactions at both sides of the plasma membrane. The specificity of integrins as cell-adhesion receptors is determined by the αβ subunit combination in the functional integrin complex. To date, at least 12 unique α-integrin subunits and 5 unique β-subunits have been identified. Two groups of integrins that are defined by the β-subunits that they contain ($\alpha\beta_1$ and $\alpha\beta_3$ integrins) include receptors for fibronectin, collagen, laminin, vitronectin, fibrinogen, von Willebrand factor, and other, as yet unidentified, extracellular matrix components. Several of these ligands contain an arginine-glycine-aspartic acid (RGD) sequence that is involved in receptor binding. Another set of integrins ($\alpha\beta_2$ integrins) present in lymphocytes mediate cell-to-cell adhesion rather than cell-to-matrix adhesion. Most cell types express various integrins simultaneously, each cell type having a characteristic complement of these receptors. Several reviews on the structural and functional properties of integrins have been published recently.[22,119,120]

TGF-β induces marked alterations in the repertoire of integrins expressed in many cell types.[121-124] The possibility that cell-adhesion receptors might be targets for TGF-β action was first suggested by the observation that TGF-β-treated cells have a higher ability to bind fibronectin and collagen.[103] Mouse thymocytes treated with TGF-β bind more readily to fibronectin-coated tissue cultures dishes, and their attachment can be prevented by short, synthetic RGD-containing peptides, indicating that adherence is mediated by integrins.[121] The mechanism by which TGF-β alters the complex balance of integrin dimers that in part share common subunits is primarily based on increasing the mRNA levels of individual subunits. TGF-β can alter the expression of all integrin subunits examined to date. These include α_1 through α_6 subunits and the β_1 subunit which combine to generate receptors for fibronectin, collagen, laminin, and other extracellular matrix molecules,[123] α_v and β_3 subunits that dimerize to form the vitronectin receptor,[124] and the α_L subunit that with the β_2 subunit constitutes LFA-1, a leukocyte receptor that binds the intercellular adhesion molecule ICAM-1 on the surface of other cells.[125]

The susceptibility of individual integrins to regulation by TGF-β is dictated by important cell determinants.[123] TGF-β can regulate the levels of integrins that are expressed by a given cell, but there is no evidence that TGF-β can induce integrins that are not being expressed by a cell in the basal state. Furthermore, cell determinants dictate which of the expressed integrin subunits will respond to TGF-β and to what extent. Although TGF-β can upregulate the expression of all the integrin subunits mentioned above, it upregulates them independently and to different extents, depending on the cell line. The expression of a given integrin subunit may not be elevated in a given cell type, or may even be strongly downregulated, as is the case of the α_3 subunit in MG-63 osteosarcoma cells.[123] The relative changes in the levels of assembled integrin dimers on the cell surface depend not only on the changes in their rate of synthesis induced by TGF-β, but also on the initial balance of the various α and β subunits, and on which of the subunits is available at rate-limiting levels to form functional integrin dimers. Through this set of upregulatory and downregulatory events, TGF-β alters the repertoire of cell-adhesion receptors and with it the ability of a cell to adhere, migrate, and home to specific locations.

VI. CONTROLLING CELL PHENOTYPE BY CONTROLLING CELL ADHESION

The finding that TGF-β controls rapidly and selectively the levels of many components of the cell adhesion apparatus is provocative for several reasons. This action of TGF-β is opposite to that of oncogenic transformation, which frequently results in a decrease or loss of cell adhesion components.[126,127] Indeed, TGF-β has been shown to repress the transformed behavior of certain tumor cells.[57,128] As mentioned above, TGF-β regulates the expression of genes that can clearly affect, positively or negatively, cell proliferation. However, it is conceivable that a normalization of the adhesive capacity by TGF-β might contribute to its growth-suppressive action in transformed cells and might similarly play a role in the growth inhibition of normal cells.

Control of cell adhesion and extracellular matrix assembly and deposition by TGF-β is particularly relevant in the control of cell phenotype by this factor. Adhesion to substrata and to other cells provides not only physical cues that affect cell morphology and position in tissues, but also transmembrane stimuli that may influence cell proliferation and, in particular, cell differentiation. Although the molecular details of the mechanism involved are unknown, there is ample evidence to indicate that changes in cell adhesion to extracellular matrices affect the expression of phenotype-determining genes.

Two recent reports have addressed the role of integrins and cell adhesion in differentiation. In one of them, it has been shown that differentiation of a human colon carcinoma cell line occurs only when cells are implanted in a three-dimensional collagen gel.[129] When an RGDT peptide is included in the gel, the cells fail to differentiate. Another peptide, RGET, which does not bind to integrins, does not block differentiation. The second report is of relevance here because it concerns the differentiation and fusion of chick myoblasts. These developmental events can be blocked with an antibody that recognizes the β_1-integrin subunit or with the peptide RGDS, which disrupts the interaction of integrins with the extracellular matrix.[130] Rat L6 myoblast differentiation can also be blocked by disturbing the extracellular matrix with the addition of exogenous fibronectin or collagen.[131] These observations and the finding that TGF-β rapidly and strongly elevates the expression of fibronectin, collagen, and proteoglycans in myoblasts suggest that inhibition of myogenic differentiation by this factor is accomplished via accumulation of an abundant extracellular matrix and modulation of cell adhesion.

There is also a very good correlation between the ability of TGF-β to elevate expression of fibronectin, collagen, and their corresponding receptors in 3T3-L1 preadipocytes, and the ability to block the adipogenic differentiation. Exogenous addition of these extracellular matrix components to cultures of 3T3 preadipocytes can prevent or delay their subsequent differentiation.[132] Soluble fibronectin or collagen added to the cultures may compete with the matrix for binding to integrins. Alternatively, oversaturation of integrins with their ligands may be incompatible with commitment to terminal differentiation.

A third example that implicates changes in cell adhesion induced by TGF-β as mediators of the overall cellular response to this factor is provided by the proliferation and colony formation of nontransformed NRK-49F fibroblasts in soft agar. The response of these cells to TGF-β is the biological effect that led to the discovery of this factor.[133,134] This effect of TGF-β can be mimicked by the addition of exogenous fibronectin, and can be specifically inhibited if TGF-β is assayed in the presence of a synthetic hexapeptide containing the RGD sequence.[103] These observations suggest that growth of fibroblasts in soft agar induced by TGF-β is mediated by the deposition of a pericellular matrix by the cells in response to this factor, and their subsequent anchored proliferation is a response to mitogens available in the culture medium.

The concept that TGF-β may exert its actions in part by controlling cell-extracellular matrix and cell-cell adhesive interactions has implications in the physiological processes in which this factor participates. Morphogenesis and other developmental events that take place in embryogenesis, during tissue growth, recycling, and repair, and in tumorigenic processes are all guided by the type of physical contacts that TGF-β regulates. Control of cell adhesion might also operate in the control of phenotype and morphogenesis by other factors of the TGF-β superfamily. The interface between the two major intercellular communication networks—growth factors and cell-adhesion components—is likely to become an important focus of attention in efforts to elucidate the molecular and cellular basis of pluricellular organism development.

REFERENCES

1. **Massagué, J.,** The TGF-β family of growth and differentiation factors, *Cell,* 49, 437, 1987.
2. **Roberts, A. B., Thompson, N. L., Heine, U., Flanders, C., and Sporn, M. B.,** Transforming growth factor-β: possible roles in carcinogenesis, *Br. J. Cancer,* 57, 594, 1988.
3. **Keski-Oja, J., Leof, E. B., Lyons, R. M., Coffey, R. J. Jr., and Moses, H. L.,** Transforming growth factors and control of neoplastic cell growth, *J. Cell. Biochem.,* 33, 95, 1987.
4. **Derynck, R., Jarrett, J. A., Chen, E. Y., Eaton, D. H., Bell, J. R., Assoian, R. K., Roberts, A. B., Sporn, M. B., and Goeddel, D. V.,** Human transforming growth factor-complementary DNA sequence and expression in normal and transformed cells, *Nature,* 316, 701, 1985.
5. **Seyedin, S. M., Thomas, T. C., Thompson, A. Y., Rosen, D. M., and Piez, K. A.,** Purification and characterization of two cartilage-inducing factors from bovine demineralized bone, *Proc. Natl. Acad. Sci. U.S.A.,* 82, 2267, 1985.
6. **Cheifetz, S., Weatherbee, J. A., Tsang, M.L.-S., Anderson, J. K., Mole, J. E., Lucas, R., and Massagué, J.,** The transforming growth factor-β systems: a complex pattern of cross-reactive ligands and receptors, *Cell,* 48, 409, 1987.
7. **Hanks, S. K., Armour, R., Baldwin, J. H., Maldonado, F., Spiess, J., and Holley, R. W.,** Amino acid sequence of the BSC-1 cell growth inhibitor (polyergin) deduced from the nucleotide sequence of the cDNA, *Proc. Natl. Acad. Sci. U.S.A.,* 85, 79, 1988.
8. **Madisen, L., Webb, T., Rose, T. M., Marquardt, H., Ikeda, T., Twardzik, D., Seyedin, S., and Purchio, A. F.,** Transforming growth factor-β2: cDNA cloning and sequence analysis, *DNA,* 7, 1, 1988.
9. **de Martin, R., Haendler, B., Hofer-Warbinek, R., Gaugitsch, H., Schlüsener, H., Seifert, J. M., Bodmer, S., Fontana, A., and Hofer, E.,** Complementary DNA for human glioblastoma-derived T cell suppressor factor, a novel member of the transforming growth factor-β gene family, *EMBO J.,* 6, 676, 1987.
10. **Derynck, R., Lindquist, P. B., Lee, A., Wen, D., Tamm, J., Graycar, J. L., Rhee, L., Mason, A. J., Miller, D. A., Coffey, R. J., Moses, H. L., and Chen, E. Y.,** A new type of transforming growth factor-β, TGF-β3, *EMBO J.,* 7, 3737, 1988.
11. **ten Dijke, P., Hansen, P., Iwata, K. K., Pieler, C., and Foulkes, J. B.,** Identification of another member of the transforming growth factor type β gene family, *Proc. Natl. Acad. Sci. U.S.A.,* 85, 4715, 1988.

12. **Lechleider, R. J., Jakowlew, S. B., Dillard, P. J., Sverha, J. P., Sporn, M. B., and Roberts, A. B.,** Complementary deoxyribonucleic acid cloning of chicken transforming growth factor-beta messenger ribonucleic acids, *J. Cell. Biochem. Suppl.,* 13B (Abstr.), E130, 1989.

13. **Jakowlew, S. B., Dillard, P. J., Sporn, M. B., and Roberts, A. B.,** Complementary deoxyribonucleic acid cloning of a messenger ribonucleic acid encoding transforming growth factor beta 4 from chicken embryo chondrocytes, *Mol. Endocrinol.,* 2, 1186, 1988.

14. **Assoian, R. K., Komoriya, A., Meyers, C. A., Miller, D. M., and Sporn, M. B.,** Transforming growth factor-β in human platelets. Identification of a major storage site, purification, and characterization, *J. Biol. Chem.,* 258, 7155, 1983.

15. **Gentry, L. E., Webb, N. R., Lim, G. J., Brunner, A. M., Ranchalis, J. E., Twardzik, D. R., Lioubin, M. N., Marquardt, H., and Purchio, A. F.,** Type 1 transforming growth factor beta: amplified expression and secretion of mature and precursor polypeptides in Chinese hamster ovary cells, *Mol. Cell. Biol.,* 7, 3418, 1987.

16. **Sharples, K., Plowman, G. D., Rose, T. M., Twardzik, D. R., and Purchio, A. F.,** Cloning and sequence analysis of simian transforming growth factor cDNA, *DNA,* 6, 239, 1987.

17. **Pircher, R., Jullien, P., and Lawrence, D. A.,** β-transforming growth factor is stored in human blood platelets as a latent high molecular weight complex, *Biochem. Biophys. Res. Commun.,* 136, 30, 1986.

18. **Miyazono, K., Hellman, U., Wernstedt, C., and Heldin, C-H.,** Latent high molecular weight complex of transforming growth factor β1. Purification from platelets and structural characterization, *J. Biol. Chem.,* 263, 6407, 1988.

19. **Wakefield, L. M., Smith, D. M., Flanders, K. C., and Sporn, M. B.,** Latent transforming growth factor-β from human platelets. A high molecular weight complex containing precursor sequences, *J. Biol. Chem.,* 263, 7646, 1988.

20. **Brunner, A. M., Gentry, L. E., Cooper, J. A., and Purchio, A. F.,** Recombinant type 1 transforming growth factor β precursor produced in Chinese hamster ovary cells is glycosylated and phosphorylated, *Mol. Cell. Biol.,* 8, 2229, 1988.

21. **Purchio, A. F., Cooper, J. A., Brunner, A. M., Lioubin, M. N., Gentry, L. E., Kovacina, K. S., Roth, R. A., and Marquardt, H.,** Identification of mannose-6-phosphate in two asparagine-linked sugar chains of recombinant transforming growth factor-β1 precursor, *J. Biol. Chem.,* 263, 14211, 1988.

22. **Ruoslahti, E. and Pierschbacher, M. D.,** New perspectives in cell adhesion: RGD and integrins, *Science,* 238, 491, 1987.

23. **Carpenter, G.,** Receptors for epidermal growth factor and other polypeptide mitogens, *Annu. Rev. Biochem.,* 56, 881, 1987.

24. **Bender, W.,** Homeotic gene products as growth factors, *Cell,* 43, 559, 1985.

25. **O'Connor-McCourt, M. D. and Wakefield, L. M.,** Latent transforming growth factor-β in serum: a specific complex with α_2-macroglobulin, *J. Biol. Chem.,* 262, 14090, 1987.

26. **Lyons, R. M., Keski-Oja, J., and Moses, H. L.,** Proteolytic activation of latent transforming growth factor-β from fibroblast conditioned medium, *J. Cell Biol.,* 106, 1659, 1988.

27. **Miyazono, K. and Heldin, C-H.,** Role for carbohydrate structures in TGF-β1 latency, *Nature,* 338, 158, 1989.

28. **Ellingsworth, L. R., Brennan, J. E., Fok, K., Rosen, D. M., Bentz, H., Piez, K. A., and Seyedin, S. M.,** Antibodies to the N-terminal portion of cartilage-inducing factor A and transforming growth factor β. Immunohistochemical localization and association with differentiating cells, *J. Biol. Chem.,* 261, 12362, 1986.

29. **Heine, U. I., Munoz, E. F., Flanders, K. C., Ellingsworth, L. R., Lam, H-Y. P., Thompson, N. L., Roberts, A. B., and Sporn, M. B.,** Role of transforming growth factor-β in the development of the mouse embryo, *J. Cell Biol.,* 105, 2861, 1987.

30. **Carrington, J. L., Roberts, A. B., Flanders, K. C., Roche, N. S., and Reddi, A. H.,** Accumulation, localization and compartmentation of transforming growth factor β during endochondral bone development, *J. Cell Biol.,* 107, 1969, 1988.

31. **Cate, R. L., Mattaliano, R. J., Hession, C., Tizard, R., Farber, N. M., Cheung, A., Ninfa, E. G., Frey, A. Z., Gash, D. J., Chow, E. B., Fisher, R. A., Bertonis, J. M., Torres, G., Wallner, B. P., Ramachandran, K. L., Ragin, R. C., Manganaro, T. F., MacLaughlin, D. T., and Donahoe, P. K.,** Isolation of the bovine and human genes for Mullerian inhibiting substance and expression of the human gene in animal cells, *Cell,* 45, 685, 1986.

32. **Mason, A. J., Hayflick, J. S., Ling, N., Esch, F., Ueno, N., Ying, S.-Y., Guillemin, R., Niall, H., and Seeburg, P. H.,** Complementary DNA sequences of ovarian follicular fluid inhibin show precursor structure and homology with transforming growth factor-β, *Nature,* 318, 659, 1985.

33. **Ying, S.-Y., Becker, A., Ling, N., Ueno, N., and Guillemin, R.,** Inhibin and beta type transforming growth factor (TGFβ) have opposite modulating effects on the follicle stimulating hormone (FSH)-induced aromatase activity of cultured rat granulosa cells, *Biochem. Biophys. Res. Commun.,* 136, 969, 1986.

34. **Ling, N., Ying, S.-Y., Ueno, N., Shimasaki, S., Esch, F., Hotta, M., and Guillemin, R.,** Pituitary FSH is released by a heterodimer of the β-subunits from the two forms of inhibin, *Nature,* 31, 779, 1986.

35. **Vale, W., Rivier, J., Vaughan, J., McClintock, R., Corrigan, A., Woo, W., Karr, D., and Spiess, J.,** Purification and characterization of an FSH releasing protein from porcine ovarian follicular fluid, *Nature,* 321, 776, 1986.

36. **Eto, Y., Tsuji, T., Takezawa, M., Takano, S., Yokogawa, Y., and Shibai, H.,** Purification and characterization of erythroid differentiation factor (EDF) isolated from human leukemia cell line THP-1, *Biochem. Biophys. Res. Commun.,* 142, 1095, 1987.

37. **Weeks, D. L. and Melton, D. A.,** A maternal mRNA localized to the vegetal hemisphere in Xenopus eggs codes for a growth factor related to TGFβ, *Cell,* 51, 861, 1987.

38. **Padgett, R. W., St. Johnston, R. D., and Gelbart, W. M.,** A transcript from a Drosophila pattern gene predicts a protein homologous to the transforming growth factor-β family, *Nature,* 325, 81, 1987.

39. **Wozney, J. M., Rosen, V., Celeste, A. J., Mitsock, L. M., Whitters, M. J., Kriz, R. W., Hewick, R. M., and Wang, E. A.,** Novel regulators of bone formation: molecular clones and activities, *Science,* 2452, 1528, 1989.

40. **Cheifetz, S., Like, B., and Massagué, J.,** Cellular distribution of type I and type II receptors for transforming growth factor-β, *J. Biol. Chem.,* 261, 9972, 1986.

41. **Cheifetz, S., Bassols, A., Stanley, K., Ohta, M., Greenberger, J., and Massagué, J.,** Heterodimeric transforming growth factor β. Biological properties and interaction with three types of cell surface receptors, *J. Biol. Chem.,* 263, 10783. 1988.

42. **Ohta, M., Anklesaria, P., Greenberger, J. S., Bassols, A., and Massagué, J.,** Two forms of transforming growth factor-β distinguished by hematopoietic progenitor cells, *Nature,* 329, 539, 1987.

43. **Segarini, P. R. and Seyedin, S. M.,** The high molecular weight receptor to transforming growth factor-β contains glycosaminoglycan chains, *J. Biol. Chem.,* 263, 8366, 1988.

44. **Cheifetz, S., Andres, J. L., and Massagué, J.,** The TGF-β receptor type III is a membrane proteoglycan. Domain structure of the receptor, *J. Biol. Chem.,* 263, 16984, 1988.

45. **Massagué, J.,** Subunit structure of a high affinity receptor for type β transforming growth factor. Evidence for a disulfide-linked glycosylated receptor complex, *J. Biol. Chem.,* 260, 7059, 1985.

46. **Segarini, P. R., Rosen, D. M., and Seyedin, S. M.,** Binding of transforming growth factor-β to cell surface proteins varies with cell type, *Mol. Endocrinol.,* 3, 261, 1989.

47. **Massagué, J., Boyd, F. T., Andres, J. L., and Cheifetz, S.,** Mediators of TGF-β action: TGF-β receptors and TGF-β-binding proteoglycans, *Ann. N.Y. Acad. Sci.,* in press.

48. **Kimchi, A., Wang, X.-F., Weinberg, R. A., Cheifetz, S., and Massagué, J.,** Absence of TGF-β receptors and growth inhibitory responses in retinoblastoma cells, *Science,* 240, 196, 1988.

49. **Cheifetz, S., Ling, N., Guillemin, R., and Massagué, J.,** A surface component on GH3 pituitary cells that recognizes transforming growth factor-β, activin, and inhibin, *J. Biol. Chem.,* 263, 17225, 1988.

50. **Like, B. and Massagué, J.,** The antiproliferative effect of type β transforming growth factor occurs at a level distal from receptors for growth-activating factors, *J. Biol. Chem.,* 261, 13426, 1986.

51. **Boyd, F. T. and Massagué, J.,** Transforming growth factor-β inhibition of epithelial cell proliferation linked to the expression of a 53-kDa membrane receptor, *J. Biol. Chem.,* 264, 2272, 1989.

52. **Ignotz, R. A., Boyd, F. T., Rauth, J., and Massagué, J.,** unpublished work.

53. **Cheifetz, S. and Massagué, J.,** TGF-β binding activity located in the core of the TGF-β receptor proteoglycan, *J. Biol. Chem.,* 264, 12025, 1989.

54. **Andres, J. L., Stanley, K., Cheifetz, S., and Massagué, J.,** Membrane-anchored and secreted forms of betaglycan, a polymorphic proteoglycan that binds transforming growth factor-β., *J. Cell Biol.,* 109, 3137, 1989.

55. **Roberts, A. B., Anzano, M. A., Lamb, L. C., Smith, J. M., and Sporn, M. B.,** New class of transforming growth factors potentiated by epidermal growth factor: isolation from nonneoplastic tissues, *Proc. Natl. Acad. Sci. U.S.A.,* 78, 5339, 1981.

56. **Tucker, R. F., Shipley, G. D., Moses, H. L., and Holley, R. W.,** Growth inhibitor from BSC-1 cells is closely related to the platelet type β transforming growth factor, *Science,* 226, 705, 1984.

57. **Roberts, A. B., Anzano, M. A., Wakefield, L. M., Roche, N. S., Stern, D. F., and Sporn, M. B.,** Type-β transforming growth factor: a bifunctional regulator of cellular growth, *Proc. Natl. Acad. Sci. U.S.A.,* 82, 119, 1985.

58. **Frater-Schröder, M., Muller, G., Birchmeier, W., and Böhlen, P.,** Transforming growth factor-β inhibits endothelial cell proliferation, *Biochem. Biophys. Res. Commun.,* 137, 295, 1986.

59. **Leof, E. B., Proper, J. A., Goustin, A. S., Shipley, G. D., DiCorleto, P. E. and Moses, H. L.,** Induction of c-sis mRNA and activity similar to platelet-derived growth factor by transforming growth factor β: a proposed model for indirect mitogenesis involving autocrine activity, *Proc. Natl. Acad. Sci. U.S.A.,* 83, 2453, 1986.

60. **Kavanaugh, W. M., Harsh, G. R., IV, Starksen, N. F., Rocco, C. M., and Williams, L. T.,** Transcriptional regulation of the A and B chain genes of platelet-derived growth factor in microvascular endothelial cells, *J. Biol. Chem.,* 263, 8470, 1988.

61. **Makela, T. P., Alitalo, R., Paulsson, Y., Westermark, B., Heldin, C.-H., and Alitalo, K.,** Regulation of platelet-derived growth factor gene expression by transforming growth factor β and phorbol ester in human leukemia cell lines, *Mol. Cell. Biol.,* 7, 3656, 1987.

62. **Chambard, J.-C. and Pouyssegur, J.,** TGF-β inhibits the growth factor-induced DNA synthesis in hamster fibroblasts without affecting the early mitogenic events, *J. Cell Physiol.,* 135, 101, 1988.

63. **Green, H. and Meuth, M.,** An established pre-adipose cell line and its differentiation in culture, *Cell,* 3, 127, 1974.

64. **Ignotz, R. A. and Massagué, J.,** Type β transforming growth factor controls the adipogenic differentiation of 3T3 fibroblasts, *Proc. Natl. Acad. Sci. U.S.A.,* 82, 8530, 1985.

65. **Sparks, R. L. and Scott, R. E.,** Transforming growth factor type β is a specific inhibitor of 3T3 T mesenchymal stem cell differentiation, *Exp. Cell Res.,* 165, 345, 1986.

66. **Torti, F. M., Torti, S. V., Larrick, J. W., and Ringold, G. M.,** Modulation of adipocyte differentiation by tumor necrosis factor and transforming growth factor beta, *J. Cell Biol.,* 108, 1105, 1989.

67. **Yaffe, D.,** Retention of differentiation potentialities during prolonged cultivation of myogenic cells, *Proc. Natl. Acad. Sci. U.S.A.,* 61, 477, 1968.

68. **Nadal-Ginard, B.,** Commitment, fusion, and biochemical differentiation of a myogenic cell line in the absence of DNA synthesis, *Cell,* 15, 855, 1978.

69. **Richler, C. and Yaffe, D.,** In vitro cultivation and differentiation of myogenic cell lines, *Dev. Biol.,* 23, 1, 1970.

70. **Massagué, J., Cheifetz, S., Endo, T., and Nadal-Ginard, B.,** Type β transforming growth factor is an inhibitor of myogenic differentiation, *Proc. Natl. Acad. Sci. U.S.A.,* 83, 8206, 1986.

71. **Ignotz, R. A., Endo, T., and Massagué, J.,** Regulation of fibronectin and type I collagen mRNA levels by transforming growth factor-β, *J. Biol. Chem.,* 262, 6443, 1987.

72. **Bassols, A. and Massagué, J.,** Transforming growth factor β regulates the expression and structure of extracellular matrix chondroitin/dermatan sulfate proteoglycans, *J. Biol. Chem.,* 263, 3039, 1988.

73. **Ewton, D. A., Spizz, G., Olson, E. N., and Florini, J. R.,** Decrease in transforming growth factor-β binding and action during differentiation in muscle cells, *J. Biol. Chem.,* 263, 4029, 1988.

74. **Olson, E. N., Sternberg, E., Hu, J. S., Spizz, G., and Wilcox, C.,** Regulation of myogenic differentiation by type β transforming growth factor, *J. Cell Biol.,* 103, 1799, 1986.

75. **Spizz, G., Hu, J.-S., and Olson, E. N.,** Inhibition of myogenic differentiation by fibroblast growth factor or type β transforming growth factor does not require persistent c-myc expression, *Dev. Biol.,* 123, 500, 1987.

76. **Allen, R. E. and Boxhorn, L. K.,** Inhibition of skeletal muscle satellite cell differentiation by transforming growth factor-beta, *J. Cell. Physiol.,* 133, 567, 1987.

77. **Seyedin, S. M., Thompson, A. Y., Bentz, H., Rosen, D. M., McPherson, J. M., Conti, A., Siegel, N. R., Galluppi, G. R., and Piez, K. A.,** Cartilage-inducing factor-A: apparent identity to transforming growth factor-β, *J. Biol. Chem.,* 261, 5693, 1986.

78. **Seyedin, S. M., Segarini, P. R., Rosen, D. M., Thompson, A. Y., Bentz, H., and Graycar, J.,** Cartilage-inducing factor-β is a unique protein structurally and functionally related to transforming growth factor-β, *J. Biol. Chem.,* 262, 1946, 1987.

79. **Centrella, M. and Canalis, E.,** Transforming and nontransforming growth factors are present in medium conditioned by fetal rat calvariae, *Proc. Natl. Acad. Sci. U.S.A.,* 82, 7335, 1985.

80. **Centrella, M., Massagué, J., and Canalis, E.,** Human platelet-derived transforming growth factor-β stimulates parameters of bone growth in fetal rat calvariae, *Endocrinology,* 119, 2306, 1986.

81. **Centrella, M., McCarthy, T. L., and Canalis, E.,** Transforming growth factor β is a bifunctional regulator of replication and collagen synthesis in osteoblast-enriched cell cultures from fetal rat bone, *J. Biol. Chem.,* 262, 2869, 1987.

82. **Noda, M., Yoon, K., Prince, C. W., Butler, W. T., and Rodan, G. A.,** Transcriptional regulation of osteopontin production in rat osteosarcoma cells by type β transforming growth factor, *J. Biol. Chem.,* 263, 13916, 1988.

83. **Noda, M. and Rodan, G. A.,** Type β transforming growth factor (TGFβ) regulation of alkaline phosphatase expression and other phenotype-related mRNAs in osteoblastic rat osteosarcoma cells, *J. Cell. Physiol;.,* 133, 426, 1987.

84. **Pfeilschiefter, J. D., Sousa, S. M., and Mundy, G. R.,** Effects of transforming growth factor-β on osteoblastic osteosarcoma cells, *Endocrinology,* 121, 212, 1987.

85. **Rosen, D. M., Stempien, S. A., Thompson, A. Y., and Seyedin, S. M.,** Transforming growth factor-beta modulated the expression of osteoblast and chondroblast phenotypes in vitro, *J. Cell. Physiol.,* 134, 337, 1988.

86. **Noda, M. and Rodan, G. A.,** Type-β transforming growth factor inhibits proliferation and expression of alkaline phosphatase in murine osteoblast-like cells, *Biochem. Biophys. Res. Commun.,* 140, 56, 1986.

87. **Wozney, J.,** Bone morphogenetic proteins, *Prog. Growth Factor Res.,* 1, 267, 1989.

88. **Greenberger, J. S., Sakakeeny, M. A., Humphries, R. K., Eaves, C. J., and Eckner, R. J.,** Demonstration of permanent factor-dependent multipotential (erythroid/neutrophil/basophil) hematopoietic progenitor cell lines, *Proc. Natl. Acad. Sci. U.S.A.,* 80, 2931, 1983.

89. **Masui, T., Wakefield, L. M., Lechner, J. F., LaVeck, M. A., Sporn, M. B., and Harris, C. C.,** Type β transforming growth factor is the primary differentiation-inducing serum factor for normal human bronchial epithelial cells, *Proc. Natl. Acad. Sci. U.S.A.,* 83, 2438, 1986.

90. **Kurokowa, M., Lynch, K., and Podolsky, D. K.,** Effect of growth factors on an intestinal epithelial cell line: transforming growth factor β inhibits proliferation and stimulates differentiation, *Biochem. Biophys. Res. Commun.,* 142, 775, 1987.

91. **Shipley, G. D., Pittelkow, M. R., Wille, J. J., Jr., Scott, R. E., and Moses, H. L.,** Reversible inhibition of normal human prokeratinocyte proliferation by type β transforming growth factor-growth inhibitor in serum-free medium, *Cancer Res.,* 46, 2068, 1986.

92. **Coffey, R. J., Jr., Bascom, C. C., Sipes, N. J., Graves-Deal, R., Weissman, B. E., and Moses, H. L.,** Selective inhibition of growth-related gene expression in murine keratinocytes by transforming growth factor β, *Mol. Cell. Biol.,* 8, 3088, 1988.

93. **Reiss, M. and Sartorelli, A. C.,** Regulation of growth and differentiation of human keratinocytes by type β transforming growth factor and epidermal growth factor, *Cancer Res.,* 47, 6705, 1987.

94. **Lin, T., Blaisdell, J., and Haskell, J. F.,** Transforming growth factor-β inhibits Leydig cell steroidogenesis in primary culture, *Biochem. Biophys. Res. Commun.,* 146, 387, 1987.

95. **Rook, A. H., Kehrl, J. H., Wakefield, L. M., Roberts, A. B., Sporn, M. B., Burlington, D. B., Lane, H. C., and Fauci, A. S.,** Effects of transforming growth factor β on the functions of natural killer cells: depressed cytolytic activity and blunting of interferon responsiveness, *J. Immunol.,* 136, 3916, 1986.

96. **Kehrl, J. H., Roberts, A. B., Wakefield, L. M., Jakowlew, S., Sporn, M. B., and Fauci, A. S.,** Transforming growth factor β is an important immunomodulatory protein for human B lymphocytes, *J. Immunol.,* 137, 3855, 1986.

97. **Tsunawaki, S., Sporn, M. B., Ding, A., and Nathan, C.,** Deactivation of macrophages by transforming growth factor-β, *Nature,* 334, 260, 1988.

98. **Hotta, M. and Baird, A.,** Differential effects of transforming growth factor type β on the growth and function of adrenocortical cells in vitro, *Proc. Natl. Acad. Sci. U.S.A.,* 83, 7795, 1986.

99. **Pertovaara, L., Sistonen, L., Box, T. J., Vogt, P. K., Keski-Oja, J., and Alitalo, K.,** Enhanced jun gene expression is an early genomic response to transforming growth factor β stimulation, *Mol. Cell. Biol.,* 9, 1255, 1989.

100. **Takehara, K., LeRoy, E. C., and Grotendorst, G. R.,** TGFβ-induction of endothelial cell proliferation: Alteration of EGF binding and EGF-induced growth-regulatory (competence) gene expression, *Cell,* 49, 415, 1987.

101. **Van Obberghen-Schilling, E., Roche, N. S., Flanders, K. C., Sporn, M. B., and Roberts, A. B.,** Transforming growth factor β1 positively regulates its own expression in normal and transformed cells, *J. Biol. Chem.,* 263, 7741, 1988.

102. **Sporn, M. B., Roberts, A. B., Shull, J. H., Smith, J. M., Ward, J. M., and Sodek, J.,** Polypeptide transforming growth factors isolated from bovine sources and used for wound healing *in vivo, Science,* 219, 1329, 1983.

103. **Ignotz, R. A. and Massagué, J.,** Transforming growth factor-β stimulates the expression of fibronectin and collagen and their incorporation into the extracellular matrix, *J. Biol. Chem.,* 261, 4337, 1986.

104. **Blatti, S. P., Foster, D. N., Rangnathan, G., Moses, H. L., and Getz, M. J.,** Induction of fibronectin gene transcription and mRNA is a primary response to growth-factor stimulation of AKR-2B cells, *Proc. Natl. Acad. Sci. U.S.A.,* 85, 1119, 1988.

105. **Dean, D. C., Newby, R. F., and Bourgeois, S.,** Regulation of fibronectin biosynthesis by dexamethasone, transforming growth factor β, and cAMP in human cell lines, *J. Cell Biol.,* 106, 2159, 1988.

106. **Raghow, R., Postlethwaite, A. E., Keski-Oja, J., Moses, H. L., and Kang, H.,** Transforming growth factor-β increases steady-state levels of type I procollagen and fibronectin messenger RNAs posttranscriptionally in cultured human dermal fibroblasts, *J. Clin. Invest.,* 79, 1285, 1987.

107. **Roberts, A. B., Sporn, M. B., Assoian, R. K., Smith, J. M., Roche, N. S., Wakefield, L. M., Heine, U. I., Liotta, L. A., Falanga, V., Kehrlk, J. H., and Fauci, A. S.,** Transforming growth factor type-β: rapid induction of fibrosis and angiogenesis *in vivo* and stimulation of collagen formation *in vitro, Proc. Natl. Acad. Sci. U.S.A.,* 83, 4167, 1986.

108. **Varga, J., Rosenbloom, J., and Jimenez, S. A.,** Transforming growth factor β (TGFβ) causes a persistent increase in steady-state amounts of type I and type III collagen and fibronectin mRNAs in normal human dermal fibroblasts, *Biochem. J.,* 247, 597, 1987.

109. **Penttinen, R. P., Kobayashi, S., and Bornstein, P.,** Transforming growth factor β increases mRNA for matrix proteins both in the presence and in the absence of changes in mRNA stability, *Proc. Natl. Acad. Sci. U.S.A.,* 85, 1105, 1988.

110. **Rossi, P., Karsenty, G., Roberts, A. B., Roche, N. S., Sporn, M. B., and de Crombrugghe, B.,** A nuclear factor 1 binding site mediates the transcriptional activation of a type I collagen promoter by transforming growth factor-β, *Cell,* 52, 405, 1988.

111. **Kim, S.-J., Jeang, K.-T., Glick, A. B., Sporn, M. B., and Roberts, A. B.,** Promoter sequences of the human transforming growth factor-β1 gene responsive to transforming growth factor-β1 autoinduction, *J. Biol. Chem.,* 264, 7041, 1989.

112. **Pearson, C. A., Pearson, D., Shibahara, S., Hofsteenge, J., and Chiquet-Ehrismann, R.,** Tenascin: cDNA cloning and induction by TGF-β, *EMBO J.,* 7, 2677, 1988.

113. **Rasmussen, S. and Rapraeger, A.,** Altered structure of the hybrid cell surface proteoglycan of mammary epithelial cells in response to transforming growth factor-β, *J. Cell Biol.,* 107, 1959, 1988.

114. **Chen, J.-K., Hoshi, H., and McKeehan, W. L.,** Transforming growth factor type β specifically stimulates synthesis of proteoglycan in human adult arterial smooth muscle cells, *Proc. Natl. Acad. Sci. U.S.A.,* 84, 5287, 1987.

115. **Laiho, M., Saksela, O., Andreasen, P. A., and Keski-Oja, J.,** Enhanced production and extracellular deposition of the endothelial-type plasminogen activator inhibitor in cultured human lung fibroblasts by transforming growth factor-β, *J. Cell Biol.,* 103, 2403, 1986.

116. **Lund, L. R., Riccio, A., Andreasen, P. A., Nielsen, L. S., Kristensen, P., Laiho, M., Saksela, O., Blasi, F., and Danø, K.,** Transforming growth factor-β is a strong and fast acting positive regulator of the level of type-1 plasminogen activator inhibitor mRNA in WI-38 human lung fibroblasts, *EMBO J.,* 6, 1281, 1987.

117. **Edwards, D. R., Murphy, G., Reynolds, J. J., Whitham, S. E., Docherty, A. J. P., Angel, P., and Heath, J. K.,** Transforming growth factor-β modulates the expression of collagenase and metalloproteinase inhibitor, *EMBO J.,* 6, 1899, 1987.

118. **Chua, C. C., Geiman, D. E., Keller, G. H., and Ladda, R. L.,** Induction of collagenase secretion in human fibroblast cultures by growth promoting factors, *J. Biol. Chem.,* 260, 5213, 1985.

119. **Hynes, R. O.,** Integrins: a family of cell surface receptors, *Cell,* 48, 549, 1987.

120. **Hemler, M. E.,** Adhesive receptors on hematopoietic cells, *Immunol. Today,* 9, 109, 1988.

121. **Ignotz, R. A. and Massagué, J.,** Cell adhesion protein receptors as targets for transforming growth factor-β action, *Cell,* 51, 189, 1987.

122. **Roberts, C. J., Birkenmeier, T. M., McQuillan, J. J., Akiyama, S. K., Yamada, S. S., Chen, W.-T., Yamada, K. M., and McDonald, J. A.,** Transforming growth factor β stimulates the expression of fibronectin and of both subunits of the human fibronectin receptor by cultured human lung fibroblasts, *J. Biol. Chem.,* 263, 4586, 1988.

123. **Heino, J., Ignotz, R. A., Hemler, M. E., Crouse, C., and Massagué, J.,** Regulation of cell adhesion receptors by transforming growth factor-β. Concomitant regulation of integrins that share a common β1 subunit, *J. Biol. Chem.,* 264, 380, 1989.

124. **Ignotz, R. A., Heino, J., and Massagué, J.,** Regulation of cell adhesion receptors by transforming growth factor-β. Regulation of vitronectin receptor and LFA-1, *J. Biol. Chem.,* 264, 389, 1989.

125. **Marlin, S. D. and Springer, T. A.,** Purified intercellular adhesion molecule-1 (ICAM-1) is a ligand for lymphocyte function-associated antigen 1 (LFA-1), *Cell,* 51, 813, 1987.

126. **Hynes, R. O.,** Fibronectin and its relation to cellular structure and behavior, in *Cell Biology of the Extracellular Matrix,* Hay, E. D., Ed., Plenum Press, New York, 1981, 295.

127. **Burridge, K.,** Substrate adhesions in normal and transformed fibroblasts: organization and regulation of cytoskeletal, membrane and extracellular matrix components at focal contacts, *Cancer Rev.,* 4, 18, 1986.

128. **Knabbe, C., Lippman, M. E., Wakefield, L. M., Flanders, K. C., Kasid, A., Derynck, R., and Dickson, R. B.,** Evidence that transforming growth factor-β is a hormonally regulated negative growth factor in human breast cancer cells, *Cell,* 48, 417, 1987.

129. **Pignatelli, M. and Bodmer, W. F.,** Genetics and biochemistry of collagen binding-triggered glandular differentiation in a human colon carcinoma cell line, *Proc. Natl. Acad. Sci. U.S.A.,* 85, 5561, 1988.

130. **Menko, A. S. and Boettiger, D.,** Occupation of the extracellular matrix receptor, integrin, is a control point for myogenic differentiation, *Cell,* 51, 51, 1987.

131. **Podleski, T. R., Greenberg, I., Schlessinger, J., and Yamada, K. M.,** Fibronectin delays the fusion of L6 myoblasts, *Exp. Cell Res.,* 122, 317, 1979.

132. **Spiegelman, B. M. and Ginty, C. A.,** Fibronectin modulation of cell shape and lipogenic gene expression in 3T3-adipocytes, *Cell,* 35, 657, 1983.

133. **DeLarco, J. E. and Todaro, G. J.,** Growth factors from murine sarcoma virus-transformed cells, *Proc. Natl. Acad. Sci. U.S.A.,* 75, 4001, 1978.

134. **Sporn, M. B. and Todaro, G. J.,** Autocrine secretion and malignant transformation of cells, *N. Engl. J. Med.,* 303, 878, 1980.

INDEX